AGROINDUSTRIAL WASTE MANAGEMENT AND NATURAL RESOURCES CONSERVATION

Environmental Challenges and Transformation into Value-Added Products

AGROINDUSTRIAL WASTE MANAGEMENT AND NATURAL RESOURCES CONSERVATION

Environmental Challenges and Transformation into Value-Added Products

José Juan Buenrostro-Figueroa, PhD

Juan Manuel Tirado-Gallegos, PhD

Mónica L. Chávez-González, PhD

A. K. Haghi, PhD

Shaheemath Suhara K. K., PhD

Editors

AAP | APPLE ACADEMIC PRESS

First edition published 2025

Apple Academic Press Inc.
1265 Goldenrod Circle, NE,
Palm Bay, FL 32905 USA

760 Laurentian Drive, Unit 19,
Burlington, ON L7N 0A4, CANADA

CRC Press
2385 NW Executive Center Drive,
Suite 320, Boca Raton FL 33431

4 Park Square, Milton Park,
Abingdon, Oxon, OX14 4RN UK

Library and Archives Canada Cataloguing in Publication

CIP data on file with Canada Library and Archives

Library of Congress Cataloging-in-Publication Data

CIP data on file with US Library of Congress

ISBN: 978-1-77491-868-5 (hbk)
ISBN: 978-1-77491-869-2 (pbk)
ISBN: 978-1-00359-530-4 (ebk)

About the Editors

José Juan Buenrostro-Figueroa, PhD
Researcher, Research Center for Food and Development, A.C., Mexico

José Juan Buenrostro-Figueroa, PhD, is Titular Researcher at the Research Center for Food and Development, A.C., Mexico. He is a food engineer. His MSc and PhD degrees were Food Science and Technology. Dr. Buenrostro has experience in bioprocess development, including microbial processes for enzyme production and recovery of bioactive compounds, valorization of agroindustrial byproducts, and extraction and characterization of bioactive compounds from plants and agroindustrial wastes. He has published 40 papers in indexed journals, 25 book chapters, six books as editor, and more than 70 contributions at scientific meetings, with over 1160 citations and h-index of 18 (Google Scholar database). He is member of the Mexican Association of Food Science (AMECA), Mexican Society for Food Protection affiliate of the International Association for Food Protection, and Food Safety Thematic Network: Agroindustrial By-products Valorization and Reduction of Food Losses and Waste (PDA). Dr. Buenrostro is a member of SNII (National System of Researches, Mexico), since 2014.

Juan Manuel Tirado-Gallegos, PhD
Researcher, School of Animal Sciences and Ecology at the Autonomous University of Chihuahua, Mexico

Juan Manuel Tirado-Gallegos, PhD, is a Research Professor at the School of Animal Sciences and Ecology at the Autonomous University of Chihuahua, Mexico. Dr. Tirado has experience in isolating, modifying, and characterizing natural polymers and developing biodegradable packaging with application in meat products. He has published 34 scientific articles, ten book chapters, and more than 40 contributions in scientific meetings. Dr. Tirado has been a member of the National System of Researchers in México since 2017. He graduated as an engineer in Food Technology from Antonio Narro Agrarian Autonomous University and earned his Master of Science in Biochemical Engineering from the Technological Institute of Veracruz,

Mexico. He earned his Doctorate in Food Biopolymers from the Research Center for Food and Development, A.C., Mexico.

Mónica L. Chávez-González, PhD
Professor and Research Scientist, School of Chemistry of the Universidad Autónoma de Coahuila, Mexico

Mónica L. Chávez-González, PhD, is a Research Professor at the School of Chemical Sciences of the Universidad Autónoma de Coahuila, Mexico. She is a pharmacobiological chemist. Her master's and doctoral program was in Food Science and Technology. She is the General Coordinator and founder of the Bioprocessing Mx-LATAM Network, whose objective is to promote R&D in the area of bioprocessing through interaction between universities, institutes, research centers, and relevant industries in Mexico and Latin America. She has been awarded the Latin American Women in Chemistry Award in the category of "Emerging Leader," granted by the American Chemical Society and the Latin American Federation of Chemical Associations (FLAQ). Her expertise is in the areas of fermentation processes, microbial biotransformation, enzyme production, industrial food waste valorization, extraction of bioactive compounds (antioxidants), and chemical characterization.

A. K. Haghi, PhD
Research Associate, Department of Chemistry, University of Coimbra, Portugal

A. K. Haghi, PhD, is a retired professor and has written, cowritten, edited, or coedited more than 1000 publications, including books, book chapters, and papers in refereed journals with over 3800 citations and h-index of 32, according to the Google Scholar database. He is currently a Research Associate at the University of Coimbra, Portugal. Professor Haghi has received several grants, consulted for several major corporations, and is a frequent speaker to national and international audiences. He is founder and former Editor-in-Chief of the *International Journal of Chemoinformatics and Chemical Engineering* and *Polymers Research Journal*. Professor Haghi has acted as an editorial board member of many international journals. He has served as a member of the Canadian Research and Development Center of Sciences & Cultures. He has supervised several PhD and MSc theses at the University of Guilan (UG) and cosupervised international doctoral projects. Professor Haghi holds a BSc in urban and environmental engineering from the University of North Carolina (USA) and holds two MSc, one in mechanical engineering from North Carolina State University (USA) and another one in applied mechanics, acoustics, and materials from the Université de Technologie de Compiègne (France). He was awarded a PhD in engineering sciences at Université de Franche-Comté (France). He is a regular reviewer for leading international journals.

Shaheemath Suhara K. K., PhD
Research Scholar, Tamil Nadu Agricultural University, Coimbatore

Shaheemath Suhara K. K. is currently pursuing a PhD in Soil and Water Conservation Engineering at Tamil Nadu Agricultural University, Coimbatore. She completed her master's degree in Soil and Water Engineering from Kerala Agricultural University with a gold medal. She has qualified the National Eligibility Test (NET) in Land and Water Management Engineering conducted by ASRB. Her research interests lie in the field of conserving natural resources for food and water security. She has research experience in hydrology, geoinformatics (GIS) and remote sensing, agricultural engineering, irrigation, and drainage engineering, as well as spatial analysis of natural resources.

Contents

Contributors

Belinda-Vallejo-Cordoba
CONAHCYT-Centro de Investigación en Alimentación y Desarrollo, Laboratorio de Calidad, Autenticidad y Trazabilidad de los Alimentos, Coordinación de Tecnología de Alimentos de Origen Animal, Hermosillo, Sonora, México

Lilia María Beltrán-Barrientos
CONAHCYT-Centro de Investigación en Alimentación y Desarrollo, Laboratorio de Calidad, Autenticidad y Trazabilidad de los Alimentos, Coordinación de Tecnología de Alimentos de Origen Animal, Hermosillo, Sonora, México

Diana Yepes Betancur
Servicio Nacional de Aprendizaje, SENA. Cra. 48 # 49-62, Rionegro, Colombia

Tanushree Bhattacharjee
Department of Chemical Engineering, Vishwakarma Institute of Technology, Bibwewadi, Pune, India

Soham Bhave
Department of Chemical Engineering, Vishwakarma Institute of Technology, Bibwewadi, Pune, India

María Esmeralda Bibián-León
Posgrado en Botánica, Colegio de Posgraduados-Montecillo, Texcoco, Mexico State, Mexico

José Juan Buenrostro-Figueroa
Laboratorio de Biotecnología y Bioingeniería, Centro de Investigación en Alimentación y Desarrollo, Delicias, Chihuahua, Mexico

Juan Carlos Bustillos-Rodríguez
Tecnológico Nacional de México, Instituto Tecnológico de Cd. Cuauhtémoc, Mexico

Adali Castañeda-Facio
Universidad Autónoma de Coahuila, Facultad de Ciencias Químicas, Blvd. Venustiano Carranza y José Cárdenas Valdés, S/N, C.P., Saltillo, Coahuila, México

Jonathan A. Díaz-Baca
Faculty of Science & Environmental Studies, Lakehead University, Thunder Bay, ON,Canada

Ángeles Domínguez-Rivera
Departamento de Biotecnología, Universidad Autónoma Metropolitana, Iztapalapa, Mexico City, Mexico

Coyolxauhqui Figueroa
Programa de Biología. Herbario UACJ- Unidad de Producción de plantas, México

Juanita D. Flores-Valdez
Universidad Autónoma de Coahuila, Facultad de Ciencias Químicas, Blvd. Venustiano Carranza y José Cárdenas Valdés, S/N, C.P., Saltillo, Coahuila, México

Aarón Fernando González-Cordova
CONAHCYT-Centro de Investigación en Alimentación y Desarrollo, Laboratorio de Calidad, Autenticidad y Trazabilidad de los Alimentos, Coordinación de Tecnología de Alimentos de Origen Animal, Hermosillo, Sonora, México

Adrián Hernández-Mendoza
CONAHCYT-Centro de Investigación en Alimentación y Desarrollo, Laboratorio de Calidad, Autenticidad y Trazabilidad de los Alimentos, Coordinación de Tecnología de Alimentos de Origen Animal, Hermosillo, Sonora, México

Claudia Carolina Hernández-Peña
Programa de Biotecnología. Departamento de Ciencias Químico Biológicas, Instituto de Ciencias, Biomédicas, Universidad Autónoma de Ciudad Juárez. Av. Benjamín Franklin #4650, Zona, PRONAF, Ciudad Juárez, Chihuahua, México

Nitzia Thalía Flores Jiménez
Centro de Tecnología de Alimentos. Universidad Autónoma de Nayarit. Ciudad de la Cultura Amado Nervo. Tepic, Nayarit. México

Balaji Kannan
Department of Physical Science and Information Technology, Agricultural Engineering College & Research Institute, Tamil Nadu Agricultural University, Coimbatore, Tamil Nadu, India

Shaheemath Suhara K. K.
Department of Soil and Water Conservation Engineering, Agricultural Engineering College & Research Institute, Tamil Nadu Agricultural University, Coimbatore, Tamil Nadu, India

Liliana Londoño-Hernández
Biotics Group. School of Basic Sciences, Technology and Engineering. Universidad Nacional Abierta y a Distancia – UNAD, Colombia

Leticia Xóchitl López-Martínez
CONAHCYT-Centro de Investigaciónen Alimentación y Desarrollo A.C. Sonora, México. Coordinación de Alimentos de Origen Vegetal, Hermosillo, Sonora, México

José Isidro Méndez-Romero
CONAHCYT-Centro de Investigación en Alimentación y Desarrollo, Laboratorio de Calidad, Autenticidad y Trazabilidad de los Alimentos, Coordinación de Tecnología de Alimentos de Origen Animal, Hermosillo, Sonora, México

Nirmala Devi M.
Department of Physical Science and Information Technology, Agricultural Engineering College & Research Institute, Tamil Nadu Agricultural University, Coimbatore, Tamil Nadu, India

Manuel Vargas-Ortiz
CONAHCYT-Centro de Investigaciónen Alimentación y Desarrollo, Laboratorio de Calidad, Autenticidad y Trazabilidad de los Alimentos, Coordinaciónde Tecnología de Alimentos de Origen Animal, Hermosillo, Sonora, México

Fernando Méndez-González
Laboratorio de Biotecnología y Bioingeniería, Centro de Investigación en Alimentación y Desarrollo, Delicias, Chihuahua, Mexico

Janani N.
Department of Soil and Water Conservation Engineering, Agricultural Engineering College & Research Institute, Tamil Nadu Agricultural University, Coimbatore, Tamil Nadu, India

Rosa Idalia Narro-Cespedes
Universidad Autónoma de Coahuila, Facultad de Ciencias Químicas, Blvd. Venustiano Carranza y José Cárdenas Valdés, S/N, C.P., Saltillo, Coahuila, México

Magali Ordóñez-García
Tecnológico Nacional de México, Instituto Tecnológico de Cd. Cuauhtémoc, Mexico

Yuridia Ortiz-Rivera
Programa de Biotecnología. Departamento de Ciencias Químico Biológicas, Instituto de Ciencias, Biomédicas, Universidad Autónoma de Ciudad Juárez. Av. Benjamín Franklin #4650, Zona, PRONAF, Ciudad Juárez, Chihuahua, México

Ginna Alejandra Ordoñez-Narvaez
Biotics Group. School of Basic Sciences, Technology and Engineering, Universidad Nacional Abierta y a Distancia – UNAD, Colombia

Ebber Addí Quintana-Obregón
CONAHCYT-Centro de Investigaciónen Alimentación y Desarrollo A.C. Sonora, México. Coordinación de Alimentos de Origen Vegetal, Hermosillo, Sonora, México

Aline Reyes-Díaz
Universidad Estatal de Sonora (UES), Unidad Académica Hermosillo, Av. Ley Federal del Trabajo, Sin Asignar en Nombre del Acuerdo, Sonora, México

Magda Piedad Valdés-Restrepo
Biotics Group. School of Basic Sciences, Technology and Engineering. Universidad Nacional Abierta y a Distancia – UNAD, Colombia

Raúl Rodríguez-Herrera
Universidad Autónoma de Coahuila, Facultad de Ciencias Químicas, Blvd. Venustiano Carranza y José Cárdenas Valdés, S/N, C.P., Saltillo, Coahuila, México

Emanuel Rivas-Robles
Universidad de Ciencias y Artes de Chiapas. Facultad de Ciencias de la Nutrición y Alimentos, Ingeniería en Agroalimentos, subsede Acapetahua, México

Aidé Saenz-Galindo
Universidad Autónoma de Coahuila, Facultad de Ciencias Químicas, Blvd. Venustiano Carranza y José Cárdenas Valdés, S/N, C.P., Saltillo, Coahuila, México

Gerardo Saucedo-Castañeda
Departamento de Biotecnología, Universidad Autónoma Metropolitana, Iztapalapa, Mexico City, Mexico

Alma Rosa Torres-Rodríguez
Universidad Autónoma de Coahuila, Facultad de Ciencias Químicas, Blvd. Venustiano Carranza y José Cárdenas Valdés, S/N, C.P., Saltillo, Coahuila, México

José Armando Ulloa
Centro de Tecnología de Alimentos. Universidad Autónoma de Nayarit. Ciudad de la Cultura Amado Nervo. Tepic, Nayarit, México

Andrea Vásquez-García
Biotics Group. School of Basic Sciences, Technology and Engineering. Universidad Nacional Abierta y a Distancia – UNAD, Colombia

Wendy Y. Villastrigo-López
Universidad Autónoma de Coahuila, Facultad de Ciencias Químicas, Blvd. Venustiano Carranza y José Cárdenas Valdés, S/N, C.P., Saltillo, Coahuila, México

Preface

This book addresses the reuse of food residue and by-product valorization of fruits, vegetables, and seeds through a survey of natural resource management and environmental issues that are at the heart of sustainability. It also provides theoretical and experimental methods used to study different aspects of valorization techniques for agri-food wastes and by-product management in parallel to conservation of nature.

This research-oriented volume presents sustainability challenges related to food wastes and by-products valorization along with emerging recovery technologies, to support circular economy concepts by managing the "natural" environment and its resources.

It also discusses the importance of green packaging for foods, fruits, vegetables, and agricultural products to limit the use of persistent plastics in the "natural" environment. These plastic packaging materials increase the risk of poor residue handling and environmental pollution as they are often discarded as waste. This book highlights the recent advancement toward the development of sustainable packaging materials, which could be biodegradable while promoting the conservation of natural resources.

Chapter 1 of this book opens up new possibilities to improve production by residue applications for controlling the negative impact of agro-industrial waste on the environment. In this chapter, shows how proper management of wastes could be very vital for the quality of human life. Hence, this chapter studies sustainable methods for innovative management of agro-industrial wastes.

Chapter 2 covers new methods for processing large quantities of agricultural wastes to obtain high value-added products in a short time. This chapter analyzes the agricultural waste used to produce biopesticides, biofertilizers, enzymes, and biosurfactants by solid-state culture. Chapter 3 reviews the possibility of the application of bioactive compounds collected from fruit and vegetable by-products for the development of functional foods.

Chapter 4 is devoted to investigate the potential of high hydrostatic pressures to improve the extraction of compounds of interest from plant tissues. It is shown that this method can be applied for utilizing waste, byproducts, or wastage generated in the value chain of fruits and vegetables. Chapter 5 reports process advancements about fruit seeds as a source of protein, including its functional properties and its enhancement by high-intensity ultrasound, and its utilization as valuable ingredients for the development of new foods.

Chapter 6 presents a detailed study on biotechnological valorization of cocoa waste, highlighting economic and environmental perspective. A case study for natural resource management is presented in Chapter 7 using Google Earth Engine. Plastic waste generation and management in agro-industrial packaging sector is discussed in Chapter 8.

In Chapter 9, ultrasound-assisted techniques for managing and reducing wastes are investigated in detail. Chapter 10 of this volume covers the application of computer vision technology-driven image-based classification for segregating mixed plastic waste.

CHAPTER 1

Agro-industrial Waste Management Practices for Sustainable Development

MAGALI ORDÓÑEZ-GARCÍA[1], JONATHAN A. DIAZ-BACA[2], and JUAN CARLOS BUSTILLOS-RODRÍGUEZ[1*]

[1]*Tecnológico Nacional de México, Instituto Tecnológico de Cd. Cuauhtémoc, Av. Tecnológico, Z.C., Mexico*

[2]*Faculty of Science and Environmental Studies, Lakehead University, Thunder Bay, ON, Canada*

*Corresponding author.

ABSTRACT

Throughout human history, our planet has presented abundance of natural resources that facilitated the development of civilizations. Nevertheless, every human activity generates waste to a greater or lesser extent, and as the population grows and demands for needs such as food and dress increase, the amount of waste generated also rises. This is causing significant harm to the environment and is becoming a serious problem for humanity. Every year, agricultural, forestry and industrial activities generate a considerable amount of waste, which causes serious environmental problems. However, these wastes contain numerous valuable compounds that can be exploited in diverse industries, making them a valuable resource. This practice opens up new possibilities to boost production by residue utilization and re-utilization while reducing the negative impact of agro-industrial waste on the environment. Adequate waste management practices are necessary to improve the quality of life worldwide, by preventing soil, water, and air pollution. Hence, this chapter highlights the investigation and implementation of current reliable and sustainable methods and approaches for managing agro-industrial wastes.

Agroindustrial Waste Management and Natural Resources Conservation: Environmental Challenges and Transformation into Value-Added Products. José Juan Buenrostro-Figueroa, Juan Manuel Tirado-Gallegos, Mónica L. Chávez-González, A. K. Haghi, and Shaheemath Suhara K. K. (Eds.)

1.1 INTRODUCTION

Population growth and economic development have led to a rise in demand for food production, which in turn has resulted in a significant increase in the generation of agro-industrial wastes.[1] Agro-industrial wastes are by-products of agriculture-based industries which are rich in bioactive compounds with great valorization potential. The type, quantity, and constituents of agro-industrial wastes mostly depend on the raw materials' characteristics, the nature of the products, the types of operations, and the processing steps involved.[2] Some examples of agro-industrial wastes are weeds, forest waste, crop residues, sawdust, leaf waste, livestock waste, food waste, crop production waste, animal production waste, and horticultural waste.[3] Unfortunately, these residues are left either untreated or unutilized and disposal is traditionally done through dumping into the land, burning, or landfilling. Improper management of these wastes leaves undesirable effects on the environment and economy of many nations.[4]

Nowadays, the accumulation of agro-industrial waste is over 2 billion tons worldwide, which makes it necessary to investigate how agro-industrial waste utilization can be advanced to the next phase to maximize benefits and to fully utilize these residues from the sector.[5] The objectives and targets set in different countries' legislations have been key drivers in improving waste management and stimulating innovation in recycling.[6] The exploitation of agro-industrial wastes could reduce the production cost of many other products, as well as reduce the generation of new wastes. This is possible due to many agro-industrial wastes present high organic load and varied nutritional composition (like high amounts of proteins, sugars, and minerals).[7]

These components can be used as feedstocks in large quantities or transformed through different waste management practices, which can reduce environmental pollution, the degradation of superficial water sources, avoid greenhouse gas generation, decrease health problems, and contribute to the sustainable and efficient use of resources, through novel approaches, eco-friendly, and sustainability.[8,9] Agro-industrial wastes also are important raw materials for the obtention of diverse products, as to produce new fuels and energy generation. Furthermore, they could be utilized for the development of functional foods for human or animal consumption. In general, the proper utilization of agro-industrial waste would promote its integral valorization and reincorporation into the food supply chain within the circular bioeconomy concept, creating revenue streams, business, and job opportunities.[10] Ultimately, an integrated approach to agro-industrial waste management could improve local and regional waste management systems and increase the environmental sustainability of the agroindustry considering the local, regional, and global context.

1.2 WASTE MANAGEMENT PRACTICES

The mass of agro-industrial waste produced in the world has been growing considerably for many decades, and it is an important situation that needs to be investigated and attended to give waste treatment or disposal alternatives.[11] This situation is caused mainly by tremendous amounts of agro-industrial food waste that usually are carelessly discarded due to the high cost associated with their treatment methods to reduce the organic load both from industries and households.[12] However, in the recent past, agro-industrial waste was considered neither a cost nor a benefit, nowadays this concept has been changing due to the emergence of several new technologies to reuse agro-industrial waste by adding value to it.[1] Currently, shortcomings of existing waste management practices are highlighted, being necessary and a conceptual framework for a centralized waste management system, should take into account the product lifecycle in reducing waste and enhancing its recovery.[13]

1.2.1 CURRENT WASTE MANAGEMENT PRACTICES

The managing practices of agro-industrial waste depend on the amount and composition of each waste, targeted product, and involved processes.[2] Several new environmentally friendly technologies such as supercritical fluid extraction (SFE), microwave-assisted extraction (MAE), and ultrasound-assisted extraction (UAE) have been studied and applied. Likewise, the energy generation from biomass, and the production of new biomaterials such as bio-nanofilms, bio-aerogels, hydrogels, and nanocomposites have been intensely studied. The production of biomaterials might be an effective path to achieve sustainable and efficient practices to manage agro-industrial wastes.[1] For instance, the new biofuels (e.g. biogas, biohydrogen, bioethanol, biodiesel) are perfect replacements for conventional fossil fuels. Biofuels are produced through diverse practices using biomass as their feedstock, including photobiological fermentation, anaerobic fermentation, enzymatic process, and microbial electrolysis, among others. The production of biofuel could help to solve the concerns associated with agro-industrial waste management.[9]

Furthermore, anaerobic digestion and fermentation are two of the most used methods for converting biomass into useful products. For instance, using microorganisms to biologically transform waste into biofuels such as bio-butanol, biogas, and bioethanol.[14] Moreover, the depolymerization of agro-industrial waste compounds into fermentable sugars to produce biochemicals has long been explored using enzymatic practices. One of the most used processes is enzymatic hydrolysis, which breaks down organic matter like biomass into simple sugars in a solvent-free and chemicals-reduced environment.[6] In the same way, green extraction and purification methods can be used to obtain natural products from agro-industrial waste, by reducing energy consumption and using renewable materials instead of traditional solvents.[15]

1.2.2 LIMITATIONS OF TRADITIONAL WASTE MANAGEMENT PRACTICES

One of the major challenges facing the world at present is the utilization of by-products, hence, at present, one major concern is the development of efficient and green technology for the exploitation of agro-industrial waste, thus contributing to their reduction, as well as long-term improvements in environmental quality and enhancement of extractability of interest compounds.[6] Although waste management efforts have significantly improved over the past decades, there are still many opportunities for improvement. Recent efforts to improve waste management practices are often limited to better treatment of waste through prominent principles like the 3Rs (Reduce, Reuse, and Recycle). Despite significant efforts being made, the full potential value of waste is not being maximized.[16]

Nowadays, one of the main challenges related to the management practices of agro-industrial waste is the reduction of costs associated with their transport. The generation of enormous waste mass requires huge economic expenses for transportation, transformation, and purification.[10] Besides, conventional practices generally involve large amounts of organic solvents, high energy consumption, and are time-consuming. The current trend is for traditional extraction methods to be gradually replaced by new, environmentally friendly ones.[1] In this sense, the heterogeneity of agro-industrial waste and its complex biochemical composition often makes it difficult to use a single production process to be successful for their complete utilization.[12]

Various processes have been proposed for addressing the aforementioned issues. Such technologies can be classified as biological or physicochemical. The biological processes are quite sensitive concerning the inlet feedstock as, potentially, microorganisms could be easily inhibited by toxic substances present in the substrate. Additionally, they need long incubation times (e.g. days or weeks). In contrast, physicochemical processes generally last from a few minutes to a few hours, but they may involve the use of toxic agents on occasion.[17]

1.3 THE PROBLEM OF AGRO-INDUSTRIAL WASTE

The agro-industrial processing and production causes the generation of large amounts of waste in various solid, liquid, or gaseous forms. Such waste can cause economic problems for its disposal, as well as environmental contamination.[18] As a result, agro-industrial waste is one of the main sources of pollution. These wastes can come from agricultural, forestry, or industrial activities. Agricultural and forestry activities include livestock slurry, manure, crop remains, and waste from pruning and the maintenance of woodlands, respectively. Industries generate organic wastes, which include the by-products of the agri-food industries such as coffee dregs, bagasse, degummed fruits and legumes, milk serum, and sludge from wool and cellulose, among others.[19]

1.3.1 THE IMPACT OF POPULATION GROWTH AND INCREASING DEMANDS ON WASTE GENERATION

The growing population has increased demand for food processing, but this generates significant waste, causing environmental and public health issues.[20] These wastes can produce adverse effects on the environment like an increase in temperature and greenhouse gases.[4]

1.3.2 THE ENVIRONMENTAL PROBLEMS CAUSED BY AGRO-INDUSTRIAL WASTES

Within agro-industrial waste, it is possible to find those of animal origin and plant origin. For instance, between 50 and 70% of seafood raw materials are discarded annually due to the low utilizable mass of the animal sources.[9]

The agro-industrial wastes are rich in nutrients and have been widely considered as a potential nutrient source for the cultivation of microalgae/duckweed. However, they commonly carry various hazardous contaminants such as heavy metals, metalloids, pathogens (e.g. bacteria, viruses, parasites), and xenobiotics (e.g., hormones, antibiotics, parasiticides), which could taint the produced biomass. Fortunately, these wastes can be eliminated using various technologies, such as anaerobic digestion.[21]

1.4 THE POTENTIAL OF AGRO-INDUSTRIAL WASTE

Agri-food industries are known to produce a considerable amount of waste such as shells, seeds, and bagasse. As a result, various industries have begun to explore ways to repurpose these wastes, to avoid their accumulation and reduce production costs and environmental pollution. Among the multiple applications that have been investigated for agro-industrial waste are the generation of energy from biomass and the generation of new materials such as bio-nanofilms and bio-nanocomposites. Many of these biomaterials have been obtained through the use of various green technologies such as SFE, MAE, and UAE.[1]

1.4.1 THE PRESENCE OF NUTRIENTS AND VALUABLE COMPOUNDS IN AGRO-INDUSTRIAL WASTES

Agro-industrial wastes are mainly composed of sugars, fibers, proteins, and minerals. The major components of agro-industrial wastes are cellulose, hemicellulose, and lignin. Because of their composition, these materials are called "lignocellulosic materials."[24] The percentage of the different components in the agro-industrial waste depends on the plant source. In general, lignocellulosic waste is inexpensive, renewable, abundant, and provides a unique natural resource.[25] It has also been proven that lignocellulosic agro-industrial waste (e.g. coconut fiber, corn cob, spent grain, spent coffee, husk, husk ash, and straw rice, soybean, and wheat bran) and not lignocellulosic by-products (e.g. eggshell and eggshell membranes) have great potential for the enzyme immobilization which contributes to generating more economical biocatalysts.[26] Despite causing pollution and health problems, agro-industrial waste is a valuable source of bioactive compounds including fatty acids, phenolic compounds, terpenes, peptides, melanin, and β-glucans.[20,22] Many of these bioactive molecules have shown antioxidant and antimicrobial properties.[20] Likewise, these bioactive compounds have proven excellent nutraceutical functions.[23]

1.4.2 THE POTENTIAL FOR REUSE AND EXPLOITATION OF RESIDUES

The utilization of agro-industrial waste can help prevent environmental pollution and ecological damage. Therefore, there has been an increased interest in exploring the best ways to use such waste.[23] Various studies have been carried out in which the high potential of agro-industrial waste has been demonstrated. These wastes are used as compost, animal feed, and biofuel production and can be used as starting substrates in biorefinery.[27] In this sense, it has been observed that the use of waste from different foods can be used as raw material to produce biodiesel and bioethanol. Among the foods used are rice, coffee, sugar cane, fruits, and vegetables.[9]

In Malaysia has been observed that wastewater and agro-industrial waste can be used as a supply of nutrients for microalgae which can be used for the production of biodiesel.[28] One of the ways to use agro-industrial waste is by producing biogas through the use of microorganisms and an anaerobic digestion reaction. This process can help reduce greenhouse gas emissions and environmental problems. To achieve this, various methods are used, and it is essential to control parameters such as digestion temperature, power supply, pH level, energy losses, gate valves, and pumps.[3] It has been shown that biogas can be used efficiently in the generation of electrical and thermal energy, which makes it an alternative renewable energy source that can be used for bioenergy generation.[29]

In addition, production of biogas, and agro-industrial waste has been used for the production of pigments. Panesar et al.[18] reviewed the waste that is used to produce microbial pigments, observing that microorganisms such as bacteria, and fungi, among others, have the ability to produce different pigments depending on the source or type of waste. It has also been shown that agro-industrial waste can be used for the production of *Saccharomyces cerevisiae* and the biomass produced can be used in food formulations due to its high protein content.[30] In a study carried out in Mexico, it was observed that waste from coconut, pineapple, and tuna fruits can be used to promote the synthesis of bioethanol by yeast *Saccharomyces cerevisiae*, observing that the highest concentration of ethanol is obtained with coconut juice.[31]

1.5 NEW TRENDS FOR MANAGING AGRO-INDUSTRIAL WASTES

Traditionally, agro-industrial wastes were viewed as problematic residues because of the environmental impact of their disposal and the expenses associated with their removal.[5] However, nowadays there is a transition from traditional methods such as incineration, composting, and animal feeding, to novel technologies focused on reusing agro-industrial waste by recovering value-added compounds.[32] In this regard, the use of agro-industrial wastes for the development of new products has evolved significantly in the last decade.[33] These new trends aimed to improve the overall sustainability of the production chain through the circular economy principles. The focus for the development of such technologies is driven by the need to gradually replace non-renewable sources with more sustainable ones.

Newly developed methods for the utilization of agro-industrial waste take green and sustainable tools as an alternate approach. The aim is to limit the use of undesired large volumes of organic solvents and reduce excessive energy consumption and time spent.[34] In this sense, the current technologies are not fully developed yet. More efficient processes are to be accomplished for the recovery, purification, and concentration of valuable compounds obtained from agro-industrial waste.[35] High-value-added products such as chemical reagents, functional and bioactive molecules, and biopolymers are the target for the design of new recovery methodologies, particularly at the industrial and commercial levels.

Another important aspect of the use of agro-industrial wastes is the generation of bioenergy. This sector is currently more advanced and developed in comparison with those focusing on the obtention of value-added products.[5] There are several companies dedicated to the production of bioenergy from lignocellulosic materials. Companies such as Lignol Innovations Inc, Verenium Corporation, and Lalleman use sustainable and renewable processes to obtain different materials from a variety of biomass feedstocks that will be used as a source of energy production (e.g. cellulosic ethanol). However, the generation of bioenergy or energy sources is outside the scope of this chapter. Much relevant literature regarding this is available for consultation.[4,36]

In the upcoming years, it is expected to have a larger number of agro-industrial wastes generated due to the increasing demand and production of commodities (e.g. extensive crops, and food products).[37] For this reason, it is a priority to develop more efficient processes that allow the transformation of agro-industrial waste into high-added-value molecules which would be a further part of high-end products such as food supplements, drugs, fertilizers, and biomaterials.[38]

1.5.1 RESEARCH AND DEVELOPMENT OF SUSTAINABLE TREATMENTS OF AGRO-INDUSTRIAL WASTE

It is important to follow specific steps to properly valorize agro-industrial waste.[34] First, it is required to have a complete characterization of the material to identify the target value-added product. Second, it is required to carry out an efficient extraction of the value-added products through adequate downstream phases.[4] Third, it is of great significance to assess the different properties and characteristics of the extracted compound which will allow tailored potential applications. Lastly, the methodologies developed at the laboratory level must be scaled to pilot and industrial levels.[5]

This section explores the latest laboratory-based techniques for repurposing agro-industrial waste. The focus is on their cost-effectiveness, utilization of green chemistry tools, and adherence to circular economy principles. The purpose of this section is to compare and contrast the different methods to better understand their benefits and limitations.

1.5.2 PHYSICOCHEMICAL TREATMENT

Various physicochemical methodologies have been developed to increase the effectiveness of treating agro-industrial waste for extracting valuable compounds. Among those methods, SFE, subcritical water extraction (SWE), UAE, MAE, and deep eutectic solvent technology, are the most studied. In general, these methods are environmentally friendly and yield high extraction rates[39–40]

1.5.2.1 SUPERCRITICAL FLUID EXTRACTION

Supercritical fluid extraction (SFE) is considered the ideal candidate for replacing solvent-based extraction methods. SFE consists of separating the component which will be extracted from a matrix using supercritical fluids as the extracting solvent.[41] These fluids are at a pressure and temperature above their critical point, which allows them to diffuse through the matrix behaving as a liquid and gas.[42] The extraction usually takes place from a solid matrix but can also be performed from liquids. Carbon dioxide (CO_2) is the most used supercritical fluid, sometimes modified by co-solvents such as methanol, ethanol, or water. SFE is commonly employed for the extraction of bioactive compounds from cereal wastes (i.e. mango peels, soybean, and brewing residues).[43–45] Oils, polyphenols, β-glucans, proteins, and nucleotides are among the types of value-added products that can be obtained through SFE. The advantages of using SFE include high yields of extraction, specificity, no extra separation steps, and preserving the chemical structure and functionality of the bioactive during the extraction process.[46–47] However, the major disadvantage is the elevated cost of the process, particularly increased by the required equipment.[39,48]

1.5.2.2 SUBCRITICAL WATER EXTRACTION

Subcritical water extraction (SWE) and subcritical water hydrolysis (SWH) are efficient water-based green techniques useful for the extraction of bioactive compounds through solubilization and for hydrolysis of large molecules (e.g. lipids, carbohydrates, and proteins) into smaller fractions, respectively.[49] SWE is designed for extracting compounds with low polarity in short periods of extraction time, around 30 min. The fundamental of the procedure consists of keeping the water in a liquid state under high pressure and a temperature range of 100–374°C. Under those conditions, water exhibits a lower dielectric constant which allows the disruption of the hydrogen bonds and behaves more like a less polar solvent (e.g. methanol or ethanol).[50] Because of those attributes, SWE can be used for extracting intracellular-entrapped compounds from different agro-industrial residues such as marine waste, oilseed by-products, or grain husks.[51–54] The use of SWE has been effective in producing saponins, phenolic compounds, and organic, amino, and fatty acids.[15] The temperature can be varied to selectively extract different types of compounds. Higher temperatures are required for less polar compounds, while lower temperatures are needed for more polar compounds.[55] However, an appropriate selection of operating conditions is crucial to maximize yield while avoiding degradation and decomposition of products.

1.5.2.3 MICROWAVE-ASSISTED EXTRACTION

The microwave-assisted extraction (MAE) is a process used to remove solutes from solid-state residues into a solvent.[56] This process involves electromagnetic, heat, mass, and momentum transfer. By using microwave radiation, which can penetrate and combine with a substrate, an accurate and controlled heat transfer is possible.[57] The microwave radiations increase the motion of the intracellular water molecules, which causes evaporation and cell wall rupture.[58] MAE is considered an economical and relatively simple technique for the extraction of value-added products from agro-industrial residues such as oilseed residues, grain brans, crop stalks and steams, and fruit peels.[58–61] Typically, MAE reduces the time extraction (e.g. ten times faster) and uses fewer volumes of solvent in comparison to more traditional methods such as Soxhlet extraction.[60,62] Furthermore, MAE has a low degradation rate on the extracted compounds.[63] However, it is important to monitor for undesired chemical reactions such as epimerization, oxidation, and polymerization during the extraction of desired products such as fatty acids, proteins, carbohydrates, betalains, and phenolic compounds.[64,65]

1.5.2.4 ULTRASOUND-ASSISTED EXTRACTION

The ultrasound-assisted extraction (UAE) is also a fast and very effective method for the extraction of value-added products. UAE is a technique that uses mechanical energy generated by ultrasound to extract various intracellular components from agro-industrial wastes such as tree pruning biomass,[66] fruit pomace, bagasse, and peels,[67,68] grains, and seed residues.[69,70] Solid particles of the treated waste should be dispersed in a proper solvent to be further ultrasonicated. The transition of ultrasonic waves creates an acoustic cavitation in the solvent which leads to the breakup of the cell wall or membrane. After the cell is collapsed, the solvent can easily solubilize and extract the intracellular components from the matrix.[58] Valuable compounds such as proteins, polysaccharides, polyols, glucosinolates, carotenoids, flavonoids, anthocyanins, and phenolic compounds can be extracted using UAE.[71–74] UAE is a relatively non-costly procedure with low instrumental requirements. A major advantage of UAE is that it does not generate heat. Thus, unlike other temperature-dependent methods, UAE is an ideal method to separate heat-labile substances without altering their properties. In contrast, a major disadvantage of UAD is its dependency on solvents, because one of the main factors influencing the efficiency of the extraction is the solvent selection. Furthermore, scaling up UAE might be problematic, as recent finding suggests.[75]

1.5.2.5 DEEP EUTECTIC SOLVENTS (DESS)

Another recently developed technology is the utilization of DESs for the extraction of valuable compounds from agro-industrial wastes. DESs are a type of ionic fluid, composed of Lewis or Brønsted acids and bases which form a eutectic mixture.[76] This eutectic mixture has lower lattice energy and lower melting points compared to each component individually.[76] DESs are a cost-effective, safe, and environmentally friendly alternative to other solvents for the treatment of different agro-industrial wastes such as coconut husks, tamarind seeds, rice straws, and oil palm trunks.[77] Their non-flammability, non-toxicity, accessibility, and ease of use make DESs a suitable choice for scaling-up processes. Usually, DESs are synthesized using bulk chemicals from natural sources, making them accessible and biodegradable. Among the many available DESs, Choline Chloride/Urea, Choline Chloride/Glycerol, Malic Acid/Glucose, and Choline Chloride/Sugars/Polyols are some commonly used mixtures.[78,79] Using DESs is an effective method for extracting important compounds such as lignin, pectin, cellulose, hemicellulose, polyphenols, and antioxidants from plant materials.[80,81] By treating lignocellulosic and pectinocellulosic biomass with DESs under mild operating conditions (temperature, pressure, and time), the lignin or pectin portions are dissolved, which frees up the cellulose chains and makes them available for further processing.[80,82] DESs are highly adaptable through varying the type, relative ratio, and components and thus have potential application in a wide variety of agro-industrial wastes. However, further research is necessary to determine the optimal biomass-to-solvent ratio for each residue.

1.5.3 BIOLOGICAL TREATMENT

The biological treatment of agro-industrial residues involves a range of methodologies, each offering sustainable and efficient solutions for agro-industrial waste management while facilitating the efficient extraction of valuable compounds.[83] Various methods such as submerged-state fermentation, solid-state fermentation (SSF), enzymatic-assisted extraction, and protein precipitation by polysaccharides are among the most explored techniques. These procedures are environmentally friendly, and selective, and yield high-quality extracted products.[84] This makes the biological treatments effective for waste valorization and the production of high-value-added products and bioactive compounds.

1.5.3.1 *SUBMERGED FERMENTATION (SMF)*

Submerged Fermentation (SmF) is a process that involves the growth of microorganisms in a free-flowing liquid medium. The process involves immersing a specific strain of microorganisms such as yeast, bacteria, or fungi in a liquid culture medium. Those microorganisms grow and multiply in this medium, metabolizing the nutrients that produce the bioactive molecules as part of their metabolic processes. Nowadays, around 90% of microbial-originated molecules are produced through SmF.[85,86] Since SSF is carried out in a liquid medium, it makes homogenization and heat transfer easier. This results in higher amounts of bioactive molecules recovered, lower production costs, and more stable molecules. However, it is important to consider factors such as the substrate composition, the addition of growth factors, and the supply of oxygen to the bioreactor. There are two common methods for performing SmF: batch-fed fermentation and continuous fermentation. In batch-fed fermentation, sterilized growth nutrients are added to the culture. On the other hand, in continuous fermentation, liquid nutrients are slowly and continuously added to the bioreactor at the same rate at which the bioactive compounds are being recovered from the system.

In terms of agro-industrial residues, SmF has been used for the production of bioactive compounds such as enzymes (lipases, laccases, inulinases, cellulases, pectinases), proteins, and pigments.[87] These bioactive compounds have been obtained through SmF using a variety of agro-industrial residues as substrates, including fruit wastes, cereal and oilseed brans, sugarcane bagasse, tofu dreg, distillers' and brewers' spent grains.[88] Various strains of yeast, bacteria, and fungi, such as *Saccharomyces*, *Kluyveromyces*, *Rheinheimera*, *Chromobacterium*, *Mucor*, and *Monascus*, have been commonly utilized for the fermentation of those agro-industrial wastes.

1.5.3.2 *SOLID-STATE FERMENTATION*

Over the last five decades, solid-state fermentation (SSF) has undergone significant development, with a particular boost since 2010.[33] Although SSF is a well-known methodology to produce enzymes or antibiotics, lately, it has been explored to produce other bioactive compounds with the purpose of agro-industrial waste valorization. Novel products obtained through SFF include organic acids, antioxidants, biosurfactants, biopolymers, and biofuels.[89–91] These biomolecules are produced mostly as metabolites generated by microorganisms.[92] The microorganisms are grown on a natural substrate which functions as a carbon and energy source in a minimal-water environment.[92]

The agro-industrial residues used as substrates in SSF can be classified into four major groups depending on their main components: lignocellulose, protein, soluble carbohydrates, and starch.[33] The selection of an appropriate microorganism strain for a determined substrate as well as the process parameters such as temperature and time are crucial factors that affect SSF process efficiency. For instance, only microorganisms capable of growing under low moisture conditions can be used (*Aspergillus*, *Penicillium*, and *Trichoderma*).[93] Moreover, precise conditions and constant regulation of the fermentation process such as heat transfer are necessary to maximize the yield and the downstream of the produced compounds.[94] Despite these challenges, this technology is extensively used because of its low water and energy consumption, and the minimal residual production.

1.5.3.3 ENZYMATIC-ASSISTED EXTRACTION (EAE)

Enzymatic-assisted extraction (EAE) is another biological approach to obtaining value-added products from agro-industrial waste. EAE utilizes specific enzymes to disrupt biomass cell walls, enabling the extraction of bioactive and industry-relevant compounds, leading to higher yields. Frequently, EAE is combined with other method methods to enhance the overall recovery of bioactive molecules.[95] For instance, microwave or ultrasound extraction is combined with EAE to improve the recovery of value-added products from agro-industrial residues.

The most used types of enzymes for the extraction of bioactive molecules are hydrolytic cellulases, hemicellulases, pectinases, and proteases.[96] In this regard, parameters such as temperature, extraction time, pH, substrate availability, and enzymatic mechanism must be studied for the adequate selection of enzyme-residue combination.[97] Moreover, when using complex matrices such as agro-industrial residues as substrate, it is recommended to use an enzymatic cocktail (i.e. a mixture of different types of enzymes) instead of a single-type medium. The incorporation of enzymes in the solid-liquid extraction method enhances the overall operation, reducing the use of organic solvents and improving the biomolecule recovery.[98] Throughout this process, different value-added products, such as carbohydrates, proteins, oils, phenols, and carotenoids, can be extracted from lignocellulosic, fruit, grain, oilseed, brewing, and dairy industry residues.[97, 99]

EAE has several advantages compared to non-biological processes. It allows for the obtention of high-quality extracts with significant bioactivity.[100] Additionally, the EAE process is generally considered to have low toxicity and minimal environmental impact. However, the EAE process can be expensive and difficult to scale up, due to the high cost of enzymatic solutions, complexity of the agro-industrial residues, and considerable water requirements.[101]

1.5.4 FUTURE PERSPECTIVES

Tight collaboration between academia and industry is essential for developing effective strategies for scaling up the production of value-added products from agro-industrial residues. It is crucial to increase the efficiency of the different extraction methods, as well as the quality of the extracted bioactive compounds from such residues. If quality meets commercial standards, the bioactive compound from agro-industrial wastes can be utilized as food additives or dietary supplements. In this regard, the development of functional foods can also benefit from the evaluation of these compounds. The effective valorization of waste streams from specific industries by connecting them to agro-based businesses can also help alleviate the issue of waste accumulation. One promising approach is the utilization of enzymatic technologies combined with physicochemical methods to produce higher quality and volume value-added products from agro-industrial waste.

1.6 CONCLUSION

The accumulation of agro-industrial waste represents a social, environmental, and economic problem that threatens various sectors. Despite this, agro-industrial waste has shown the potential to be an important source of bioactive compounds. Numerous technological advances have permitted that agro-industrial waste has multiple applications, like, biofuel and pigment production, among others. The focus on the development of technologies is driven by the need to gradually replace non-renewable sources with more sustainable ones. This technology is based on low water and energy consumption, and minimal residual production. Effective and novel practices of agro-industrial waste treatment are currently an active field of research interest.

KEYWORDS

- **environment**
- **exploitation**
- **reuse**
- **technology**
- **value-added**

REFERENCES

1. Freitas, L. C.; Rodrigues Barbosa, J.; Caldas da Costa, A. L.; Figueiredo Bezerra, F. W.; Holanda Pinto, R. H.; Nunes de Carvalho Junior, R. From Waste to Sustainable Industry: How can Agro-industrial Wastes Help in the Development of New Products? *Resour. Conserv. Recycl.* **2021,** *169,* 105466.
2. Senthilkumar, K.; Naveen Kumar, M.; Chitra Devi, V.; Saravanan, K.; Easwaramoorthi, S. Agro-industrial Waste Valorization to Energy and Value Added Products for Environmental Sustainability. In *Biomass Valorization to Bioenergy*; Praveen Kumar, R.; Bharathiraja, B.; Kataki, R.; Moholkar, V. S., Eds; Springer Singapore: 2020; pp 1–9.
3. Devi, M. K.; Manikandan, S.; Oviyapriya, M.; Selvaraj, M.; Assiri, M. A.; Vickram, S.; Subbaiya, R.; Karmegam, N.; Ravindran, B.; Chang, S. Recent Advances in Biogas Production Using Agro-Industrial Waste: A Comprehensive Review Outlook of Techno-Economic Analysis. *Bioresour. Technol.* **2022,** *363,* 127871.
4. Yaashikaa, P.; Kumar, P. S.; Varjani, S. Valorization of Agro-industrial Wastes for Biorefinery Process and Circular Bioeconomy: A Critical Review. *Bioresour. Technol.* **2022,** *343,* 126126.
5. Singh, R.; Das, R.; Sangwan, S.; Rohatgi, B.; Khanam, R.; Peera, S. P. G.; Das, S.; Lyngdoh, Y. A.; Langyan, S.; Shukla, A. Utilisation of Agro-industrial Waste for Sustainable Green Production: A Review. *Environ. Sustain.* **2021,** *4* (4), 619–636.
6. Radenkovs, V.; Juhnevica-Radenkova, K.; Górnaś, P.; Seglina, D. Non-waste Technology Through the Enzymatic Hydrolysis of Agro-industrial by-products. *Trends Food Sci. Technol.* **2018,** *77,* 64–76.
7. Sadh, P. K.; Duhan, S.; Duhan, J. S. Agro-industrial Wastes and Their Utilization using Solid State Fermentation: A Review. *Bioresour. Bioprocess.* **2018,** *5* (1), 1–15.
8. Diaz-Vazquez, D.; Carrillo-Nieves, D.; Orozco-Nunnelly, D. A.; Senés-Guerrero, C.; Gradilla-Hernández, M. S. An Integrated Approach for the Assessment of Environmental Sustainability in Agro-industrial Waste Management Practices: The Case of the Tequila Industry. *Front. Environ. Sci.* **2021,** *9,* 682093.
9. Nair, L. G.; Agrawal, K.; Verma, P. An Overview of Sustainable Approaches for Bioenergy Production from Agro-industrial Wastes. *Energy Nexus.* **2022,** *6,* 100086.

10. Gómez-García, R.; Campos, D. A.; Aguilar, C. N.; Madureira, A. R.; Pintado, M. Valorisation of Food Agro-industrial by-products: From the Past to the Present and Perspectives. *J. Environ. Manage.* **2021,** *299,* 113571.
11. Giusti, L. A Review of Waste Management Practices and Their Impact on Human Health. *Waste manage.* **2009,** *29* (8), 2227–2239.
12. Kumar, V.; Sharma, N.; Umesh, M.; Selvaraj, M.; Al-Shehri, B. M.; Chakraborty, P.; Duhan, L.; Sharma, S.; Pasrija, R.; Awasthi, M. K. Emerging Challenges for the Agro-industrial Food Waste Utilization: A Review on Food Waste Biorefinery. *Bioresour. Technol.* **2022,** 127790.
13. Esmaeilian, B.; Wang, B.; Lewis, K.; Duarte, F.; Ratti, C.; Behdad, S. The Future of Waste Management in Smart and Sustainable Cities: A Review and Concept Paper. *Waste manage.* **2018,** *81,* 177–195.
14. Manikandan, S.; Vickram, S.; Sirohi, R.; Subbaiya, R.; Krishnan, R. Y.; Karmegam, N.; Sumathijones, C.; Rajagopal, R.; Chang, S. W.; Ravindran, B. Critical Review of Biochemical Pathways to Transformation of Waste and Biomass Into Bioenergy. *Bioresour. Technol.* **2023,** 128679.
15. Zuin, V. G.; Ramin, L. Z. Green and Sustainable Separation of Natural Products from Agro-industrial Waste: Challenges, Potentialities, and Perspectives on Emerging Approaches. *Chem. Chem. Technol.* **2018,** 229–282.
16. Romero-Hernández, O.; Romero, S. Maximizing the Value of Waste: From Waste Management to the Circular Economy. *Thunderbird Int. Bus. Rev.* **2018,** *60* (5), 757–764.
17. Basso, D.; Patuzzi, F.; Castello, D.; Baratieri, M.; Rada, E. C.; Weiss-Hortala, E.; Fiori, L. Agro-industrial Waste to Solid Biofuel Through Hydrothermal Carbonization. *Waste Manage.* **2016,** *47,* 114–121.
18. Panesar, R.; Kaur, S.; Panesar, P. S. Production of Microbial Pigments Utilizing Agro-industrial Waste: A Review. *Curr. Opin. Food Sci.* **2015,** *1,* 70–76.
19. Yusuf, M. Agro-industrial Waste Materials and Their Recycled Value-added Applications. In *Handbook of Ecomaterials*; Torres Martínez, L. M.; Vasilievna Kharissova, O.; Ildusovich Kharisov, B., Eds; Springer Cham, 2017; Vol. 1, pp 1–11.
20. Leyva-López, N.; Lizárraga-Velázquez, C. E.; Hernández, C.; Sánchez-Gutiérrez, E. Y. Exploitation of Agro-industrial Waste as Potential Source of Bioactive Compounds for Aquaculture. *Foods.* **2020,** *9* (7), 843.
21. Markou, G.; Wang, L.; Ye, J.; Unc, A. Using Agro-industrial Wastes for the Cultivation of Microalgae and Duckweeds: Contamination Risks and Biomass Safety Concerns. *Biotechnol. Adv.* **2018,** *36* (4), 1238–1254.
22. Vodnar, D. C.; Călinoiu, L. F.; Mitrea, L.; Precup, G.; Bindea, M.; Păcurar, A. M.; Ştefănescu, B. E. A New Generation of Probiotic Functional Beverages Using Bioactive Compounds from Agro-industrial Waste. In *Functional and Medicinal Beverages*; Grumezescu, A. M.; Holban, A. M., Eds; Academic Press, 2019; pp 483–528.
23. Beltrán-Ramírez, F.; Orona-Tamayo, D.; Cornejo-Corona, I.; González-Cervantes, J. L. N.; de Jesús Esparza-Claudio, J.; Quintana-Rodríguez, E. Agro-industrial Waste Revalorization: the Growing Biorefinery. In *Biomass for Bioenergy-recent Trends and Future Challenges*, El-Fatah Abomohra, A., Ed. 2019; pp 83–102.
24. Mussatto, S. I.; Ballesteros, L. F.; Martins, S.; Teixeira, J. A. Use of Agro-industrial Wastes in Solid-State Fermentation Processes. *Industrial Waste* **2012,** 274.
25. Anwar, Z.; Gulfraz, M.; Irshad, M. Agro-industrial Lignocellulosic Biomass a Key to Unlock the Future Bio-energy: A Brief Review. *JRRAS.* *7* (2), 163–173.

26. Girelli, A. M.; Astolfi, M. L.; Scuto, F. R. Agro-industrial Wastes as Potential Carriers for Enzyme Immobilization: A Review. *Chemosphere.* **2020,** *244,* 125368.
27. Taurisano, V.; Anzelmo, G.; Poli, A.; Nicolaus, B.; Di Donato, P. Re-use of Agro-Industrial Waste: Recovery of Valuable Compounds by Eco-friendly Techniques. *Int. J. Perform. Eng.* **2014,** *10* (4), 419.
28. Jayakumar, S.; Yusoff, M. M.; Rahim, M. H. A.; Maniam, G. P.; Govindan, N. The Prospect of Microalgal Biodiesel Using Agro-industrial and Industrial Wastes in Malaysia. *Renew. Sust. Energ. Rev.* **2017,** *72,* 33–47.
29. Dalpaz, R.; Konrad, O.; da Silva Cyrne, C. C.; Barzotto, H. P.; Hasan, C.; Guerini Filho, M. Using Biogas for Energy Cogeneration: An Analysis of Electric and Thermal Energy Generation from Agro-industrial Waste. *Sustain. Energy Technol. Assess.* **2020,** *40,* 100774.
30. Bacha, U.; Nasir, M.; Khalique, A.; Anjum, A. A.; Jabbar, M. A. Comparative Assessment of Various Agro-industrial Wastes for *Saccharomyces Cerevisiae* Biomass Production and its Quality Evaluation as Single Cell Protein. *J. Anim. Plant Sci.* **2011,** *21* (4), 844–849.
31. Domínguez-Bocanegra, A. R.; Torres-Muñoz, J. A.; López, R. A. Production of Bioethanol from Agro-industrial Wastes. *Fuel* **2015,** *149,* 85–89.
32. Naik, B.; Kumar, V.; Rizwanuddin, S.; Chauhan, M.; Gupta, A. K.; Rustagi, S.; Kumar, V.; Gupta, S. Agro-industrial Waste: A Cost-effective and Eco-friendly Substrate to Produce Amylase. *Food Prod. Process. Nutr.* **2023,** *5* (1), 30.
33. Yafetto, L. Application of Solid-state Fermentation by Microbial Biotechnology for Bioprocessing of Agro-industrial Wastes from 1970 to 2020: A Review and Bibliometric Analysis. *Heliyon.* 2022, 8, e09173.
34. Fernández-Ochoa, Á.; Leyva-Jiménez, F. J.; Pimentel-Moral, S.; del Carmen Villegas-Aguilar, M.; Alañón, M. E.; Segura-Carretero, A.; de la Luz Cádiz-Gurrea, M. Revalorisation of Agro-Industrial Wastes into High Value-Added Products. In *Sustainable Bioconversion of Waste to Value Added Products*; Inamuddin; Khan, A., Eds; 2021; pp 229–245.
35. Caballero, E.; Soto, C. Valorization of Agro-industrial Waste into Bioactive Compounds: Techno-economic Considerations. In *Biorefinery: Integrated Sustainable Processes for Biomass Conversion to Biomaterials, Biofuels, and Fertilizers*; Schmidt, J. E.; Bastidas-Oyanedel, J.-R., Eds; 2019; pp 235–252.
36. Lee, S. Y.; Sankaran, R.; Chew, K. W.; Tan, C. H.; Krishnamoorthy, R.; Chu, D.-T.; Show, P.-L. Waste to Bioenergy: A Review on the Recent Conversion Technologies. *BMC Energy* **2019,** *1* (1), 1–22.
37. Peng, X.; Jiang, Y.; Chen, Z.; Osman, A. I.; Farghali, M.; Rooney, D. W.; Yap, P.-S. Recycling Municipal, Agricultural and Industrial Waste into Energy, Fertilizers, Food and Construction Materials, and Economic Feasibility: A Review. *Environ. Chem. Lett.* **2023,** *21* (2), 765–801.
38. Almaraz-Sánchez, I.; Amaro-Reyes, A.; Acosta-Gallegos, J. A.; Mendoza-Sánchez, M. Processing Agroindustry By-Products for Obtaining Value-Added Products and Reducing Environmental Impact. *J. Chem.* **2022,** 3656932, 13.
39. Fărcaş, A. C.; Socaci, S. A.; Nemeş, S. A.; Salanţă, L. C.; Chiş, M. S.; Pop, C. R.; Borşa, A.; Diaconeasa, Z.; Vodnar, D. C. Cereal Waste Valorization through Conventional and Current Extraction Techniques—An Up-to-Date Overview. *Foods* **2022,** *11* (16), 2454.

40. Mishra, K.; Siwal, S. S.; Nayaka, S. C.; Guan, Z.; Thakur, V. K. Waste-to-chemicals: Green Solutions for Bioeconomy Markets. *Sci. Total Environ.* **2023,** 164006.
41. Herrero, M.; Mendiola, J. A.; Cifuentes, A.; Ibáñez, E. Supercritical Fluid Extraction: Recent Advances and Applications. *J. Chromatogr. A.* **2010,** *1217* (16), 2495–2511.
42. Rostagno, M. A.; Villares, A.; Guillamón, E.; García-Lafuente, A.; Martínez, J. Sample Preparation for the Analysis of Isoflavones from Soybeans and Soy Foods. *J. Chromatogr. A.* **2009,** *1216* (1), 2–29.
43. Alvarez, M. V.; Cabred, S.; Ramirez, C. L.; Fanovich, M. A. Valorization of an Agroindustrial Soybean Residue by Supercritical Fluid Extraction of Phytochemical Compounds. *J. Supercrit. Fluids.* **2019,** *143,* 90–96.
44. del Pilar Sanchez-Camargo, A.; Gutierrez, L.-F.; Vargas, S. M.; Martinez-Correa, H. A.; Parada-Alfonso, F.; Narvaez-Cuenca, C.-E. Valorisation of Mango Peel: Proximate Composition, Supercritical Fluid Extraction of Carotenoids, and Application as an Antioxidant Additive for an Edible Oil. *J. Supercrit. Fluids.* **2019,** *152,* 104574.
45. Olivares-Galván, S.; Marina, M.; García, M. Extraction of Valuable Compounds from Brewing Residues: Malt Rootlets, Spent Hops, and Spent Yeast. *Trends Food Sci Technol.* **2022,** *127,* 181–197.
46. Fraguela-Meissimilly, H.; Bastías-Monte, J. M.; Vergara, C.; Ortiz-Viedma, J.; Lemus-Mondaca, R.; Flores, M.; Toledo-Merma, P.; Alcázar-Alay, S.; Gallón-Bedoya, M. New Trends in Supercritical Fluid Technology and Pressurized Liquids for the Extraction and Recovery of Bioactive Compounds from Agro-Industrial and Marine Food Waste. *Molecules* **2023,** *28* (11), 4421.
47. Sodhi, A. S.; Sharma, N.; Bhatia, S.; Verma, A.; Soni, S.; Batra, N. Insights on Sustainable Approaches for Production and Applications of Value Added Products. *Chemosphere* **2022,** *286,* 131623.
48. Li, J.; Pettinato, M.; Campardelli, R.; De Marco, I.; Perego, P. High-Pressure Technologies for the Recovery of Bioactive Molecules from Agro-Industrial Waste. *Appl. Sci.* **2022,** *12* (7), 3642.
49. Ziero, H. D. D.; Buller, L. S.; Mudhoo, A.; Ampese, L. C.; Mussatto, S. I.; Carneiro, T. F. An Overview of Subcritical and Supercritical Water Treatment Of Different Biomasses for Protein and Amino Acids Production and Recovery. *J. Environ. Chem. Eng.* **2020,** *8* (5), 104406.
50. Ko, M.-J.; Nam, H.-H.; Chung, M.-S. Subcritical Water Extraction Of Bioactive Compounds from *Orostachys japonicus* A. Berger (Crassulaceae). *Sci. Rep.* **2020,** *10* (1), 10890.
51. Awaluddin, S.; Thiruvenkadam, S.; Izhar, S.; Hiroyuki, Y.; Danquah, M. K.; Harun, R. Subcritical Water Technology for Enhanced Extraction of Biochemical Compounds from *Chlorella vulgaris. Biomed Res. Int.* **2016,** 5816974, 10 pages
52. Bodoira, R.; Velez, A.; Maestri, D.; Herrera, J. Bioactive Compounds Obtained from Oilseed by-products with Subcritical Fluids: Effects on *Fusarium verticillioides* Growth. *Waste Biomass Valorization.* **2020,** *11,* 5913–5924.
53. Draszewski, C. P.; Silveira, N. M.; Brondani, M.; Cruz, A. d. S.; Rezzadori, K.; Mayer, F. D.; Abaide, E. R.; Mazutti, M. A.; Tres, M. V.; Zabot, G. L. Use of Rice Husk Hydrolyzed by Subcritical Water to Obtain Silica from Agro-industrial Waste. *Environ. Eng. Sci.* **2022,** *39* (12), 938–946.

54. Yoshida, H.; Terashima, M.; Takahashi, Y. Production of Organic Acids and Amino Acids from Fish Meat by Sub-critical Water Hydrolysis. *Biotechnol. Prog.* **1999,** *15* (6), 1090–1094.
55. Gharib-Bibalan, S. High Value-added Products Recovery from Sugar Processing by-Products and Residuals by Green Technologies: Opportunities, Challenges, and Prospects. *Food Eng. Rev.* **2018,** *10,* 95–111.
56. Veggi, P. C.; Martinez, J.; Meireles, M. A. A. Fundamentals of Microwave Extraction. In *Microwave-assisted Extraction for Bioactive Compounds: Theory and Practice*; Chemat, F.; Cravotto, G., Eds; Springer Science & Business Media, 2012; Vol. 4, pp 15–52.
57. Eskilsson, C. S.; Björklund, E. Analytical-scale Microwave-Assisted Extraction. *J. Chromatogr. A.* **2000,** *902* (1), 227–250.
58. Amirabbasi, S.; Elhamirad, A. H.; Saeediasl, M. R.; Armin, M.; Ziaolhagh, S. H. R. Optimization of Polyphenolic Compounds Extraction Methods from Okra Stem. *J. Food Meas. Charact.* **2021,** *15* (1), 717–734.
59. Diaconeasa, Z.; Iuhas, C. I.; Ayvaz, H.; Mortas, M.; Farcaş, A.; Mihai, M.; Danciu, C.; Stanilă, A. Anthocyanins from Agro-Industrial Food Waste: Geographical Approach and Methods of Recovery—A Review. *Plants* **2022,** *12* (1), 74.
60. Hakimi, M. I.; Ilham, Z.; Abdul Kohar, R. A. Enhancement of Agro-industrial Copra Residue Oil Yield Using Microwave-Assisted Extraction. Waste Biomass Valor. **2019,** *10,* 2681–2688.
61. Ordoñez-Torres, A.; Torres-León, C.; Hernández-Almanza, A.; Flores-Guía, T.; Luque-Contreras, D.; Aguilar, C. N.; Ascacio-Valdés, J. Ultrasound-microwave-assisted Extraction of Polyphenolic Compounds from Mexican "Ataulfo" Mango Peels: Antioxidant Potential and Identification by HPLC/ESI/MS. *Phytochem Anal.* **2021,** *32* (4), 495–502.
62. Cardoso-Ugarte, G.; Sosa-Morales, M.; Ballard, T.; Liceaga, A.; San Martín-González, M. Microwave-assisted Extraction of Betalains From Red Beet (*Beta vulgaris*). *LWT - Food Sci.* **2014,** *59* (1), 276–282.
63. Görgüç, A.; Özer, P.; Yılmaz, F. M. Microwave-assisted Enzymatic Extraction of Plant Protein with Antioxidant Compounds from the Food Waste Sesame Bran: Comparative Optimization Study and Identification of Metabolomics using LC/Q-TOF/MS. *J. Food Process. Preserv.* **2020,** *44* (1), e14304.
64. Mena-García, A.; Rodríguez-Sánchez, S.; Ruiz-Matute, A. I.; Sanz, M. L. Exploitation of Artichoke Byproducts to Obtain Bioactive Extracts Enriched in Inositols and Caffeoylquinic Acids by Microwave Assisted Extraction. *J. Chromatogr. A.* **2020,** *1613,* 460703.
65. Zin, M. M.; Anucha, C. B.; Bánvölgyi, S. Recovery of Phytochemicals via Electromagnetic Irradiation (Microwave-Assisted-Extraction): Betalain and Phenolic Compounds in Perspective. *Foods* **2020,** *9* (7), 918.
66. Martínez-Patiño, J. C.; Gullón, B.; Romero, I.; Ruiz, E.; Brnčić, M.; Žlabur, J. Š.; Castro, E. Optimization of Ultrasound-Assisted Extraction of Biomass from Olive Trees Using Response Surface Methodology. *Ultrason. Sonochem.* **2019,** *51,* 487–495.
67. Backes, E.; Pereira, C.; Barros, L.; Prieto, M.; Genena, A. K.; Barreiro, M. F.; Ferreira, I. C. Recovery of Bioactive Anthocyanin Pigments from *Ficus carica* L. Peel by Heat, Microwave, and Ultrasound Based Extraction Techniques. *Food Res. Int.* **2018,** *113,* 197–209.

68. Dos Santos, S. S.; Paraíso, C. M.; Rodrigues, L. M.; Madrona, G. S. Agro-Industrial Waste as a Source of Bioactive Compounds: Ultrasound-assisted Extraction from Blueberry (*Vaccinium myrtillus*) and Raspberry (*Rubus idaeus*) Pomace. *Acta Sci. Technol.* **2021,** *43,* 1–8.
69. Phongthai, S.; Lim, S. T.; Rawdkuen, S. Ultrasonic-Assisted Extraction of Rice Bran Protein Using Response Surface Methodology. *J. Food Biochem.* **2017,** *41* (2), e12314.
70. Segatto, M. L.; Stahl, A. M.; Zanotti, K.; Zuin, V. G. Green and Sustainable Extraction of Proteins from Agro-industrial Waste: An Overview and a Closer Look to Latin America. *Curr. Opin. Green Sustain.* **2022,** 100661.
71. Chen, H.-m.; Fu, X.; Luo, Z.-G. Properties and Extraction of Pectin-enriched Materials from Sugar Beet Pulp by Ultrasonic-Assisted Treatment Combined with Subcritical Water. *Food Chem.* **2015,** *168,* 302–310.
72. Gómez-Cruz, I.; Contreras, M. D. M.; Carvalheiro, F.; Duarte, L. C.; Roseiro, L. B.; Romero, I.; Castro, E. Recovery of Bioactive Compounds from Industrial Exhausted Olive Pomace Through Ultrasound-Assisted Extraction. *Biology* **2021,** *10* (6), 514.
73. Pagliari, S.; Giustra, C. M.; Magoni, C.; Celano, R.; Fusi, P.; Forcella, M.; Sacco, G.; Panzeri, D.; Campone, L.; Labra, M. Optimization of Ultrasound-Assisted Extraction of Naturally Occurring Glucosinolates from by-products of *Camelina sativa* L. and their Effect on Human Colorectal Cancer Cell Line. *Front. Nutr.* **2022,** *9,* 901944.
74. Serna-Jiménez, J.; Torres-Valenzuela, L.; Villarreal, A. S.; Roldan, C.; Martín, M.; Siles, J.; Chica, A. Advanced Extraction of Caffeine and Polyphenols from Coffee Pulp: Comparison of Conventional and Ultrasound-assisted Methods. *LWT* **2023,** *177,* 114571.
75. Eom, S. J.; Kim, Y. E.; Kim, J.-E.; Park, J.; Kim, Y. H.; Song, K.-M.; Lee, N. H. Production of *Undaria pinnatifida* Sporophyll Extract using Pilot-scale Ultrasound-assisted Extraction: Extract Characteristics and Antioxidant and Anti-inflammatory Activities. *Algal Res.* **2020,** *51,* 102039.
76. Smith, E. L.; Abbott, A. P.; Ryder, K. S. Deep Eutectic Solvents (DESs) and Their Applications. *Chem. Rev.* **2014,** *114* (21), 11060–11082.
77. De, D.; Sai, M. S. N.; Aniya, V.; Satyavathi, B. Strategic Biorefinery Platform for Green Valorization of Agro-industrial Residues: A Sustainable Approach Towards Biodegradable Plastics. *J. Clean. Prod.* **2021,** *290,* 125184.
78. García-Roldán, A.; Piriou, L.; Jauregi, P. Natural Deep Eutectic Solvents as a Green Extraction of Polyphenols from Spent Coffee Ground with Enhanced Bioactivities. *Front. Plant Sci.* **2023,** *13,* 1072592.
79. Gómez, A. V.; Biswas, A.; Tadini, C. C.; Furtado, R. F.; Alves, C. R.; Cheng, H. N. Use of Natural Deep Eutectic Solvents for Polymerization and Polymer Reactions. *J. Braz. Chem. Soc.* **2019,** *30,* 717–726.
80. Orejuela-Escobar, L. M.; Landázuri, A. C.; Goodell, B. Second Generation Biorefining in Ecuador: Circular bioeconomy, Zero Waste Technology, Environment and Sustainable Development: The Nexus. *J. Bioresour. Bioprod.* **2021,** *6* (2), 83–107.
81. Procentese, A.; Raganati, F.; Olivieri, G.; Russo, M. E.; Rehmann, L.; Marzocchella, A. Deep eutectic Solvents Pretreatment of Agro-industrial Food Waste. *Biotechnol. Biofuels.* **2018,** *11,* 1–12.
82. Shrestha, S.; Kognou, A. L. M.; Zhang, J.; Qin, W. Different Facets of Lignocellulosic Biomass Including Pectin and Its Perspectives. Waste Biomass Valor. **2021,** *12,* 4805–4823.

83. Lemes, A. C.; Egea, M. B.; Oliveira Filho, J. G. D.; Gautério, G. V.; Ribeiro, B. D.; Coelho, M. A. Z. Biological Approaches for Extraction of Bioactive Compounds from Agro-industrial by-Products: A Review. *Front. Bioeng. Biotechnol.* **2022,** *9,* 802543.
84. Ravindran, R.; Hassan, S. S.; Williams, G. A.; Jaiswal, A. K. A Review on Bioconversion of Agro-industrial Wastes to Industrially Important Enzymes. *Bioengineering* **2018,** *5* (4), 93.
85. Mazutti, M. A.; Skrowonski, A.; Boni, G.; Zabot, G. L.; Silva, M. F.; de Oliveira, D.; Di Luccio, M.; Filho, F. M.; Rodrigues, M. I.; Treichel, H. Partial Characterization of Inulinases Obtained by Submerged and Solid-state Fermentation Using Agroindustrial Residues as Substrates: A Comparative Study. *Appl. Biochem. Biotechnol.* **2010,** *160,* 682–693.
86. Singhania, R. R.; Sukumaran, R. K.; Patel, A. K.; Larroche, C.; Pandey, A. Advancement and Comparative Profiles in the Production Technologies using Solid-state and Submerged Fermentation for Microbial Cellulases. *Enzyme Microb. Technol.* **2010,** *46* (7), 541–549.
87. Al Mousa, A. A.; Hassane, A. M.; Gomaa, A. E.-R. F.; Aljuriss, J. A.; Dahmash, N. D.; Abo-Dahab, N. F. Response-surface Statistical Optimization of Submerged Fermentation for Pectinase and Cellulase Production by *Mucor Circinelloides* and *M. hiemalis*. *Fermentation* **2022,** *8* (5), 205.
88. Grewal, J.; Wołącewicz, M.; Pyter, W.; Joshi, N.; Drewniak, L.; Pranaw, K. Colorful Treasure from Agro-industrial Wastes: A Sustainable Chassis for Microbial Pigment Production. *Front. Microbiol.* **2022,** *13,* 832918.
89. Begum, W.; Saha, B.; Mandal, U. A Comprehensive Review on Production of Bio-Surfactants by Bio-degradation of Waste Carbohydrate Feedstocks: An Approach Towards Sustainable Development. *RSC Adv.* **2023,** *13* (36), 25599–25615.
90. Ranganathan, S.; Dutta, S.; Moses, J.; Anandharamakrishnan, C. Utilization of Food Waste Streams for the Production of Biopolymers. *Heliyon* **2020,** 6, e04891.
91. Szymczak, T.; Cybulska, J.; Podleśny, M.; Frąc, M. Various Perspectives on Microbial Lipase Production Using Agri-food Waste and Renewable Products. *Agriculture* **2021,** *11* (6), 540.
92. Pandey, A.; Soccol, C. R.; Mitchell, D. New Developments in Solid State Fermentation: I-Bioprocesses and Products. *Process Biochem.* **2000,** *35* (10), 1153–1169.
93. Prabhu, G.; Bhat, D.; Bhat, R. M.; Selvaraj, S. A Critical Look at Bioproducts co-cultured Under Solid State Fermentation and Their Challenges and Industrial Applications. Waste Biomass Valor. **2022,** *13* (7), 3095–3111.
94. Cerda, A.; Artola, A.; Barrena, R.; Font, X.; Gea, T.; Sánchez, A. Innovative Production of Bioproducts from Organic Waste Through Solid-State Fermentation. *Front. Sustain. Food Syst.* **2019,** *3,* 63.
95. Casas-Godoy, L.; Campos-Valdez, A. R.; Alcázar-Valle, M.; Barrera-Martínez, I. Comparison of Extraction Techniques for the Recovery of Sugars, Antioxidant and Antimicrobial Compounds from Agro-Industrial Wastes. *Sustainability* **2022,** *14* (10), 5956.
96. Radenkovs, V.; Juhnevica-Radenkova, K.; Kviesis, J.; Lazdina, D.; Valdovska, A.; Vallejo, F.; Lacis, G. Lignocellulose-degrading Enzymes: A Biotechnology Platform for Ferulic acid Production from Agro-industrial Side Streams. *Foods* **2021,** *10* (12), 3056.

97. Barcelos, M. C.; Ramos, C. L.; Kuddus, M.; Rodriguez-Couto, S.; Srivastava, N.; Ramteke, P. W.; Mishra, P. K.; Molina, G. Enzymatic Potential for the Valorization of Agro-Industrial by-products. *Biotechnol. Lett.* **2020,** *42,* 1799–1827.
98. Hernández Becerra, E.; De Jesús Pérez López, E.; Zartha Sossa, J. W. Recovery of Biomolecules from Agroindustry by Solid-liquid Enzyme-assisted Extraction: A Review. *Food Anal. Methods* **2021,** *14,* 1744–1777.
99. Gonçalves, D. A.; González, A.; Roupar, D.; Teixeira, J. A.; Nobre, C. How Prebiotics have been Produced from Agro-industrial Waste: An Overview of the Enzymatic Technologies Applied and the Models used to Validate Their Health Claims. *Trends Food Sci. Technol.* 2023.
100. Lima, A. R.; Cristofoli, N. L.; da Costa, A. M. R.; Saraiva, J. A.; Vieira, M. C. Comparative Study of the Production of Cellulose Nanofibers from Agro-industrial Waste Streams of *Salicornia ramosissima* by Acid and Enzymatic Treatment. *Food Bioprod. Process.* **2023,** *137,* 214–225.
101. Vilas-Franquesa, A.; Casertano, M.; Tresserra-Rimbau, A.; Vallverdú-Queralt, A.; Torres-León, C. Recent Advances in Bio-based extraction Processes for the Recovery of Bound Phenolics from Agro-industrial by-products and Their Biological Activity. *Crit. Rev. Food Sci. Nutr.* **2023,** 1–25.

CHAPTER 2

From Farm to Industry: Valorization of Agricultural Waste by Solid-State Culture for Sustainable Development

ÁNGELES DOMÍNGUEZ-RIVERA[1], MARÍA ESMERALDA BIBIÁN-LEÓN[2], GERARDO SAUCEDO-CASTAÑEDA[1], JOSÉ JUAN BUENROSTRO-FIGUEROA[3], and FERNANDO MÉNDEZ-GONZÁLEZ[3*]

[1]*Departamento de Biotecnología, Universidad Autónoma Metropolitana, Iztapalapa, Mexico City, Mexico*

[2]*Posgrado en Botánica, Colegio de Posgraduados-Montecillo, Texcoco, Mexico State, Mexico*

[3]*Laboratorio de Biotecnología y Bioingeniería, Centro de Investigación en Alimentación y Desarrollo, Delicias, Chihuahua, Mexico*

*Corresponding author.

ABSTRACT

Food production and processing generates a large amount of waste. A small fraction of agricultural waste is used locally to produce composts and biofuels. However, agricultural waste can be used as substrates in solid-state culture (SSC) processes to obtain biopesticides, biofertilizers, enzymes, and biosurfactants. These products are obtained in bioreactors with different configurations, ranging from laboratory-scale to industrial-scale. The adaptation and implementation of these biotechnological processes allow the comprehensive use of agricultural waste, contributing to agricultural sustainability. Therefore, this chapter analyzes the bioreactors, microorganisms, and agricultural waste used to produce biopesticides, biofertilizers, enzymes, and biosurfactants by SSC.

Agroindustrial Waste Management and Natural Resources Conservation: Environmental Challenges and Transformation into Value-Added Products. José Juan Buenrostro-Figueroa, Juan Manuel Tirado-Gallegos, Mónica L. Chávez-González, A. K. Haghi, and Shaheemath Suhara K. K. (Eds.)

2.1 INTRODUCTION

SSC is an aerobic or anaerobic process in which the microorganism grows on a solid matrix (substrate or support).[1] The carbon source, nutrients, and water required for culture are in the solid matrix. Interparticle spaces allow gas exchange and heat removal.[2] The SSC simulates the natural conditions for microorganism growth,[3] allowing the obtaining of various products from fungi, bacteria, and actinomycetes.[4] Among the advantages of SSC is the possibility of using agricultural waste (AW) as a substrate,[5] allowing its management. Worldwide, more than 5×10^9 t/year of AW are produced,[6] almost all of them are organic waste, and although they are generated in large quantities, only a small fraction is used to obtain compost[7] and energy sources such as biogas[8] and bioethanol.[9] The AW composition allows for other products, such as biopesticides, biofertilizers, enzymes, and biosurfactants, improving their management and opening new markets and income sources. Bioreactors with different configurations can be used for their production on different scales. The characteristics and operating conditions of the bioreactor essentially depend on the type of product, microorganism used, and level of technology in the region. Therefore, this chapter will first describe the types of bioreactors used in SSC and, subsequently, the microorganisms and substrates used to produce biopesticides, biofertilizers, enzymes, and biosurfactants.

2.2 SSC BIOREACTORS

The SSC bioreactors (aerated or not) present two configurations: static bed and agitated.[10] Due to their capacity, these can be classified as laboratory-scale (from a few milliliters to <2 L capacity),[2] bench-scale (from 2 to <5 L capacity),[2,11] pilot-scale (from 5 to <50 L capacity),[12–14] and industrial-scale (>50 L capacity). Among the different designs and operating parameters of SSC bioreactors are those that promote growth and metabolic product formation from specific microorganisms.[15] Therefore, to recognize the selection criteria of SSC bioreactors, their characteristics will be described in the following subsections.

2.2.1 STATIC BED BIOREACTORS

Tray (TB) and packed bed columns (PBCB) are classic examples of static bed bioreactors.[10] The TB (Fig. 2.1) is a thin layer of substrate spread over a rectangular surface.[16] The trays are fabricated with different materials, such as wood, plastic, and metal.[17] In its classic presentation (trays without perforation), heat removal is carried out by conduction and evaporation,[18,19] and the gas exchange is carried out by diffusion and natural convection.[17,20] Therefore, heat and mass transport rates depend on the culture bed height.[21] A proper bed height from 2 to 4 cm can be used to avoid overheating and the anaerobic zones.[17,22,23] Additionally, the TB bottom can be perforated, allowing for supply forced aeration through the bed (when airflow is supplied).[16] TBs are incubated in temperature-regulated rooms[16] or closed chambers,[24] where the trays are arranged one above the other a few centimeters apart. Closed chambers allow for monitoring and controlling operating conditions (temperature, moisture, temperature, gaseous atmosphere composition, etc.).[18,21,23] In addition, in aerobic cultures, forced aeration can be supplied to incubation chambers for improving heat removal and gas exchange.[23] TB scale-up is possible by increasing the number of trays inside the incubation chamber, reducing the distance between them.[25] The TB characteristics allow a controlled process, but they require intensive labor[26] and a large area for the process.[16]

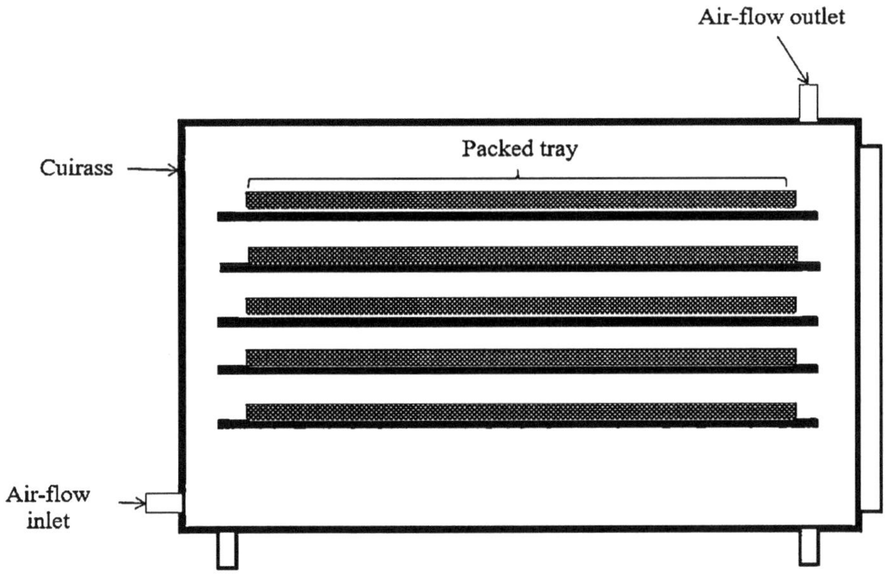

FIGURE 2.1 Tray bioreactor.

On the other hand, PBCBs are regularly cylindrical systems[10] when forced aeration can be supplied[27] (Fig. 2.2). Forced aeration allows for heat removal and gas exchange (O_2 supply and CO_2 removal) by convective mechanisms.[17,28] It is possible to modify the aeration flow rate and its moisture content during the process to control the culture temperature.[16,29] The PBCB´s design facilitates the use of systems to monitor and control the culture conditions.[30–32] In PBCB, using high-porosity substrates such as straw, bagasse, bran, and husk, facilitates maintaining the operating conditions during the culture because they improve the air distribution.[13,29,33,34] The above makes it possible to operate PBCB at a pilot scale.[28,29,35] The lab-scales PBCB are regularly fabricated in glass[17,36] and larger scales in metal.[37] However, cheaper materials such as polyvinyl chloride can be used to manufacture PBCB at different scales.[2] Additionally, PBCB can be divided into axial sections to avoid substrate shrinkage and increase working height.[30,38]

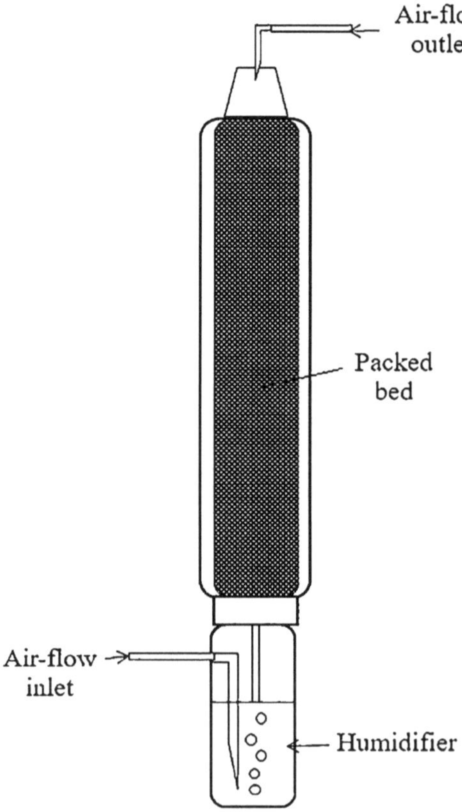

FIGURE 2.2 Packed bed column bioreactor.

2.2.2 AGITATED BIOREACTORS

Agitated bioreactors have two mixing frequencies: intermittent and continuous.[39] Agitation avoids substrate agglomeration and shrinkage, contributing to heat removal and gas exchange.[25,40] Bag bioreactors (BB) used to produce biopesticides, biofertilizers, and enzymes operate with intermittent mixing, carried out by operators.[41] The BB is generally fabricated with plastic, such as high-density polypropylene.[17] In BB manufactured with porous plastics, gas exchange is carried out by diffusivity through micropores.[42] However, diffusivity transport causes CO_2 accumulation within the system.[43] Therefore, the most BB used industrially are low capacity ($\leq$1 kg of substrate).[10,44] In some cases, the bags are partially closed or plugs with greater porosity than the bags are used to facilitate gas exchange.[44] The BB process presents low cost and technical simplicity,[45,46] but it requires labor-intensive and large areas for the process, and it lacks monitoring and control instrumentation.[10,41]

The rotating drum (RDB) or internal stirred (ISB) are other agitated bioreactors[10] (Fig. 2.3). Both are mixed mechanically and can operate under continuous or intermittent agitation.[15] Forced aeration can supplied to RDB and ISB, improving the heat and mass transport.[11,47] These agitated bioreactors are used in controlled processes at an industrial scale (up to 25 t) under low aseptic conditions.[48,49] The RDB is cylindrical and rotates on its horizontal axis; the smooth movement causes soft mechanical damage to the microorganism or the substrate.[25]

On the other hand, ISBs are horizontal[11] and vertical[40] and have a propeller system for mixing the substrate. The propeller movement can cause mechanical damage to the microbial cells and the substrate, but it contributes to the uniformity of the bed.[39,40] Therefore, the intensity and frequency of agitation must be appropriately adjusted to allow microbial growth.

The bioreactors described above have different characteristics that promote using certain microorganisms and substrates to obtain microbial products. In addition to the biological aspect, the selection of the bioreactor type can consider the region's economic and technological level. As mentioned, AW can be used as a substrate in SSC processes to obtain biopesticides, biofertilizers, enzymes, and biosurfactants. The following sections will describe the characteristics of these products and some recent processes for obtaining them.

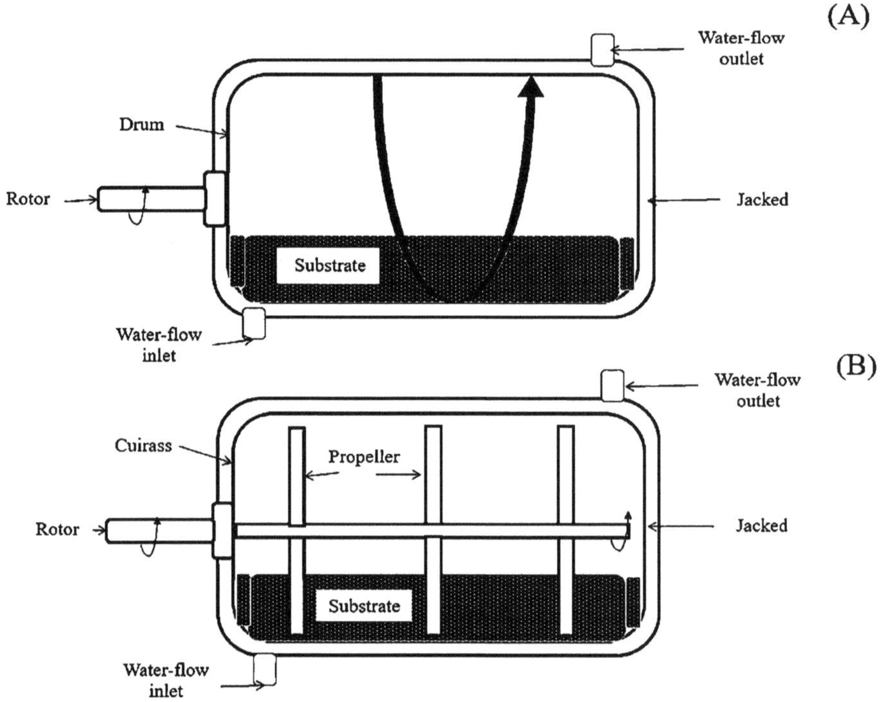

FIGURE 2.3 Agitated bioreactors: (A) rotating drum bioreactor and (B) internal stirred bioreactor.

2.3 BIOPESTICIDES AND BIOFERTILIZERS PRODUCTION

Biopesticides and biofertilizers have the potential to be used instead of chemicals, improving agricultural sustainability. Bacteria and fungi propagules are used as active principles in commercial formulations.[15,50–52] Microorganisms used as biological control agents can reduce the pest populations that include organisms such as insects, weeds, nematodes, bacteria, and fungi.[51,53–58] On the other hand, biofertilizers include microorganisms for nitrogen-fixing and phosphate, sulfur, and potassium solubilization.[59] Their application improves quality soil and crop growth.

The propagules produced (regularly spores) by SSC include simple and highly technological processes. Plastic containers[60] and bags[44] are bioreactors in simple and low-cost processes. More technical processes include TB, PBCB, and ISB with monitoring and control systems,[14,17,24] improving production yields but increasing initial investment and cost maintenance. Static bed bioreactors are preferred for fungal spore production since they avoid mycelial damage.[15] Some spore production processes use AW as substrate (Table 2.1), which achieves production yields similar to those obtained in first-generation substrates. The spores produced on AW can be recovered by mechanical[61] or fluidized bed[62] systems for fabricating commercial formulations.

2.4 ENZYMES PRODUCTION

Enzymes are molecules produced by microorganisms to catalyze metabolic reactions. Enzymes are widely used in industrial processes to manufacture fur, pharmaceutical, food, textile, chemical, and energy products.[73] SSC processes allow for high enzyme concentrations[74] with higher physicochemical stability than those produced in liquid culture.[75] AW such as fruit peels, bagasse, straw, pulp, and leaves (Table 2.2) have been successfully used to produce enzymes. Among the enzymes produced by SSC on AW are:

Amylases: Amylases hydrolyze starch, making them useful for producing food, textiles, paper, and medicines.[76] Grain residues such as wheat bran and soybean hulls can be used for amylase production using microorganisms such as *B. subtilis, B. amyloliquefaciens*, *A. oryzae*, and *A. tamarii*.[77–80] Amylase production in wheat bran is improved by the addition of potato peels (Table 2.2). Similarly, wastes such as vine shoot and sugarcane bagasse are mixed with chitin, potato flour, and olive oil to promote amylase production by *Trichoderma* fungi.[68,81] On the other hand, high amylase activity (>850 U/g) can be obtained from pulp[11] and fruit peels[12] using *M. phaseolina* and *T. virens* (respectively).

Cellulases: These enzymes catalyze the hydrolysis of cellulose and hemicellulose chains. The cellulases are classified into endoglucanase, exoglucanase, and β-glucosidase.[82] Cellulases occupy the third place of the most used enzymes at an industrial level due to their important applications in bioenergy, paper, textile, and food industries.[73] Microorganisms such as *A. oryzae*, *A. fumigatus*, *A. niger*, and *A. flavus* produce cellulases on grain and fruit residues,[74] and Basidiomycotas such as *Pleurotus ostreatus* are effective with forest residues.[83] Additionally, some working groups evaluated co-cultures as a strategy to improve cellulase productivity by SSC.[84–86] In co-cultures, microorganisms with different growth rates and environmental tolerance (temperature, pH, humidity, etc.) are used, which allow for reduced lag time and increased substrate hydrolysis and process operability under variable environmental conditions.

TABLE 2.1 Spore Production from Agricultural Waste by SSC.

Microorganism	Substrate	Bioreactor	Spore production (spore × 10^9/gdm)	Reference
Metarhizium anisopliae	Hemp and nutrient	Packed bed column	8	[29]
	Broken rice and wheat bran	Tray	1.7	[24]
	Wheat bran	Packed bed column	5.6	[29]
	Recycled parboiled rice	Plastic Bag	1.4–2.3	[63]
	Rice bran	Packed bed column	3.4	[29]
Lasiodiploidia pseudotheobromae	Rice bran	Plastic Bag	1.8–3.1	[64]
Beauveria bassiana	Vine shoots, wheat bran, olive pomace	Packed bed (Zymotis-2)	7.5	[35]
	Potato residues and sugarcane bagasse	Packed bed column	9.8	[65]
	Potato peel	Packed bed column	1.3	[66]
	Beer draff	Packed bed column	2.5	[67]
Trichoderma harzianum	Sugarcane bagasse, wheat bran and nutrient	Packed bed column	1.08	[68]
	Potato peel	Packed bed column	6.4	[66]
	Beer draff	Packed bed column	7.5	[66]
Trichoderma asperellum	Sugarcane bagasse, wheat bran and nutrient	Packed bed column	1.23–1.30	[68]
	Vine shoots, jatropha cake, olive pomace and olive oil	Packed bed column	8.5	[69]
	Vine shoots, wheat bran, olive pomace	Packed bed	5.4	[35]
Tichoderma longibranchiatum	Sugarcane bagasse, wheat bran, and nutrient	Packed bed column	1.31	[68]
Trichoderma yunnanese	Sugarcane bagasse, wheat bran and nutrient	Packed bed column	1.2	[68]
Coniothyrium minitans	Hemp and nutrient	Packed bed column	12.8	[70]
Bacillus amyloliquefaciens	Corncobs	Bag	94	[71]
Bacillus subtilis	Soybean curd residue and nutrient	Flask	2.75	[72]

*dm: dry mass

TABLE 2.2 Enzyme Production from Agricultural Waste by SSC.

Enzyme	Microorganism	Substrate	Bioreactor	Productivity	Reference
Amylase	*Bacillus subtilis*	Wheat bran	Flask	82.6 U/gdm	[77]
	Macrophomia phaseolina	Apple pomace	Flask	3309 U/g, 5 d	[108]
	Trichoderma asperellum	Vine shoots, potatoes flour, jatropha, olive pomace, and olive oil	Packed bed column	46.3 U/g, 2 d	[81]
	Trichoderma longibranchiatum	Sugarcane bagasse, wheat bran, chitin, potato flour, and olive oil	Packed bed column	28.3 U/g, 2 d	[68]
	Trichoderma yunnanense	Sugarcane bagasse, wheat bran, chitin, potato flour, and olive oil	Packed bed column	33.68 U/g, 3 d	[68]
α-amylase	*Bacillus amyloliquefaciens*	Wheat bran and potato peel	Flask	1.3 U/ug, 4 d	[78]
	Aspergillus tamarii	Wheat bran and potato peel	Flask	519.40 U/g, 5 d	[79]
	Trichoderma virens	Watermelon peel	Flask	880 U/g, 3 d	[109]
	Aspergillus oryzae	Soybean hulls and flour milling byproduct	Tray	45.900 U/g, 6 d	[80]
Cellulase	*Trichoderma asperellum*	Vine shoots, potatoes flour, jatropha, olive pomace, and olive oil	Packed bed column	34.3 U/g, 2d	[81]
	Trichoderma reesei and *Penicillium citrinum*	Wheat bran	Flask	22.3 U/gds, 6 d	[86]
	Trichoderma longibrachiatum and *Aspergillus fumigatus*	Banana stem	Flask	61 U/gdm, 9 d	[85]
	Cladosporium cladosporioides and *Penicillium funiculosum*	Corn straw	Flask	1.5 U/mL, 5 d	[84]
	Pleorotus ostreatus	Forestal waste	Flask	21 (cellulase) and 77.1 U/g (hemicellulase), 15 d	[83]
	T. yunnanense	Sugarcane bagasse, wheat bran, chitin, potato flour, and olive oil	Packed bed column	38.32 U/g, 3 d (endoglucanase)	[68]

TABLE 2.2 *(Continued)*

Enzyme	Microorganism	Substrate	Bioreactor	Productivity	Reference
	Trichoderma harzianum	Sugarcane bagasse, wheat bran, chitin, potato flour, and olive oil	Packed bed column	31.7 U/g, 3 d (endoglucanase)	[68]
	Aspergillus flavus	Rice straw	Tray	12,5 UI/gdm (FPasa), 235 UI/gdm (CMCasa), 190 UI/gdm (β-glucosidasa), 5 d	[104]
	Aspergillus niger	Green tea waste	Flask	65,85 IU/g (endoglucanase), 4 d 1.016 IU/g (β-glucosidase), 5 d	[110]
	Myceliophthora termophila	Sugarcane bagasse and wheat bran	Packed bed column	110 U/gms, 4 d	[13]
Lipase	*Aspergillus niger* KLC-33	Coconut pulp	Flask	10.83 U/mL, 7 d	[91]
	Aspergillus niger	Rice husk, cottonseed cake and red gram husk	Tray	24,38 U/gdm, 2 d	[111]
	Aspergillus niger	Rice bran and *Jathropa* seed cake	Tray	282 U/mL, 5 d	[105])
	Aspergillus niger	Guava tree leaves	–	132,72 U/mg, 4 d	[92]
	Bacillus amyloliquefaciens	Rice bran	Flask	41,20 U/mL, 2 d	[112]
	Trichoderma asperellum	Vine shoots, potatoes flour, jatropha, olive pomace, and olive oil	Packed bed column	30.6 U/g, 2d	[81]
	Thermomucor indicae	Sugarcane bagasse and soybean oil	Bag	15 U/g, 3 d	[88]
Xylanase	*Aspergillus flavus*	Rice straw	Tray	180 IU/gdm, 5 d	[104]
	Aspergillus niger	Rice husk	Flask	8 U/mL, 6 d	[113]
	Pichia kudriavzevii	Wheat straw	Flask	273,02 IU/mL, 7 d	[94]
	Trichoderma harzianum	Pine sawdust and culture media	Stirred	306 U/g, 2 d	[11]
	Myceliophthora termophila	Sugarcane bagasse and wheat bran	Packed bed column	1100 U/gms, 4 d	[13]
Laccase	*Bacillus* sp.	Fruit juice waste	Flask	1645 IU/g	[114]
	Funalia trogii	Rice straw	Packed bed column	2912.34 U/g, 14 d	[115]
	Trametes hirsuta	Pineapple leaves	Flask	6513.8 U/mL, 21 d	[116]

TABLE 2.2 *(Continued)*

Enzyme	Microorganism	Substrate	Bioreactor	Productivity	Reference
	Trametes versicolor	Corn cane	Tray	6.885 U/mL, 13 d	[106]
	Fomes fomentarius	Corn straw	Flask	260 U/mL, 10 d	[117]
	Penicillium sp.	Barley bran	Flask	36.29 U/g, 1.8 d	[118]
	Trametes versicolor	Pineapple Crown leaves	Flask	60.73 U/g, 7 d	[98]
	Trametes versicolor	Dry pineapple crown	Flask	77.88 U/g, 7 d	[99]
Pectinase	*Aspergillus* sp.	Coffee waste (Pulp and husk)	Tray	29.9 UI/g	[101]
n-demethylase	*Rhizopus oryzae*	Sorghum and coffee waste	Tray	18.76 U/g, 1 d	[119]
Inulinase	*Nothophoma anigozanthi*	Banana plant shoot poder	Flask	2.58 U/gdm, 3 d	[103]

*dm: dry mass

TABLE 2.3 Biosurfactant Production from Agricultural Waste by SSC.

Biosurfactant class	Microorganism	Substrate	Bioreactor	Production yield	Reference
Sophorolipids	*Aspergillus niger*	Wheat bran, corncob, and sugarcane molasses	Flask	6.67 ± 0.06 UE	[125]
	Starmerella bombicola	Wheat straw, sunflower oil, sugar beet molasses	Aerated Flask	0.179 g/gdm	[132]
Rhamnolipids	*Pseudomonas aeruginosa*	Soybean meal	Flask	19.68 g/kgdm	[122]
	Pseudomonas aeruginosa	Sugarcane bagasse and sunflower seed meal	Flask	4685 g/L-IS (IS, impregnant solution)	[121]
Surfactin	*Bacillus subtilis*	Rapeseed meal and oat	Flask	4 g/kgdm	[133]
	Bacillus pumilus	Okara and sugarcane bagasse	Flask	3.3 g/kgdm	[134]
Lipopeptides	*Bacillus subtilis*	Aleppo pine seeds and confectionery effluent	Flask	27.6 g/kg	[135]
Unidentified	*Aspergillus niger*	Banana stalks	Flask	5.5 g/L	[136]

*dm: means dry mass

**UE means units of emulsifying activity

Lipases: Lipases catalyze the hydrolysis of fats and oils, converting them into free fatty acids, diglycerides, monoglycerides, and glycerol.[87] These enzymes are helpful in producing biodiesel.[88,89] However, lipases also have important applications in producing drugs, foods, cosmetics, textiles, and detergents,[74] which are considered the most important industrial enzymes.[87] Additionally, lipases can be used in water and soil remediation processes.[90] Lipases are usually produced on oily residues or oil-enriched fibrous substrates.[74] Some authors used fibrous substrates (such as rice husk and bran) mixed with seeds to produce lipases using *A. niger* (Table 2.2). The seeds provide a source of lipids to the culture. Similarly, oil-rich substrates such as coconut pulp can be exploited to produce lipases.[91] Additionally, small lipase concentrations (<1 U/g) can be obtained from pruning waste such as guava leaves.[92]

Xylanases: Xylanases hydrolyze the 1,4-α-D-xylosidic bonds contained in hemicellulose molecules.[74] By contributing to the degradation of hemicellulose, microbial xylanases are helpful in the paper industry.[93] Bacteria and filamentous fungi are effective microorganisms for producing xylanases on fibrous waste (Straw, bran, bagasse, husk, etc.).[74] However, Poomani et al.[94] obtained ~6885 IU/mL from wheat straw using the yeast *Pichia kudriavzevii*. The production of xylanases from agricultural fibrous waste is improved by applying an alkaline pretreatment, eliminating part of the lignin from the substrate.[95]

Laccases: These enzymes (Benzenediol: oxygen oxidoreductase) can catalyze the oxidation of compounds such as polyphenols, methoxyphenols, aromatic amines, and ascorbate.[96] They can be applied for biobleaching, bioremediation of xenobiotics, discoloration of textile dyes, biosensors, food industry, degradation of plastics, and lignin degradation.[97] Basidiomycota fungi are commonly used to produce laccases from grains[74] and fruits[98] waste. However, Ascomycota fungi and bacteria are also helpful in producing these enzymes (Table 2.2). The addition of laccase-inducing substances such as farnesol (sesquiterpene alcohol) is an effective strategy to improve the productivity of the process.[99]

Pectinases: Pectinases catalyze depolymerization or de-esterification reactions for the degradation of pectins. Among its applications are the clarification of juices and beverages and the processing (production, extraction, clarification, and purification) of oils and bioactive substances.[100] Pectinases have been obtained by SSC from citrus residues using *A. niger* and *A.*

oryzae.[74] However, using *Aspergillus* sp. Núñez-Pérez et al.[101] produced pectinases from coffee waste (pulp and husk).

Inulinase: 1-D-fructan fructanohydrolase (inulinase) is an enzyme that catalyzes the breakdown of inulin to fructose by hydrolyzing β-2,1 glycosidic bonds endothermically.[102] Recently, Tippanavar et al.[103] produced inulinase from banana plant shoot powder using *Nothophoma anigozanthi*.

The enzyme production by SSC involves the use of bioreactors with different configurations. Simple processes with low cost can be carried out in BB[88] and trays.[80,104–106] However, the production of amylases, cellulases, and lipases (produced by *Trichoderma*) is enhanced by forced aeration supply.[68,81] Therefore, aerated bioreactors such as the PBCB or the one presented by Maiga et al.[107] and Hamrouni et al.[35] (Zymotis-2) are a useful option for enzyme production from AW. On the other hand, López-Ramírez et al.[11] demonstrated that bioreactors with internal agitation operated at 1 rpm are useful for producing enzymes (xylanase) from *Trichoderma harzianum*. Therefore, both bioreactor types (static bed and stirred) could be incorporated into controlled enzyme production processes to use AW.

2.5 BIOSURFACTANTS PRODUCTION

Biosurfactants (BS) are amphipathic molecules produced by archaea,[120] bacteria,[121–123] yeasts,[124] and filamentous fungi.[125] These compounds have a hydrophobic fraction formed by fatty acids and a hydrophilic fraction formed by amino acids, peptides, (poly) saccharides, or alcohols.[126] BS molecules include glycolipids, lipopeptides, lipoproteins, and polymeric surfactants.[120,127] Due to their physicochemical properties, BS reduce the interfacial tension between two immiscible liquids.[127] Therefore, they have many applications in oil, medical, food, mining, cosmetics, textile, chemical, and agricultural processes.[128] Among their applications are hydrocarbon removal, metal capture, detergent manufacturing, and bioactive compound separation.[4,129] Furthermore, BS present higher biodegradability and physicochemical stability (under extreme pH, salinity, and temperature) and a lower risk of toxicity than chemically synthesized surfactants.[130] However, the high production cost of BS limits their application instead of chemical surfactants. BS production on AW as a substrate was developed as a strategy to reduce costs. Using these substrates allows for valorizing waste and reducing production costs, improving the BS´s commercial competitiveness.[127]

Biosurfactants such as sophorolipids can be produced on substrates that contain easily assimilated carbon sources (soluble sugars).[122,131] However, surfactant production is enhanced by the combination of hydrophobic and hydrophilic substrates. Lipid carbon source promotes the hydrophobic fraction formation. On the other hand, soluble sugars (hydrophilic substrates) contribute to synthesizing the hydrophilic fraction. Therefore, vegetable oils and molasses have been mixed with AW, such as wheat bran, corn cobs, wheat straw, and sugarcane bagasse, to produce sophorolipids, rhamnolipids, and surfactin (Table 2.3). Studies in liquid culture show that oily waste can be used for BS production.[124,128] Therefore, these could be used in SSC, increasing the process profitability. Additionally, milling residues such as soybean flour have nutritional requirements to produce BS.[122] The BS production processes in SSC have been developed at a laboratory scale in flasks with and without forced aeration supply (Table 2.3). However, Pedraza-Segura et al.[120] estimate that in a pilot-scale ISB (batch or continuous), it is possible to reach a sophorolipid concentration of ~100 mg/gdm. It has been observed that unicellular microorganisms grow satisfactorily in ISB;[14] therefore, their implementation could allow BS production on a large scale. On the other hand, BS production by filamentous fungi could be carried out in static bed bioreactors to avoid damage to mycelium .[15]

2.6 CONCLUDING REMARKS

Worldwide, food production and processing generate a large amount of AW. Traditionally, compost and biofuels are produced from them. However, obtaining these products requires long process times (>30 d), and they are overwhelmed by the enormous waste generated. Therefore, it is necessary to implement new strategies that allow for processing high quantities of AW in less time. Among them, it has been shown that AW can be used to obtain biopesticides, biofertilizers, enzymes, and biosurfactants through SSC. However, the biosynthesis of many of these products has been achieved only in laboratory-scale bioreactors. However, packed bed columns and stirred bioreactors can operated on a pilot scale for the controlled production of microbial propagules and enzymes. Therefore, although the design of large-scale production processes is an emerging field, there are encouraging expectations. Consolidating these processes to obtain products with high-added value from AW would improve agrarian sustainability.

KEYWORDS

- **bioprocesses**
- **solid-state bioreactors**
- **biopesticides**
- **biofertilizers**
- **enzyme production**
- **biosurfactants**

REFERENCES

1. Viniegra-González, G. Solid State Fermentation: Definition, Characteristics, Limitations and Monitoring. In *Advances in Solid State Fermentation*; Roussos, S., Lonsane, B. K., Raimbault, M., Viniegra-González, G., Eds.; Springer Netherlands: Dordrecht, 1997; pp 5–22. https://doi.org/10.1007/978-94-017-0661-2_2.
2. Méndez-González, F.; Figueroa-Montero, A.; Saucedo-Castañeda, G.; Loera, O.; Favela-Torres, E. Addition of Spherical-style Packing Improves the Production of Conidia by *Metarhizium robertsii* in Packed Column Bioreactors. *J. Chem. Technol. Biotechnol.* **2022,** *97* (6), 1517–1525. https://doi.org/10.1002/jctb.6993.
3. Vandenberghe, L. P. S.; Pandey, A.; Carvalho, J. C.; Letti, L. A. J.; Woiciechowski, A. L.; Karp, S. G.; Thomaz-Soccol, V.; Martínez-Burgos, W. J.; Penha, R. O.; Herrmann, L. W.; Rodrigues, A. O.; Soccol, C. R. Solid-State Fermentation Technology and Innovation for the Production of Agricultural and Animal Feed Bioproducts. *Syst. Microbiol. Biomanufac.* **2021,** *1* (2), 142–165. https://doi.org/10.1007/s43393-020-00015-7.
4. Costa, J. A. V.; Treichel, H.; Kumar, V.; Pandey, A. Advances in Solid-State Fermentation. In Current Developments in Biotechnology and Bioengineering; Pandey, A., Larroche, C., Soccol, C., Eds.; Elsevier, 2018; pp 1–17. https://doi.org/10.1016/B978-0-444-63990-5.00001-3.
5. Singhania, R. R.; Patel, A. K.; Soccol, C. R.; Pandey, A. Recent Advances in Solid-State Fermentation. *Biochem. Eng. J.* **2009,** *44* (1), 13–18. https://doi.org/10.1016/j.bej.2008.10.019.
6. Vassilev, N.; de Oliveira Mendes, G. Solid-State Fermentation and Plant-Beneficial Microorganisms. In Current Developments in Biotechnology and Bioengineering; Pandey, A., Larroche, C., Soccol, C., Eds.; Elsevier, 2018; pp 435–450. https://doi.org/10.1016/B978-0-444-63990-5.00019-0.
7. Bernal, M. P.; Sánchez-Monedero, M. A.; Paredes, C.; Roig, A. Carbon Mineralization from Organic Wastes at Different Composting Stages during Their Incubation with Soil. *Agric. Ecosyst. Environ.* **1998,** *69* (3), 175–189. https://doi.org/10.1016/S0167-8809(98)00106-6.
8. Xu, S. Y.; Lam, H. P.; Karthikeyan, O. P.; Wong, J. W. C. Optimization of Food Waste Hydrolysis in Leach Bed Coupled with Methanogenic Reactor: Effect of PH and

Bulking Agent. *Bioresour. Technol.* **2011,** *102* (4), 3702–3708. https://doi.org/10.1016/j.biortech.2010.11.095.
9. Akbas, M. Y.; Stark, B. C. Recent Trends in Bioethanol Production from Food Processing Byproducts. *J. Ind. Microbiol. Biotechnol.* **2016,** *43* (11), 1593–1609. https://doi.org/10.1007/s10295-016-1821-z.
10. Méndez-González, F.; Loera-Corral, O.; Saucedo-Castañeda, G.; Favela-Torres, E. Bioreactors for the Production of Biological Control Agents Produced by Solid-State Fermentation. In *Current Developments in Biotechnology and Bioengineering*; Pandey, A., Larroche, C., Soccol, C. R., Eds.; Elsevier, 2018; pp 109–121. https://doi.org/10.1016/B978-0-444-63990-5.00007-4.
11. Lopez-Ramirez, N.; Volke-Sepulveda, T.; Gaime-Perraud, I.; Saucedo-Castañeda, G.; Favela-Torres, E. Effect of Stirring on Growth and Cellulolytic Enzymes Production by *Trichoderma harzianum* in a Novel Bench-Scale Solid-State Fermentation Bioreactor. *Bioresour. Technol.* **2018,** *265* (April), 291–298. https://doi.org/10.1016/j.biortech.2018.06.015.
12. Carboué, Q.; Rébufa, C.; Dupuy, N.; Roussos, S.; Bombarda, I. Solid State Fermentation Pilot-Scaled Plug Flow Bioreactor, Using Partial Least Square Regression to Predict the Residence Time in a Semicontinuous Process. *Biochem. Eng. J.* **2019,** *149,* 107248. https://doi.org/10.1016/j.bej.2019.107248.
13. Lopes-Perez, C.; Casciatori, F. P.; Thoméo, J. C. Improving Enzyme Production by Solid-State Cultivation in Packed-Bed Bioreactors by Changing Bed Porosity and Airflow Distribution. *Bioproc. Biosyst. Eng.* **2021,** *44* (3), 537–548. https://doi.org/10.1007/s00449-020-02466-7.
14. Martínez-Ramírez, C.; Esquivel-Cote, R.; Ferrera-Cerrato, R.; Martínez-Ruiz, J. A.; Rodríguez-Serrano, G.; Saucedo-Castañeda, G. Solid-State Culture of *Azospirillum brasilense*: A Reliable Technology for Biofertilizer Production from Laboratory to Pilot Scale. *Bioproc. Biosyst. Eng.* **2021,** *44* (7), 1525–1538. https://doi.org/10.1007/s00449-021-02537-3.
15. Méndez-González, F.; Castillo-Minjarez, J. M.; Loera, O.; Favela-Torres, E. Current Developments in the Resistance, Quality, and Production of Entomopathogenic Fungi. *World J. Microbiol. Biotechnol.* **2022,** *38* (7), 115. https://doi.org/10.1007/s11274-022-03301-9.
16. Lonsane, B. K.; Ghildyal, N. P.; Budiatman, S.; Ramakrishna, S. V. Engineering Aspects of Solid State Fermentation. *Enzyme Microb. Technol.* **1985,** *7* (6), 258–265. https://doi.org/10.1016/0141-0229(85)90083-3.
17. Méndez-González, F.; Loera, O.; Saucedo-Castañeda, G.; Favela-Torres, E. Forced Aeration Promotes High Production and Productivity of Infective Conidia from *Metarhizium robertsii* in Solid-State Fermentation. *Biochem. Eng. J.* **2020,** *156,* 107492. https://doi.org/10.1016/j.bej.2020.107492.
18. Figueroa-Montero, A.; Esparza-Isunza, T.; Saucedo-Castañeda, G.; Huerta-Ochoa, S.; Gutiérrez-Rojas, M.; Favela-Torres, E. Improvement of Heat Removal in Solid-State Fermentation Tray Bioreactors by Forced Air Convection. *J. Chem. Technol. Biotechnol.* **2011,** *86* (10), 1321–1331. https://doi.org/10.1002/jctb.2637.
19. Diniz da Silva, M. P.; Dutra, R. S.; Casciatori, F. P.; Grajales, L. M. A Two-Phase Model for Simulation of Water Transfer during Lipase Production by Solid-State Cultivation in a Tray Bioreactor Using Babassu Residues as Substrate. *Chem. Eng. Process.* **2022,** *177,* 108981. https://doi.org/10.1016/j.cep.2022.108981.

20. Raghava Rao, K. S. M. S.; Gowthaman, M. K.; Ghildyal, N. P.; Karanth, N. G. A Mathematical Model for Solid State Fermentation in Tray Bioreactors. *Bioprocess Eng.* **1993,** *8* (5–6), 255–262. https://doi.org/10.1007/BF00369838.
21. da Cunha, L. P.; Casciatori, F. P.; de Cenço Lopes, I.; Thoméo, J. C. Production of Conidia of the Entomopathogenic Fungus *Metarhizium anisopliae* ICB 425 in a Tray Bioreactor. *Bioproc. Biosyst. Eng.* **2019,** *42* (11), 1757–1768. https://doi.org/10.1007/s00449-019-02172-z.
22. Xie, L.; Chen, H. M.; Yang, J. Bin. Conidia Production by *Beauveria bassiana* on Rice in Solid-State Fermentation Using Tray Bioreactor. *Adv. Mat. Res.* **2012,** 610–613, 3478–3482. https://doi.org/10.4028/www.scientific.net/AMR.610-613.3478.
23. Ye, S. D.; Ying, S. H.; Chen, C.; Feng, M. G. New Solid-State Fermentation Chamber for Bulk Production of Aerial Conidia of Fungal Biocontrol Agents on Rice. *Biotechnol. Lett.* **2006,** *28* (11), 799–804. https://doi.org/10.1007/s10529-006-9004-z.
24. Dallastra, E. D. G.; Dias, A. C. P.; de Morais, P. B.; da Silva, J. F. M.; Casciatori, F. P.; Grajales, L. M. Development of a Novel Pilot-Scale Tray Bioreactor for Solid-State Fermentation Aiming at Process Intensification. *Chem. Eng. Process.* **2023,** *193,* 109526. https://doi.org/10.1016/j.cep.2023.109526.
25. Brand, D.; Soccol, C. R.; Sabu, A.; Roussos, S. Production of Fungal Biological Control Agents through Solid State Fermentation: A Case Study on *Paecelomyces lilacinus* Againts Root-Knot Nematodes. *Micol. Apl. Int.* **2010,** *22* (1), 31–48.
26. Pandey, A. Aspects of Fermenter Design for Solid-State Fermentations. *Process Biochem.* **1991,** *26* (6), 355–361. https://doi.org/10.1016/0032-9592(91)85026-K.
27. Mitchell, D. A.; Lonsane, B. K.; Reanud, R.; Almanza, S.; Maratray, J.; Desgranes, C.; Crooke, P. S.; Hong, K.; Tanner, R. D.; Malaney, G. C. General Principles of Bioreactor Design and Operation for SSC. In *Solid State Cultivation*; Doelle, H., Mitchell, D. A., Rolz, C. E., Eds.; Elsevier Applied Science: London, 1993; pp 115–136.
28. Lopes-Perez, C.; Casciatori, F. P.; Thoméo, J. C. Strategies for Scaling-up Packed-Bed Bioreactors for Solid-State Fermentation: The Case of Cellulolytic Enzymes Production by a Thermophilic Fungus. *Chem. Eng. J.* **2019,** *361* (December 2018), 1142–1151. https://doi.org/10.1016/j.cej.2018.12.169.
29. Van Breukelen, F. R.; Haemers, S.; Wijffels, R. H.; Rinzema, A. Bioreactor and Substrate Selection for Solid-State Cultivation of the Malaria Mosquito Control Agent *Metarhizium anisopliae*. *Process Biochem.* **2011,** *46* (3), 751–757. https://doi.org/10.1016/j.procbio.2010.11.023.
30. da Cunha, L. P.; Casciatori, F. P.; Vicente, I. V.; Garcia, R. L.; Thoméo, J. C. *Metarhizium anisopliae* Conidia Production in Packed-Bed Bioreactor Using Rice as Substrate in Successive Cultivations. *Process Biochem.* **2020,** *97* (July), 104–111. https://doi.org/10.1016/j.procbio.2020.07.002.
31. Méndez-González, F.; Chávez-Escalante, G.; Loera, O.; Favela-Torres, E. What Does Respirometric Analysis Tell Us about *Metarhizium robertsii*. *J. Appl. Biotechnol. Bioeng.* **2022,** *9* (4), 94–96. https://doi.org/10.15406/jabb.2022.09.00292.
32. Henrique, J. P.; Casciatori, F. P.; Thoméo, J. C. Automatic System for Monitoring Gaseous Concentration in a Packed-Bed Solid-State Cultivation Bioreactor. *Chem. Eng. Sci.* **2022,** *259,* 117793. https://doi.org/10.1016/j.ces.2022.117793.
33. Casciatori, F. P.; Laurentino, C. L.; Taboga, S. R.; Casciatori, P. A.; Thoméo, J. C. Structural Properties of Beds Packed with Agro-Industrial Solid by-Products Applicable

for Solid-State Fermentation: Experimental Data and Effects on Process Performance. *Chem. Eng. J.* **2014,** *255,* 214–224. https://doi.org/10.1016/j.cej.2014.06.040.
34. Casciatori, F. P.; Thoméo, J. C. Heat Transfer in Packed-Beds of Agricultural Waste with Low Rates of Air Flow Applicable to Solid-State Fermentation. *Chem. Eng. Sci.* **2018,** *188,* 97–111. https://doi.org/10.1016/j.ces.2018.05.024.
35. Hamrouni, R.; Dupuy, N.; Karachurina, A.; Mitropoulou, G.; Kourkoutas, Y.; Molinet, J.; Maiga, Y.; Roussos, S. Biotechnological Potential of Zymotis-2 Bioreactor for the Cultivation of Filamentous Fungi. *Biotechnol. J.* **2022,** *17* (1). https://doi.org/10.1002/biot.202100288.
36. Saucedo-Castañeda, G.; Gutiérrez-Rojas, M.; Bacquet, G.; Raimbault, M.; Viniegra-González, G. Heat Transfer Simulation in Solid Substrate Fermentation. *Biotechnol. Bioeng.* **1990,** *35* (8), 802–808. https://doi.org/10.1002/bit.260350808.
37. Lopes, R. B.; Faria, M.; Glare, T. R. A Nonconventional Two-Stage Fermentation System for the Production of Aerial Conidia of Entomopathogenic Fungi Utilizing Surface Tension. *J. Appl. Microbiol.* **2019,** *126* (1), 155–164. https://doi.org/10.1111/jam.14137.
38. Manan, M. A.; Webb, C. Newly Designed Multi-Stacked Circular Tray Solid-State Bioreactor: Analysis of a Distributed Parameter Gas Balance during Solid-State Fermentation with Influence of Variable Initial Moisture Content Arrangements. *Bioresour. Bioproc.* **2020,** *7* (1). https://doi.org/10.1186/s40643-020-00307-9.
39. Lonsane, B. K.; Saucedo-Castañeda, G.; Raimbault, M.; Roussos, S.; Viniegra-González, G.; Ghildyal, N. P.; Ramakrishna, M.; Krishnaiah, M. M. Scale-up Strategies for Solid State Fermentation Systems. *Process Biochem.* **1992,** *27* (5), 259–273.
40. Durand, A. Bioreactor Designs for Solid State Fermentation. *Biochem. Eng. J.* **2003,** *13* (2–3), 113–125.
41. Méndez-González, F.; Loera, O.; Favela-Torres, E. Conidia Production of *Metarhizium anisopliae* in Bags and Packed Column Bioreactors. *Curr. Biotechnol.* **2018,** *7* (1), 65–69. https://doi.org/10.2174/2211550105666160926123350.
42. Jenkins, N. E.; Goettel, M. S. Methods for Mass-Production of Microbial Control Agents of Grasshoppers and Locusts. *Mem. Ent. Soc. Can.* **2011,** *129* (S171), 37–48. https://doi.org/10.4039/entm129171037-1.
43. Arias Barreto, A.; Abalone, R.; Gastón, A.; Ochandio, D.; Cardoso, L.; Bartosik, R. Validation of a Heat, Moisture and Gas Concentration Transfer Model for Soybean (Glycine Max) Grains Stored in Plastic Bags (Silo Bags). *Biosyst. Eng.* **2017,** *158,* 23–37. https://doi.org/10.1016/j.biosystemseng.2017.03.009.
44. Jaronski, S. T. Mass Production of Entomopathogenic Fungi—State of the Art. In *Mass Production of Beneficial Organisms*; Morales-Ramos, J., Rojas, G., Shapiro-Ilan, Eds.; Elsevier, 2023; pp 317–357. https://doi.org/10.1016/B978-0-12-822106-8.00017-8.
45. Swanson, D. Economic Feasibility of Two Technologies for Production of Mycopesticide in Magadascar. *Mem. Ent. Soc. Can.* **1997,** *129* (S171), 101–113. https://doi.org/10.4039/entm129171101-1.
46. Cherry, A. J.; Jenkins, N. E.; Heviefo, G.; Bateman, R.; Lomer, C. J. Operational and Economic Analysis of a West African Pilot-Scale Production Plant for Aerial Conidia of *Metarhizium* Spp. for Use as a Mycoinsecticide against Locusts and Grasshoppers. *Biocontrol. Sci. Technol.* **1999,** *9* (1), 35–51. https://doi.org/10.1080/09583159929893.
47. Desobgo, S. C. Z.; Mishra, S. S.; Behera, S. K.; Panda, S. K. Scaling-up and Modelling Applications of Solid-State Fermentation and Demonstration in Microbial Enzyme

Production Related to Food Industries. In *Microbial Enzyme Technology in Food Applications*; Ray, R., Rosell, C., Eds.; [2016] | Series: Food biology series | A science publishers book.; CRC Press: Boca Raton, FL : CRC Press, 2017; pp 452–468. https://doi.org/10.1201/9781315368405-29.

48. Xue, M.; Liu, D.; Zhang, H.; Qi, H.; Lei, Z. A Pilot Process of Solid State Fermentation from Sugar Beet Pulp for the Production of Microbial Protein. *J. Ferment. Bioeng.* **1992,** *73,* 203–205.
49. Durand, A.; Chereau, D. A New Pilot Reactor for Solid-State Fermentation: Application to the Protein Enrichment of Sugar Beet Pulp. *Biotechnol. Bioeng.* **1988,** *31* (5), 476–486. https://doi.org/10.1002/bit.260310513.
50. Deans, C.; Krischik, V. The Current State and Future Potential of Microbial Control of Scarab Pests. *Appl. Sci.* **2023,** *13* (2), 766. https://doi.org/10.3390/app13020766.
51. Mattedi, A.; Sabbi, E.; Farda, B.; Djebaili, R.; Mitra, D.; Ercole, C.; Cacchio, P.; Del Gallo, M.; Pellegrini, M. Solid-State Fermentation: Applications and Future Perspectives for Biostimulant and Biopesticides Production. *Microorganisms* **2023,** *11* (6), 1408. https://doi.org/10.3390/microorganisms11061408.
52. Triolet, M.; Guillemin, J.; Andre, O.; Steinberg, C. Fungal-based Bioherbicides for Weed Control: A Myth or a Reality? *Weed Res.* **2020,** *60* (1), 60–77. https://doi.org/10.1111/wre.12389.
53. Roberts, J.; Florentine, S.; Fernando, W. G. D.; Tennakoon, K. U. Achievements, Developments and Future Challenges in the Field of Bioherbicides for Weed Control: A Global Review. *Plants* **2022,** *11* (17), 2242. https://doi.org/10.3390/plants11172242.
54. Mathulwe, L. L.; Malan, A. P.; Stokwe, N. F. Mass Production of Entomopathogenic Fungi, *Metarhizium robertsii* and *Metarhizium pinghaens*, for Commercial Application Against Insect Pests. *J. Vis. Exp.* **2022,** No. 181. https://doi.org/10.3791/63246.
55. Woo, S. L.; Hermosa, R.; Lorito, M.; Monte, E. Trichoderma: A Multipurpose, Plant-Beneficial Microorganism for Eco-Sustainable Agriculture. *Nat. Rev. Microbiol.* **2023,** *21* (5), 312–326. https://doi.org/10.1038/s41579-022-00819-5.
56. Hegedus, D. D.; Khachatourians, G. G. The Impact of Biotechnology on Hyphomycetous Fungal Insect Biocontrol Agents. *Biotechnol. Adv.* **1995,** *13* (3), 455–490. https://doi.org/10.1016/0734-9750(95)02006-O.
57. Ons, L.; Bylemans, D.; Thevissen, K.; Cammue, B. P. A. Combining Biocontrol Agents with Chemical Fungicides for Integrated Plant Fungal Disease Control. *Microorganisms* **2020,** *8* (12), 1930. https://doi.org/10.3390/microorganisms8121930.
58. Sharma, A.; Kaushik, N.; Sharma, A.; Djébali, N. Development of Novel Liquid Formulation of Bacillus Siamensis with Antifungal and Plant Growth Promoting Activity. *ACS Agric. Sci. Technol.* **2023,** *3* (1), 55–66. https://doi.org/10.1021/acsagscitech.2c00224.
59. Gautam, K.; Sirohi, C.; Singh, N. R.; Thakur, Y.; Jatav, S. S.; Rana, K.; Chitara, M.; Meena, R. P.; Singh, A. K.; Parihar, M. Microbial Biofertilizer: Types, Applications, and Current Challenges for Sustainable Agricultural Production. In *Biofertilizers*; Rakshit, A., Singh, V., Parihar, M., Singh, H., Singh, A., Eds.; Elsevier, 2021; pp 3–19. https://doi.org/10.1016/B978-0-12-821667-5.00014-2.
60. Grzywacz, D.; Moore, D.; Rabindra, R. J. Mass Production of Entomopathogens in Less Industrialized Countries. In Mass Production of Beneficial Organisms; Morales-Ramos, J., Rojas, G., Shapiro-Ilan, D., Eds.; Elsevier, 2014; pp 519–561. https://doi.org/10.1016/B978-0-12-391453-8.00015-7.

61. Diego-Nava, F.; Granados-Echegoyen, C.; Ruíz-Vega, J.; Aquino-Bolaños, T.; Pérez-Pacheco, R.; Díaz-Ramos, A.; Alonso-Hernández, N.; Arroyo-Balán, F.; López-Hernández, M. B. Functional and Quality Assessment of a Spore Harvester for Entomopathogenic Fungi for Biopesticide Production. *AgriEngineering* **2023,** *5* (2), 801–813. https://doi.org/10.3390/agriengineering5020049.
62. Velozo, S. G. M.; Velozo, M. R.; Domingues, M. M.; Becchi, L. K.; Carvalho, V. R. de; Passos, J. R. de S.; Zanuncio, J. C.; Serrão, J. E.; Stephan, D.; Wilcken, C. F. From the Dual Cyclone Harvest Performance of Single Conidium Powder to the Effect of *Metarhizium anisopliae* on the Management of Thaumastocoris Peregrinus (Hemiptera: Thaumastocoridae). *PLoS One* **2023,** *18* (3), e0283543. https://doi.org/10.1371/journal.pone.0283543.
63. Barra-Bucarei, L.; Vergara, P.; Cortes, A. Conditions to Optimize Mass Production of *Metarhizium anisopliae* (Metschn.) Sorokin 1883 in Different Substrates. *Chil. J. Agric. Res.* **2016,** *76* (4), 448–454. https://doi.org/10.4067/S0718-58392016000400008.
64. Adetunji, C. O.; Oloke, J. K. Effect of Wild and Mutant Strain of *Lasiodiplodia pseudotheobromae* Mass Produced on Rice Bran as a Potential Bioherbicide Agents for Weeds under Solid State Fermentation **2013,** *1* (02), 18–23. https://doi.org/10.7324/JABB.2013.1204.
65. Dalla Santa, H. S.; Sousa, N. J.; Brand, D.; Dalla Santa, O. R.; Pandey, A.; Sobotka, M.; Páca, J.; Soccol, C. R. Conidia Production of *Beauveria* sp. by Solid-State Fermentation for Biocontrol of *Ilex paraguariensis* Caterpillars. *Folia Microbiol. (Praha)* **2004,** *49* (4), 418–422. https://doi.org/10.1007/BF02931603.
66. Sala, A.; Vittone, S.; Barrena, R.; Sánchez, A.; Artola, A. Scanning Agro-Industrial Wastes as Substrates for Fungal Biopesticide Production: Use of *Beauveria bassiana* and *Trichoderma harzianum* in Solid-State Fermentation. *J. Environ. Manage* **2021,** *295,* 113113. https://doi.org/10.1016/j.jenvman.2021.113113.
67. Sala, A.; Barrena, R.; Meyling, N. V.; Artola, A. Conidia Production of the Entomopathogenic Fungus *Beauveria bassiana* Using Packed-Bed Bioreactor: Effect of Substrate Biodegradability on Conidia Virulence. *J. Environ. Manage* **2023,** *341,* 118059. https://doi.org/10.1016/j.jenvman.2023.118059.
68. De la Cruz-Quiroz, R.; Robledo-Padilla, F.; Aguilar, C. N.; Roussos, S. Forced Aeration Influence on the Production of Spores by *Trichoderma* Strains. *Waste Biomass Valor.* **2017,** *8* (7), 2263–2270. https://doi.org/10.1007/s12649-017-0045-4.
69. Rayhane, H.; Josiane, M.; Gregoria, M.; Yiannis, K.; Nathalie, D.; Ahmed, M.; Sevastianos, R. From Flasks to Single Used Bioreactor: Scale-up of Solid State Fermentation Process for Metabolites and Conidia Production by *Trichoderma asperellum*. *J. Environ. Manage* **2019,** *252,* 109496. https://doi.org/10.1016/j.jenvman.2019.109496.
70. De Vrije, T.; Antoine, N.; Buitelaar, R. M.; Bruckner, S.; Dissevelt, M.; Durand, A.; Gerlagh, M.; Jones, E. E.; Lüth, P.; Oostra, J.; Ravensberg, W. J.; Renaud, R.; Rinzema, A.; Weber, F. J.; Whipps, J. M. The Fungal Biocontrol Agent *Coniothyrium minitans*: Production by Solid-State Fermentation, Application and Marketing. *Appl. Microbiol. Biotechnol.* **2001,** *56* (1–2), 58–68. https://doi.org/10.1007/s002530100678.
71. Berikashvili, V.; Sokhadze, K.; Kachlishvili, E.; Elisashvili, V.; Chikindas, M. L. *Bacillus amyloliquefaciens* Spore Production Under Solid-State Fermentation of Lignocellulosic Residues. *Probiotics Antimicrob. Proteins* **2018,** *10* (4), 755–761. https://doi.org/10.1007/s12602-017-9371-x.

72. Mizumoto, S.; Shoda, M. Medium Optimization of Antifungal Lipopeptide, Iturin A, Production by *Bacillus subtilis* in Solid-State Fermentation by Response Surface Methodology. *Appl. Microbiol. Biotechnol.* **2007,** *76* (1), 101–108. https://doi.org/10.1007/s00253-007-0994-9.
73. Chapman, J.; Ismail, A.; Dinu, C. Industrial Applications of Enzymes: Recent Advances, Techniques, and Outlooks. *Catalysts* **2018,** *8* (6), 238. https://doi.org/10.3390/catal8060238.
74. Soccol, C. R.; Costa, E. S. F. da; Letti, L. A. J.; Karp, S. G.; Woiciechowski, A. L.; Vandenberghe, L. P. de S. Recent Developments and Innovations in Solid State Fermentation. *Biotechnol. Res. Innov.* **2017,** *1* (1), 52–71. https://doi.org/10.1016/j.biori.2017.01.002.
75. Abu Yazid, N.; Barrena, R.; Komilis, D.; Sánchez, A. Solid-State Fermentation as a Novel Paradigm for Organic Waste Valorization: A Review. *Sustainability* **2017,** *9* (2), 224. https://doi.org/10.3390/su9020224.
76. Almanaa, T. N.; Vijayaraghavan, P.; Alharbi, N. S.; Kadaikunnan, S.; Khaled, J. M.; Alyahya, S. A. Solid State Fermentation of Amylase Production from *Bacillus subtilis* D19 Using Agro-Residues. *J. King. Saud Univ. Sci.* **2020,** *32* (2), 1555–1561. https://doi.org/10.1016/j.jksus.2019.12.011.
77. Pranay, K.; Padmadeo, S. R.; Prasad, B. Production of Amylase from *Bacillus subtilis* sp. Strain KR1 under Solid State Fermentation on Different Agrowastes. *Biocatal. Agric Biotechnol.* **2019,** *21,* 101300. https://doi.org/10.1016/j.bcab.2019.101300.
78. Mojumdar, A.; Deka, J. Recycling Agro-Industrial Waste to Produce Amylase and Characterizing Amylase–Gold Nanoparticle Composite. *Int. J. Recycl. Org. Waste Agric.* **2019,** *8* (S1), 263–269. https://doi.org/10.1007/s40093-019-00298-4.
79. Premalatha, A.; Vijayalakshmi, K.; Shanmugavel, M.; Rajakumar, G. S. Optimization of Culture Conditions for Enhanced Production of Extracellular A-amylase Using Solid-state and Submerged Fermentation from *Aspergillus tamarii* MTCC5152. *Biotechnol. Appl. Biochem.* **2023,** *70* (2), 835–845. https://doi.org/10.1002/bab.2403.
80. Melnichuk, N.; Braia, M. J.; Anselmi, P. A.; Meini, M.-R.; Romanini, D. Valorization of Two Agroindustrial Wastes to Produce Alpha-Amylase Enzyme from *Aspergillus oryzae* by Solid-State Fermentation. *Waste Manag.* **2020,** *106,* 155–161. https://doi.org/10.1016/j.wasman.2020.03.025.
81. Hamrouni, R.; Molinet, J.; Dupuy, N.; Taieb, N.; Carboue, Q.; Masmoudi, A.; Roussos, S. The Effect of Aeration for 6-Pentyl-Alpha-Pyrone, Conidia and Lytic Enzymes Production by *Trichoderma asperellum* Strains Grown in Solid-State Fermentation. *Waste Biomass Valor.* **2020,** *11* (11), 5711–5720. https://doi.org/10.1007/s12649-019-00809-4.
82. Kuhad, R. C.; Deswal, D.; Sharma, S.; Bhattacharya, A.; Jain, K. K.; Kaur, A.; Pletschke, B. I.; Singh, A.; Karp, M. Revisiting Cellulase Production and Redefining Current Strategies Based on Major Challenges. *Renew. Sustain. Energy Rev.* **2016,** *55,* 249–272. https://doi.org/10.1016/j.rser.2015.10.132.
83. Zamora Zamora, H. D.; Silva, T. A. L.; Varão, L. H. R.; Baffi, M. A.; Pasquini, D. Simultaneous Production of Cellulases, Hemicellulases, and Reducing Sugars by *Pleurotus ostreatus* Growth in One-Pot Solid State Fermentation Using Alstroemeria Sp. Waste. *Biomass Convers Biorefin.* **2023,** *13* (6), 4879–4892. https://doi.org/10.1007/s13399-021-01723-3.
84. Castro-Ochoa, L. D.; Hernández-Leyva, S. R.; Medina-Godoy, S.; Gómez-Rodríguez, J.; Aguilar-Uscanga, M. G.; Castro-Martínez, C. Integration of Agricultural Residues

as Biomass Source to Saccharification Bioprocess and for the Production of Cellulases from Filamentous Fungi. *3 Biotech* **2023,** *13* (2), 43. https://doi.org/10.1007/s13205-022-03444-4.

85. Legodi, L. M.; La Grange, D. C.; van Rensburg, E. L. J. Production of the Cellulase Enzyme System by Locally Isolated *Trichoderma* and *Aspergillus* Species Cultivated on *Banana pseudostem* during Solid-State Fermentation. *Fermentation* **2023,** *9* (5), 412. https://doi.org/10.3390/fermentation9050412.
86. Lodha, A.; Pawar, S.; Rathod, V. Optimised Cellulase Production from Fungal Co-Culture of *Trichoderma reesei* NCIM 1186 and *Penicillium citrinum* NCIM 768 under Solid State Fermentation. *J. Environ. Chem. Eng.* **2020,** *8* (5), 103958. https://doi.org/10.1016/j.jece.2020.103958.
87. Kapoor, M.; Gupta, M. N. Lipase Promiscuity and Its Biochemical Applications. *Process Biochem.* **2012,** *47* (4), 555–569. https://doi.org/10.1016/j.procbio.2012.01.011.
88. Ferrarezi, A. L.; Hideyuki Kobe Ohe, T.; Borges, J. P.; Brito, R. R.; Siqueira, M. R.; Vendramini, P. H.; Quilles, J. C.; da Costa Carreira Nunes, C.; Bonilla-Rodriguez, G. O.; Boscolo, M.; Da-Silva, R.; Gomes, E. Production and Characterization of Lipases and Immobilization of Whole Cell of the Thermophilic *Thermomucor indicae seudaticae* N31 for Transesterification Reaction. *J. Mol. Catal. B Enzym.* **2014,** *107,* 106–113. https://doi.org/10.1016/j.molcatb.2014.05.012.
89. Aguieiras, E. C. G.; Cavalcanti-Oliveira, E. D.; Cammarota, M. C.; Freire, D. M. G. Solid-State Fermentation for the Production of Lipases for Environmental and Biodiesel Applications. In *Current Developments in Biotechnology and Bioengineering*; Pandey A., Larroche, C., Soccol, C., Eds.; Elsevier, 2018; pp 123–168. https://doi.org/10.1016/B978-0-444-63990-5.00008-6.
90. Hubbe, M. A.; Lavoine, N.; Lucia, L. A.; Dou, C. Formulating Bioplastic Composites for Biodegradability, Recycling, and Performance: A Review. *Bioresources* **2020,** *16* (1), 2021–2083. https://doi.org/10.15376/biores.16.1.Hubbe.
91. Suyanto, E.; Soetarto, E. S.; Cahyanto, M. N. Production and Optimization of Lipase by Aspergillus Niger Using Coconut Pulp Waste in Solid State Fermentation. *J. Phys. Conf. Ser.* **2019,** *1374* (1), 012005. https://doi.org/10.1088/1742-6596/1374/1/012005.
92. Ali, U.; Anwar, Z.; Hasan, S.; Zafar, M.; Ain, N. ul; Afzal, F.; Khalid, W.; Rahim, M. A.; Mrabti, H. N.; AL-Farga, A.; Eljeam, H. A. R. A. Bioprocessing and Screening of Indigenous Wastes for Hyper Production of Fungal Lipase. *Catalysts* **2023,** *13* (5), 853. https://doi.org/10.3390/catal13050853.
93. Chilakamarry, C. R.; Mimi Sakinah, A. M.; Zularisam, A. W.; Sirohi, R.; Khilji, I. A.; Ahmad, N.; Pandey, A. Advances in Solid-State Fermentation for Bioconversion of Agricultural Wastes to Value-Added Products: Opportunities and Challenges. *Bioresour. Technol.* **2022,** *343,* 126065. https://doi.org/10.1016/j.biortech.2021.126065.
94. Poomani, M. S.; Mariappan, I.; Muthan, K.; Subramanian, V. A Thermotolerant Yeast from Cow's Rumen Utilize Lignocellulosic Biomass from Wheat Straw for Xylanase Production and Fermentation to Ethanol. *Biocatal. Agric. Biotechnol.* **2023,** *50,* 102741. https://doi.org/10.1016/j.bcab.2023.102741.
95. Sun, S.; Sun, S.; Cao, X.; Sun, R. The Role of Pretreatment in Improving the Enzymatic Hydrolysis of Lignocellulosic Materials. *Bioresour. Technol.* **2016,** *199,* 49–58. https://doi.org/10.1016/j.biortech.2015.08.061.
96. Cruz-Vázquez, A.; Tomasini, A.; Armas-Tizapantzi, A.; Marcial-Quino, J.; Montiel-González, A. M. Extracellular Proteases and Laccases Produced by *Pleurotus ostreatus*

PoB: The Effects of Proteases on Laccase Activity. *Int. Microbiol.* **2022,** *25* (3), 495–502. https://doi.org/10.1007/s10123-022-00238-9.

97. Kudanga, T.; Le Roes-Hill, M. Laccase Applications in Biofuels Production: Current Status and Future Prospects. *Appl. Microbiol. Biotechnol.* **2014,** *98* (15), 6525–6542. https://doi.org/10.1007/s00253-014-5810-8.
98. Backes, E.; Kato, C. G.; da Silva, T. B. V.; Uber, T. M.; Pasquarelli, D. L.; Bracht, A.; Peralta, R. M. Production of Fungal Laccase on Pineapple Waste and Application in Detoxification of Malachite Green. *J. Environ. Sci. Health, Part B* **2022,** *57* (2), 90–101. https://doi.org/10.1080/03601234.2022.2025739.
99. Backes, E.; Kato, C. G.; de Oliveira Junior, V. A.; Uber, T. M.; dos Santos, L. F. O.; Corrêa, R. C. G.; Bracht, A.; Peralta, R. M. Overproduction of Laccase by *Trametes versicolor* and *Pycnoporus sanguineus* in Farnesol-Pineapple Waste Solid Fermentation. *Fermentation* **2023,** *9* (2), 188. https://doi.org/10.3390/fermentation9020188.
100. Cavalieri de Alencar Guimarães, N.; Glienke, N. N.; Silva Galeano, R. M.; Ruller, R.; Zanoelo, F. F.; Masui, D. C.; Giannesi, G. C. Polygalacturonase from *Aspergillus japonicus* (PGAj): Enzyme Production Using Low-Cost Carbon Source, Biochemical Properties and Application in Clarification of Fruit Juices. *Biocatal. Agric. Biotechnol.* **2022,** *39,* 102233. https://doi.org/10.1016/j.bcab.2021.102233.
101. Núñez Pérez, J.; Chávez Arias, B. S.; de la Vega Quintero, J. C.; Zarate Baca, S.; Pais-Chanfrau, J. M. Multi-Objective Statistical Optimization of Pectinolytic Enzymes Production by an *Aspergillus* sp. on Dehydrated Coffee Residues in Solid-State Fermentation. *Fermentation* **2022,** *8* (4), 170. https://doi.org/10.3390/fermentation8040170.
102. Bender, J. P.; Mazutti, M. A.; Oliveira, Dé. De; Luccio, M. Di; Treichel, H. Inulinase Production by *Kluyveromyces marxianus* NRRL Y-7571 Using Solid State Fermentation. *Appl. Biochem. Biotechnol.* **2006,** *132* (1–3), 951–958. https://doi.org/10.1385/ABAB:132:1:951.
103. Tippanavar, N.; Bhat, D.; Rebello, O.; Prabhu, G.; Selvaraj, S.; Bhat, R. M. Bio-Utilization of Agricultural Residue Banana Plant Shoot through Solid State Fermentation for Production of Inulinase Using Newly Isolated *Nothophoma anigozanthi* JAM. *Biomass Convers Biorefin.* **2023,** *14*, 14755–14769. https://doi.org/10.1007/s13399-023-03835-4.
104. Singh, A.; Bajar, S.; Devi, A.; Bishnoi, N. R. Adding Value to Agro-Industrial Waste for Cellulase and Xylanase Production via Solid-State Bioconversion. *Biomass Convers Biorefin.* **2023,** *13* (9), 7481–7490. https://doi.org/10.1007/s13399-021-01503-z.
105. Putri, D. N.; Khootama, A.; Perdani, M. S.; Utami, T. S.; Hermansyah, H. Optimization of *Aspergillus niger* Lipase Production by Solid State Fermentation of Agro-Industrial Waste. *Energy Rep.* **2020,** *6,* 331–335. https://doi.org/10.1016/j.egyr.2019.08.064.
106. Perdani, M. S.; Margaretha, G.; Sahlan, M.; Hermansyah, H. Solid State Fermentation Method for Production of Laccase Enzyme with Bagasse, Cornstalk and Rice Husk as Substrates for Adrenaline Biosensor. *Energy Rep.* **2020,** *6,* 336–340. https://doi.org/10.1016/j.egyr.2019.08.065.
107. Maïga, Y.; Carboué, Q.; Hamrouni, R.; Tranier, M.-S.; Ben Menadi, Y.; Roussos, S. Development and Evaluation of a Disposable Solid-State Culture Packed-Bed Bioreactor for the Production of Conidia from *Trichoderma asperellum* Grown Under Water Stress. *Waste Biomass Valor.* **2021,** *12* (6), 3223–3231. https://doi.org/10.1007/s12649-020-01210-2.

108. Qureshi, A. S.; Khushk, I.; Ali, C. H.; Chisti, Y.; Ahmad, A.; Majeed, H. Coproduction of Protease and Amylase by Thermophilic *Bacillus* Sp. BBXS-2 Using Open Solid-State Fermentation of Lignocellulosic Biomass. *Biocatal. Agric. Biotechnol.* **2016,** *8,* 146–151. https://doi.org/10.1016/j.bcab.2016.09.006.
109. Abdel-Mageed, H. M.; Barakat, A. Z.; Bassuiny, R. I.; Elsayed, A. M.; Salah, H. A.; Abdel-Aty, A. M.; Mohamed, S. A. Biotechnology Approach Using Watermelon Rind for Optimization of α-Amylase Enzyme Production from *Trichoderma virens* Using Response Surface Methodology under Solid-State Fermentation. *Folia. Microbiol. (Praha)* **2022,** *67* (2), 253–264. https://doi.org/10.1007/s12223-021-00929-2.
110. Saldaña-Mendoza, S. A.; Palacios-Ponce, A. S.; Ruiz, H. A.; Ascacio-Valdés, J. A.; Aguilar, C. N. Revalorization of Green Tea Waste through the Production of Cellulases by Solid-State Fermentation Using a Aspergillus Niger 28A. *Biomass Convers Biorefin.* **2024,** 14, 16711–16724. https://doi.org/10.1007/s13399-023-03919-1.
111. Nema, A.; Patnala, S. H.; Mandari, V.; Kota, S.; Devarai, S. K. Production and Optimization of Lipase Using Aspergillus Niger MTCC 872 by Solid-State Fermentation. *Bull Natl. Res. Cent.* **2019,** *43* (1), 82. https://doi.org/10.1186/s42269-019-0125-7.
112. Mazhar, H.; Ullah, I.; Ali, U.; Abbas, N.; Hussain, Z.; Ali, S. S.; Zhu, H. Optimization of Low-Cost Solid-State Fermentation Media for the Production of Thermostable Lipases Using Agro-Industrial Residues as Substrate in Culture of *Bacillus amyloliquefaciens*. *Biocatal. Agric. Biotechnol.* **2023,** *47,* 102559. https://doi.org/10.1016/j.bcab.2022.102559.
113. Afiqah Razali, S.; Rasit, N.; Kuan Ooi, C. Statistical Analysis of Xylanase Production from Solid State Fermentation of Rice Husk Associated Fungus *Aspergillus niger*. *Mater Today Proc.* **2021,** *39,* 1082–1087. https://doi.org/10.1016/j.matpr.2020.06.366.
114. Sondhi, S.; Saini, K. Response Surface Based Optimization of Laccase Production from *Bacillus* sp. MSK-01 Using Fruit Juice Waste as an Effective Substrate. *Heliyon* **2019,** *5* (5), e01718. https://doi.org/10.1016/j.heliyon.2019.e01718.
115. Wang, L.; An, N.; Gao, J.; Xue, H.; Li, G. The Feasibility of Sodium Hydroxide Pretreatment of Rice Straw for Solid Substrate Preparation to Enhance Laccase Production by Solid State Fermentation. *BMC Biotechnol.* **2023,** *23* (1), 16. https://doi.org/10.1186/s12896-023-00789-3.
116. Chablé-Villacis, R.; Olguin-Maciel, E.; Toledano-Thompson, T.; Alzate-Gaviria, L.; Ruiz, H. A.; Tapia-Tussell, R. Enzymatic Hydrolysis Assisted with Ligninocellulolytic Enzymes from *Trametes hirsuta* Produced by Pineapple Leaf Waste Bioconversion in Solid-State Fermentation. *Biomass Convers Biorefin.* **2023,** *13* (10), 9095–9106. https://doi.org/10.1007/s13399-021-01851-w.
117. Riffat, A.; Anwar, Z.; Zafar, M.; Nadeem, F.; Mehmood, T. Optimization of Physicochemical Parameters and Characterization Laccase Enzyme Produced by a Novel Strain of *Fomes fomentarius* through Solid-State Fermentation. *Biomass Convers Biorefin.* **2024,** 14, 3999–4006. https://doi.org/10.1007/s13399-022-02621-y.
118. Javed, F.; Ali, Z.; Ali, S.; Ahmed, N.; Alam, M. K.; Mahmood, Y.; Wali, A. Barley Bran, a Novel Agricultural Waste for the Improved Production of an Extracellular Lacsasse from a Soil-Inhabited *Penicillium* spp. *J. Microbiol., Biotechnol. Food Sci.* **2022,** e3631. https://doi.org/10.55251/jmbfs.3631.
119. Peña-Lucio, E. M.; Londoño-Hernández, L.; Ascacio-Valdes, J. A.; Chavéz-González, M. L.; Bankole, O. E.; Aguilar, C. N. Use of Coffee Pulp and Sorghum Mixtures in the

Production of N-Demethylases by Solid-State Fermentation. *Bioresour. Technol.* **2020,** *305,* 123112. https://doi.org/10.1016/j.biortech.2020.123112.
120. Pedraza-Segura, L.; Rodríguez-Durán, L. V.; Saucedo-Castañeda, G.; de Jesús Cázares-Marinero, J. Application of Microorganisms in Biosurfactant Production. In *Bioprospecting of Microorganism-Based Industrial Molecules*; Singh S., Upadhyay S. K., Eds.; Wiley, 2021; pp 6–30. https://doi.org/10.1002/9781119717317.ch2.
121. El-Housseiny, G. S.; Aboshanab, K. M.; Aboulwafa, M. M.; Hassouna, N. A. Rhamnolipid Production by a Gamma Ray-Induced *Pseudomonas aeruginosa* Mutant under Solid State Fermentation. *AMB Express* **2019,** *9* (1), 7. https://doi.org/10.1186/s13568-018-0732-y.
122. Dabaghi, S.; Ataei, S. A.; Taheri, A. Production of Rhamnolipid Biosurfactants in Solid-State Fermentation: Process Optimization and Characterization Studies. *BMC Biotechnol.* **2023,** *23* (1), 2. https://doi.org/10.1186/s12896-022-00772-4.
123. Vedaraman, N.; Venkatesh, N. Production of Surfactin by *Bacillus subtilis* mtcc 2423 from Waste Frying Oils. *Braz. J. Chem. Eng.* **2011,** *28* (2), 175–180. https://doi.org/10.1590/S0104-66322011000200001.
124. Van Bogaert, I. N. A.; Zhang, J.; Soetaert, W. Microbial Synthesis of Sophorolipids. *Process Biochem.* **2011,** *46* (4), 821–833. https://doi.org/10.1016/j.procbio.2011.01.010.
125. Colla, L. M.; Rizzardi, J.; Pinto, M. H.; Reinehr, C. O.; Bertolin, T. E.; Costa, J. A. V. Simultaneous Production of Lipases and Biosurfactants by Submerged and Solid-State Bioprocesses. *Bioresour. Technol.* **2010,** *101* (21), 8308–8314. https://doi.org/10.1016/j.biortech.2010.05.086.
126. Oiza, N.; Moral-Vico, J.; Sánchez, A.; Oviedo, E. R.; Gea, T. Solid-State Fermentation from Organic Wastes: A New Generation of Bioproducts. *Processes* **2022,** *10* (12), 2675. https://doi.org/10.3390/pr10122675.
127. Domínguez Rivera, Á.; Martínez Urbina, M. Á.; López y López, V. E. Advances on Research in the Use of Agro-Industrial Waste in Biosurfactant Production. *World J. Microbiol. Biotechnol.* **2019,** *35* (10), 155. https://doi.org/10.1007/s11274-019-2729-3.
128. Van Bogaert, I. N. A.; Zhang, J.; Soetaert, W. Microbial Synthesis of Sophorolipids. *Process Biochem.* **2011,** *46* (4), 821–833. https://doi.org/10.1016/j.procbio.2011.01.010.
129. Nitschke, M.; Silva, S. S. e. Recent Food Applications of Microbial Surfactants. *Crit. Rev. Food. Sci. Nutr.* **2018,** *58* (4), 631–638. https://doi.org/10.1080/10408398.2016.1208635.
130. Bjerk, T. R.; Severino, P.; Jain, S.; Marques, C.; Silva, A. M.; Pashirova, T.; Souto, E. B. Biosurfactants: Properties and Applications in Drug Delivery, Biotechnology and Ecotoxicology. *Bioengineering* **2021,** *8* (8), 115. https://doi.org/10.3390/bioengineering8080115.
131. Dierickx, S.; Castelein, M.; Remmery, J.; De Clercq, V.; Lodens, S.; Baccile, N.; De Maeseneire, S. L.; Roelants, S. L. K. W.; Soetaert, W. K. From Bumblebee to Bioeconomy: Recent Developments and Perspectives for Sophorolipid Biosynthesis. *Biotechnol. Adv.* **2022,** *54,* 107788. https://doi.org/10.1016/j.biotechadv.2021.107788.
132. Jiménez-Peñalver, P.; Gea, T.; Sánchez, A.; Font, X. Production of Sophorolipids from Winterization Oil Cake by Solid-State Fermentation: Optimization, Monitoring and Effect of Mixing. *Biochem. Eng. J.* **2016,** *115,* 93–100. https://doi.org/10.1016/j.bej.2016.08.006.

133. Szmigiel, I.; Kwiatkowska, D.; Łukaszewicz, M.; Krasowska, A. Xylan Decomposition in Plant Cell Walls as an Inducer of Surfactin Synthesis by *Bacillus subtilis*. *Biomolecules* **2021,** *11* (2), 239. https://doi.org/10.3390/biom11020239.
134. Slivinski, C. T.; Mallmann, E.; de Araújo, J. M.; Mitchell, D. A.; Krieger, N. Production of Surfactin by *Bacillus pumilus* UFPEDA 448 in Solid-State Fermentation Using a Medium Based on Okara with Sugarcane Bagasse as a Bulking Agent. *Process Biochem.* **2012,** *47* (12), 1848–1855. https://doi.org/10.1016/j.procbio.2012.06.014.
135. Bouassida, M.; Mnif, I.; Hammami, I.; Triki, M.-A.; Ghribi, D. *Bacillus subtilis* SPB1 Lipopeptide Biosurfactant: Antibacterial Efficiency against the Phytopathogenic Bacteria *Agrobacterium tumefaciens* and Compared Production in Submerged and Solid State Fermentation Systems. *Food Sci. Biotechnol.* **2023,** *32* (11), 1595–1609. https://doi.org/10.1007/s10068-023-01274-5.
136. Asgher, M.; Arshad, S.; Qamar, S. A.; Khalid, N. Improved Biosurfactant Production from *Aspergillus niger* through Chemical Mutagenesis: Characterization and RSM Optimization. *SN Appl. Sci.* **2020,** *2* (5), 966. https://doi.org/10.1007/s42452-020-2783-3.

CHAPTER 3

Innovation in the Sustainable Development of Functional Foods and Nutraceuticals from Fruit and Vegetable Waste

LILIANA LONDOÑO-HERNÁNDEZ[1*], DIANA YEPES-BETANCOURT[2], MAGDA PIEDAD VALDÉS-RESTREPO[1], ANDREA VÁSQUEZ-GARCÍA[1], and GINNA ALEJANDRA ORDOÑEZ NARVÁEZ[1]

[1]*Biotics Group, School of Basic Sciences, Technology and Engineering. Universidad Nacional Abierta y a Distancia—UNAD, Colombia*

[2]*Servicio Nacional de Aprendizaje, SENA, Rionegro, Colombia*

*Corresponding author.

ABSTRACT

Increasing consumer demand for healthy and safe foods that offer benefits beyond basic nutrition and improve their health has driven research into the development of functional foods. According to the World Health Organization (WHO), functional and nutraceutical foods are those that have been scientifically proven to provide beneficial effects on the health of the target population groups. Consequently, in recent years, there has been a trend toward the incorporation of bioactive compounds in food matrices to obtain functional foods. Bioactive compounds are substances derived from the secondary metabolism of plants, animals, and microorganisms, which have active properties such as antioxidants, antimicrobial, anticarcinogenic, among others. The market for bioactive compounds is experiencing constant expansion, which highlights the importance of seeking alternative sources to obtain them. In this context, fruit and vegetable residues are presented as

Agroindustrial Waste Management and Natural Resources Conservation: Environmental Challenges and Transformation into Value-Added Products. José Juan Buenrostro-Figueroa, Juan Manuel Tirado-Gallegos, Mónica L. Chávez-González, A. K. Haghi, and Shaheemath Suhara K. K. (Eds.)

a potential source of bioactive compounds, both for the amount of residues produced annually and for the presence of phytochemicals with biological activity. For this reason, this chapter will show the trends and uses of fruit and vegetable wastes for the development of functional foods and nutraceuticals.

3.1 INTRODUCTION

Due to the cultural changes brought about by technological evolution, access to information, high urbanization, changes in lifestyles, and food consumption habits, these have undergone a variation showing a trend toward healthy, safe products with greater health benefits. Likewise, after the situation derived from the COVID-19 pandemic, the demand for foods that improve the immune system is evident. This has prompted research into the development of functional foods. According to the World Health Organization (WHO), functional and nutraceutical foods are those that have been scientifically proven to have beneficial effects on the health of the target population groups.

The term "Functional foods" was initially proposed in Japan in the 1980s, given the need to improve the quality of life of the population and reduce the occurrence of chronic noncommunicable diseases such as diabetes, hypertension, cancer, among others, which had been on the rise after the end of World War II. The concept of "functional foods" was derived from the research carried out with the purpose of recognizing which were the other functions of foods, in addition to nourishing. In this sense, it was defined that foods should have three functions, the main one of nourishing, the second function of satisfying the consumer, and a third physiological function. This last function is where functional foods are framed.[60,91]

Over the years and with the research developed, the concept of functional foods has been standardized worldwide. Likewise, the identification and characterization of food compounds, such as fiber, vitamins, minerals, among others, has allowed the understanding of the functioning and mechanisms of action of each of the components in the metabolism and physiology, generating innovations in the development of functional foods.[107] It has been found that the functionalities of foods depend on different compounds that present biological activities such as antioxidants, anticancer, antiviral, antimicrobial, among others. Consequently, in recent years, there has been a trend toward the incorporation of bioactive compounds in food matrices to obtain functional foods. Bioactive compounds are substances derived from the secondary metabolism of plants, animals, and microorganisms, such as carotenoids, phenols, prebiotics, and probiotics, which have demonstrated biological activity. The market for bioactive compounds is constantly expanding, for example, the prebiotics market is expected to exceed USD 8.5 billion by 2024,[34] which reflects the importance of this type of compounds in the world's economic dynamics. In coherence with the above, it is necessary to search for economic and abundant sources of bioactive compounds, such as fruit and vegetable residues.

Fruits and vegetables are one of the most consumed foods worldwide. According to FAO statistics, the annual production of fruits and vegetables is around 1957 million tons, of which about 7% is destined for fresh consumption, while more than 50% is industrialized to produce juices, jams, compotes, pickles, sauces, among others. During processing, between 30%–40% of byproducts such as seeds, pulp, and peels are produced, which are not used and are sometimes improperly discarded, causing environmental, economic, and social problems. Fruit and vegetable byproducts are characterized by their high content of fiber, vitamins, minerals, and phytochemicals with biological activity. In this sense, bioactive compounds derived from fruit and vegetable byproducts are presented as a valuable source of ingredients for the formulation of functional foods, contributing to the reduction of food waste and the sustainable use of resources.

Despite the recognized benefits of bioactive compounds, the incorporation, formulation, and production of functional foods still face technological challenges. Bioactive compounds are sensitive to factors such as temperature, pH, acidity, among others; exposure to these conditions during food production or consumption can result in the loss of their biological activity and degradation of their functionality. Therefore, for the development of functional foods, maintaining the integrity of the compounds throughout the manufacturing process and during the shelf life of the product has become a challenge for the food industry.

One of the most important strategies for the protection of bioactive compounds is encapsulation, whereby the compounds are coated with a protective layer that acts as a physical barrier between the ingredients and the external environment. The capsule formed around the compound allows the structure and function of the bioactive ingredients to be preserved, as well as ensuring their controlled release during digestion. Encapsulation has additional benefits such as improved solubility and reduced off-flavors of functional ingredients. Other innovative technologies, such as nanotechnology, have also been proposed to achieve more precise and efficient encapsulation. These emerging technologies address current limitations, leading to innovation in functional foods.

Considering the above, this chapter will address the use of bioactive compounds obtained from fruit and vegetable by-products for the development of functional foods. In this context, different strategies for the extraction of bioactive compounds present in fruit and vegetable by-products will be explored, as well as the techniques used to protect and preserve the biological activity of the compounds during their incorporation into food matrices. Also, advances in research on the health benefits associated with the consumption of functional foods made from fruit and vegetable by-products will be discussed.

3.2 FRUIT AND VEGETABLE BY-PRODUCTS

Fruits and vegetables have some similarities at a nutritional level; they are sources of vitamins, antioxidants, minerals, and dietary fibers. However, they are highly perishable, which is why one of the main challenges of agricultural research and policies is to increase the production of the fruit and vegetable sector and find uses for the waste generated, with the aim of feeding the 9100 million people expected in 2050.[30] The World Health Organization (WHO) recommends a minimum fruit consumption of 400 grams per person or five portions of 80 grams each; however, most people's intake falls short of the minimum consumption for a healthy diet,[58] also recommended daily intake according to age based on health models; a more rigorous recommendation suggests an intake of 600 grams per day for adults and adolescents, 480 grams per day for children aged 5–14 years, and 330 grams per day for children aged 0–4 years[66] and although two billion tons of fruits and vegetables are produced per year, enough to cover the requirements of the world population, 22% is lost from postharvest to distribution. Therefore, availability is decreased and global requirements are not met ("Innovation in fruit and vegetable supply chains,"[44]). According to FAO,[31] in 2017 about 3.9 million deaths worldwide occurred because of lack of fruit and vegetable consumption, because insufficient intake of fruits and vegetables has caused globally 14% of deaths from gastrointestinal cancer, 11% of deaths from ischemic heart disease, and 9% from stroke.[2]

Fruits and vegetables, being perishable and with high moisture content, have a short shelf life and in days or almost immediately are discarded by some particularity (bruising or color change) in postharvest, where the greatest losses are found, which is why when a food is discarded as waste is considered waste, according to the United Nations Organization is considered waste to "any material that has no direct use value and is discarded by its owner."[92] Baysal and Ülkü[15] affirm that one-third of the food produced annually is wasted, and food losses and waste (FWL) have become a worldwide concern, since it is estimated that 1300 tons of food are eliminated not because they cannot be consumed, but because they do not meet some quality or aesthetic parameter, such as shape and size; this situation is occurring at the production, harvest, and postharvest levels, and is that fruits and vegetables generate the largest losses with 45% in production and 30% generated by the consumer. This is because there is a part of the fruits and vegetables that are not valued according to the nature such as leaves, stems, and skins.[41,78] In addition to this situation, when food is wasted, other resources that were used in its production (fertilizers, water, energy, human capital) are implicitly wasted, and greenhouse gases are generated,[16] becoming a social, cultural, economic, and environmental problem, which is why a disruptive technology is sought in front of a sustainable strategy; for example, those fruits and vegetables that are discarded for not meeting consumer expectations can be left on the shelf at a lower price and thus can contribute to improving people's nutrition.

Globally, it is estimated that fruits and vegetables generate 46% of food waste; in European Union households, approximately 29% of fresh fruits and vegetables are wasted;[27] 35% are wasted in supermarkets, for example, bananas are discarded because of bruises or brown marks.[68] According to Zhang et al.,[114] fruit peels produce between 15 and 60% of residues; in turn, Leong and Chang[55] mention that fruits such as mango contribute residues between 30% and 50%, orange between 30 and 50%, pineapple between 40 and 50%, and banana 20%. In the field there are losses that reduce farmers' income, due to poor agricultural management, low harvesting technology, poor storage infrastructure, and precarious transportation; in addition, pollution is generated due to the use of pesticides for the management and control of pests and diseases or to improve quality and increase production to meet the demand for food, but it must be recognized that pesticides are essential for agricultural production, and without them production would be reduced by 30%–40%.[25]

Fruit and vegetable waste generates two types of waste: solid waste (hulls, seeds, pods, stones, etc.) and liquid waste, the latter being the product of leachates from the decomposition of the material, and due to its high moisture content and microbial load, it causes significant environmental damage, since the organic matter in the waste produces compounds such as methane (CH_4) and carbon dioxide (CO_2), both of which are greenhouse gases that have a negative impact on the environment and contribute to climate change.[57] Processing generates between 30 and 50% of residues and is closely linked to geographical location, harvesting season, transportation and processing methods, generating losses during basic unit operations such as washing, peeling, filtration, and others. In turn, fruits rich in enzymes undergo enzymatic browning, which makes the generation of residues inevitable,[115] since this chemical reaction is not desirable either for agro-industrial processes or for minimally processed products.

Most fruits and vegetables are seasonal, being available for a short time, which is why postharvest handling must be improved to ensure availability for longer periods.[58] Fruit and vegetable residues derived from agroindustrial processes are considered specialized waste, which must be managed with a holistic approach and the volume of waste generated, its nature and characterization must be identified in order to know the specific properties of the substances that are extracted from the residues; once the material has been identified, the best extraction technique must be evaluated. That is why they are mostly used for animal feed and compost; however, these residues contain bioactive compounds such as polyphenols, flavonoids, antioxidants, minerals, essential oils, pigments, fibers, vitamins, polysaccharides, and sugars, that is, a proper use of these residues can help reduce the carbon footprint; this is how the extraction of bioactive compounds can be used in the development of nutraceutical compounds, functional foods, or food additives.[78] Incineration and open burning have been used to dispose of the residues, generating secondary problems;[90] however, fruit residues can be used as feedstock for energy production such as bioethanol (fermentation), biogas, biomethane, and carbon dioxide (anaerobic digestibility [AD]), biohydrogen, among others,[55] in addition, thermal technologies such as incineration to produce electricity and heat, gasification and pyrolysis to produce gas;[115] this is because fruits contain high contents of soluble and insoluble sugars, such as sucrose, fructose, and glucose that can be used for bioethanol fermentation.

Traditional methods and emerging technologies are used for the extraction of bioactive compounds from fruit and vegetable residues, depending on the part of the plant to be processed, such as leaves stems, peels, pulp, and others. Traditional methods use high amounts of solvents and energy for extraction; however, alcohol has been considered safe for the extraction of food additives, and emerging technologies include microwave-assisted extraction (MAE), supercritical fluid extraction (SFE), pressurized liquid extraction (ELE), ultrasound-assisted extraction (UAE), high hydrostatic pressure (HHP), pulsed electric field (PEF), enzyme-assisted extraction (EAE), and ionic liquid extraction (ILE).[115] Therefore, fruits and vegetables are a valuable food resource not only for their nutritional contribution but also for the residues generated, being considered a raw material with ample potential to produce or extract compounds that can be incorporated into other processes, contributing not only to food security but also to the environment.

3.3 BIOACTIVE COMPOUNDS

According to the Food and Agriculture Organization of the United Nations (FAO), worldwide losses and wastage of fruits and vegetables are estimated to be the highest among all food types, reaching up to 60%[37] and by-products from fruit and vegetable processing comprise 25 to 30% of the total staple food group.[51,76] Bioactive compounds can be found as secondary metabolites of fruit and vegetable by-products, usually produced in small quantities,[21,101] although these metabolites can be used as functional ingredients,[19,103] most of them are not extracted, which increases the environmental impact generated by these by-products.[76,78] Through conventional or "green" technologies, fruit and vegetable by-products have been converted into animal feed or soil fertilizers.[73] However, techniques for obtaining bioactive compounds from these by-products are currently being optimized.[84] Most of the by-products are seed, husk, pomace, and stem,[33,81] which have a high content of bioactive components such as phenolic compounds, carotenoids, dietary fibers, vitamins, enzymes, and oils, among others.[95] Table 3.1 shows the main bioactive compounds obtained from fruit and vegetable residues.

3.3.1 PHENOLIC COMPOUNDS

Phenolic compounds are a group of plant secondary metabolites with wide distribution and varied structures.[56] These secondary metabolites are crucial in plant defense mechanism as signaling molecules to protect plants against UV radiation,[26] oxidative stress, and other adverse effects such as drought, salinity, temperature extremes, and nutritional imbalances.[28] Phenolic compounds gained attention because of their functional and therapeutic potential.[109]

TABLE 3.1 Bioactive Compounds From Fruit and Vegetable Waste.

By product/waste	Bioactive compounds	Reference
Mango (*Mangifera indica* L.) kernel	Saturated fatty acid	[11, 18, 87]
Apple pomace	Polyphenols compounds, carotenoids, and anthocyanins	[4]
Mandarin (*Citrus reticulata*) peel	Rutin	[59]
Cashew nut testa	Polyphenols compounds	[49, 104]
Rapeseed meal byproduct	Gallic, protocatechuic, vanillic, chlorogenic, neochlorogenic, caffeic, syringic, 4- hydroxybenzoic, cryptochlorogenic, p-coumaric, ferulic, sinapic, and cinnamic acids	[73]
Pineapple (peel, pulp's leftovers)	Phenolic, anthocyanins, yellow flavonoids, b-carotene, and lycopene	[22]
Acerola (seed)	Phenolic, anthocyanins, yellow flavonoids, b-carotene, and lycopene	[22]
Cruciferous vegetable (kale) waste	Phenolic compounds	[53]
Winery wastes	Polyphenols, anthocyanins	[13]
Mango seed	Polyphenols compounds	[80]
Grape pomace	Polyphenols, anthocyanins	[13]
Mango peel, spent coffee ground, malta peel, grape pomace, grape seed, grapefruit solid waste, hawthorn seed, olive leaf extract, coconut shell power, jabuticaba peel, pomegranate peels, soyabean germ	Polyphenols compounds	[51]
Mango peel	Coumaric acid, gallic acid, acid ferulic, epicatechin	[94]
Grape pomace, orange peel, potato peels, spinach waste	Phenols, flavonoids, and anthocyanins	[76]
Tobacco industrial waste	Polyphenols, flavonoids, rutin	[12]
Pomegranate peels	Ellagic acid, gallic acid	[84]
Jack fruit peel	Pectin	[84]
Vine pruning residue	Ellagic acid	[84]
Carrot waste	Carotenoids	[79]
Mango (*Mangifera indica* L.) peel	Carotenoids	[96]
Pomegranate (*Punica granatum* L.) peel waste	Catechin, rutin, acid gallic	[42]

TABLE 3.1 *(Continued)*

By product/waste	Bioactive compounds	Reference
Pomegranate (*Punica granatum* L.) seed and peel wastes	Linolenic acid, Linoleic acid, Phenolics, Tocopherols, Punicic acid	[8, 10, 14, 108]
Passion fruit (*Passiflora edulis*) and blackberry (*Rubus glaucus*) seed	Fatty acids, sterols, tocopherols, and diacylglycerols	[11, 87]
Carrot peels	Carotenoids	[84]
Papaya seeds	Flavonoids	[84]
Tamarind seeds (*Tamarindus indica*), sweet	Fatty acid antioxidants (9, 12- octadecadienoic acid, cis- 10- heptadecenoic, and n-hexadecanoic acid	[88]

3.3.2 ESSENTIAL OILS

They are volatile substances formed by the secondary metabolism of aromatic plants;[82] their constituents are usually of low molecular weight.[97] However, some natural factors, such as geographical location, elevations,[32] physiological variations, environmental conditions, geographical variations, genetic factors, and plant evolution, can alter the chemical composition and yield of these oils.[97] Over the years, their pharmaceutical and therapeutic potential as antioxidants, anticancer, antiprotozoal, antimicrobial, anti-inflammatory, phytotoxic, and neuroprotective, in addition to improving energy and mental clarity, has been demonstrated.[20]

3.3.3 DIETARY FIBERS

Dietary fibers are composed of major carbohydrates that form a polymer that provides structural support to plants and helps maintain the shape of plant parts because it is widely distributed.[7] Such dietary fibers are abundantly present in fruits and vegetables and comprise cellulose, hemicellulose, and lignin as the major polymeric group present in fruits and vegetables.[37] Benefits associated with their adequate consumption include regulation of intestinal transit, and prevention or treatment of diabetes, cardiovascular disease, and colon cancer.[43,63]

3.3.4 CAROTENOIDS

Carotenoids are a class of natural tetraterpene pigment compounds that give yellow, red, and orange color to plants,[21] they are the most widely distributed in nature, they can be found in fruits and vegetables, algae, fungi, and bacteria.[84] They are an integral part of the plant's photosynthetic machinery and help increase its ability to survive by acting as photoprotectors.[65] Carotenoids have important properties as antioxidants and are an important dietary source of vitamin A. β-carotene, α-carotene, lycopene, and lutein are the common carotenoids present in fruits and vegetables and are part of the human diet.[99,104]

3.3.5 TOCOPHEROLS

Tocopherols are a group of vitamin E-containing organic compounds that play an essential role in human nutrition.[104] These lipid-synthesized soluble compounds are antioxidants.[47] There are different isoforms of tocopherols depending on the number and position of methyl groups on the aromatic ring.[62] Tocopherol is commonly found in vegetable oils, oilseeds, nuts, cereals, whole grains, margarine, etc.[39]

3.4 FUNCTIONAL FOODS

In recent years, changes in consumer lifestyles have focused the attention of researchers on the development of healthy foods, and the term "functional foods" has become a key concept in the food industry. The definition of "functional foods" was given around 1980 in Japan and has evolved over the years. The development of foods with functions beyond nutrition originated as an innovative alternative to solve the problems caused by chronic noncommunicable diseases such as diabetes, hypertension, and cancer, which had been on the rise after the end of World War II. In 1991, the World Health Organization (WHO) defined them as "foods that, in addition to their basic nutritional values, contain biologically active components that may have beneficial effects on health, reducing the risk of chronic diseases or improving general health." In 1999, the Scientific Committee on Food of the European Union (EU) defined functional foods as "foods which, in addition to their basic nutritional values, contain biologically active components that may have beneficial effects on health, and which can be consumed as part of a normal, balanced diet." Around 2003, the European Food Safety Agency (EFSA) defined functional foods as "foods which, in addition to their basic nutritional values, contain biologically active components that may have beneficial effects on health, and which can be consumed as part of a normal, balanced diet." Although there is no single definition of functional foods, it is generally recognized that they must have a property beyond nutrition, and their benefits must be scientifically proven, identifying the biologically active compound that fulfills the function.[23,29,36]

In view of the above, functional foods can be defined as those that, in addition to providing essential nutrients for the body, offer additional, scientifically proven health benefits. These benefits derive from the bioactive compounds present in foods, which have biological activities such as antioxidants, antimicrobials, antivirals, among others, and can contribute to the prevention of chronic diseases, strengthening the immune system, the digestive system, among other benefits.[29]

In this sense, the search for biologically active compounds, the understanding of the mechanisms of action of these compounds, and the development of functional foods arose from the need to develop functional foods to improve the quality of life and reduce the incidence of chronic diseases. The proposal for these foods is that they not only satisfy a basic nutritional need but can also offer other specific health benefits.

The definition of functional foods is based on three essential functions that foods must fulfill[5]:

1. Nutritional function: functional foods must provide all the basic and essential nutrients for the organism.
2. Satisfaction function: functional foods must meet the tastes and needs of consumers to ensure their acceptance and consumption.
3. Physiological function: functional foods must have a specific health effect. This function makes the difference between a conventional food and a functional food.

The physiological function is generally associated with a bioactive compound with some proven biological activity, which, in an adequate concentration, can help to improve health and prevent diseases. The functionality of bioactive compounds occurs through different mechanisms of action that have been studied in recent years to improve the understanding of how foods can act to improve health. Among these mechanisms, the following stand out[35,61,72,86,116]:

a. Regulation of cellular processes, modulating some of the metabolic pathways or through the interaction between cells and tissues.
b. Antioxidant activity: antioxidant compounds help neutralize free radicals, reducing oxidative stress in the body, which translates into prevention of some chronic noncommunicable diseases.

c. Modulation of the intestinal microbiota, mainly by strengthening the growth of probiotic microorganisms in the intestine.
d. Reduction of inflammation, inhibiting the production pathways of inflammatory molecules in the body.
e. Regulation of metabolism and body weight, influencing satiety and appetite through the release of process-regulating hormones.
f. Improved cardiovascular function, reducing cholesterol levels in the blood

Over the years, the concept of functional foods has been expanding and integrating into cultures and industry practices around the world. Public awareness of the health benefits and disease prevention derived from a healthy diet has led to an increased demand for this type of food. Likewise, scientific and technological advances in the food industry have made it possible to know which compounds present in foods exert a physiological effect and their mechanisms of action. Thanks to this, functional foods now include a wide range of products, from yogurts with probiotics to fermented vegetables, making the incorporation of bioactive ingredients such as fibers, prebiotics, probiotics, paraprobiotics, among others, the common denominator in food formulation.[102]

3.4.1 *INNOVATION IN THE DEVELOPMENT OF FUNCTIONAL FOODS*

Food production and agroindustry in general has faced one of the most important sustainability challenges: the valorization of waste from the food chain worldwide. The production of functional foods and nutraceuticals is an alternative development, which contributes not only to supply nutritional needs but also to the use of promising fruits and vegetables, adaptation and utilization of horticultural waste, potential in general.[75]

This need has arisen among other things due to the depletion of resources and especially the limitations in the production of raw materials, which is why the generation of technological, cultural, and/or social innovations has been proposed as part of the optimization, which has allowed the inclusion of new products and supplies the demand in order to fulfill an important role within the circular economy.[85]

The objective of innovation in the generation of functional foods and nutraceuticals is to respond to recent trends of more conscious consumers, who contemplate a holistic concept, where the products, in addition to having nutritional value and biologically active compounds beneficial to prevent diseases, must be of organic origin and or be part of a process of utilization with low environmental impact. In addition to the above, the use of emerging technologies has allowed the conservation of food surpluses, the recovery of compounds, safety and overall quality of the new food product.

An evaluation of the factors affecting nutritional and bioactive compounds and the use of nonthermal technologies has shown that the bioavailability of food components can be maintained, their functional and technological properties improved, and recovery yields of agricultural products increased. However, optimization of operating parameters is vital to avoid degradation of macromolecules and oxidation of labile compounds.[38]

For the extraction of bioactive compounds, there is evidence of innovation in the use of combined methods to obtain better results; traditional analytical techniques are the key to obtaining residual extracts or "cocktails" of biochemical compounds with high potential applications, and subsequently using emerging methods efficiently will allow obtaining the specific compound of interest.[48] Table 3.2 presents some of the most used traditional and emerging methods.

TABLE 3.2 List of Traditional and Innovative Techniques That are Combined for The Extraction of Bioactive Compounds Used in Functional Foods and Nutraceuticals.

Waste source	Bioactive compounds	Conventional/emerging methods	Reference
Orange peels	α-Carotene	Soxhlet/ultrasound-assisted extraction (UAE)	[52]
Tomato waste	Lycopene	Hydrodistillation/supercritical fluids – FSC	[50]
Grape and must residues	polyphenols Anthocyanins	Distillation/pulsed Electrical Pulses Soxhlet/ultrasound-assisted extraction (UAE)	[13] [40]
Cocoa residue	Total phenols	Soxhlet/microwave ultrasound – UMAE	[93]
Carrot waste	Carotenoids	Chromatographic separation/ enzyme-assisted extraction	[79]
Papaya seeds	Flavonoids	Percolation/extraction with supercritical fluid – FSC	[84]
Passion fruit and blackberry seed	Tocopherols, sterols, fatty acids	Soxhlet/supercritical Fluid extraction – FSC	[11] [87]
Pomegranate seed and peel	Linoleic acid Phenolics Tocopherols	Soxhlet/Supercritical Fluid Extraction – FSC	[8] [87]

The techniques used for the extraction of bioactive compounds can be distinguished into two categories: conventional methods (maceration, distillation, steam, Soxhlet, and solvent extraction), although widely used, conventional techniques have important limitations: (i) low extraction rates; (ii) the need for expensive, high purity, and often hazardous organic solvents when target compounds have low polarity; (iii) low selectivity; (iv) risk of loss of volatile compounds; and (v) high energy costs associated with evaporation of large volumes of liquids for solvent recovery and recycling.[111] Green extraction methods (ultrasound-assisted, microwave, supercritical fluid, subcritical water, pressurized liquid, and enzyme-assisted), the main challenges derived from emerging technologies are related to their scaling up and implementation in comprehensive valorization strategies. All extractive techniques have the following objectives: (a) to extract the desired target compounds, minimizing their degradation, and without altering their properties; (b) to obtain the extract in the most stable and pure form; (c) to increase the extraction yield of the desired compounds; and finally, to reduce costs and processing time.[103]

By using techniques such as ultrasound-assisted extraction (UAE), a higher yield has been found, for example, for obtaining phenolic compounds, soluble and hydrolyzable characteristics, condensed tannins and total polyphenols, in fruit by-products such as soursop, which include the seed, peel, columella, and pulp. This technique has resulted in better yields in optimal conditions for UAE (37%) compared to traditional analytical methods; additionally, contents of gallic acid, coumaric acid, and chlorogenic acid are presented with higher concentration in their extracts, concluding that obtaining these bioactive compounds by UAE is efficient on a commercial scale.[3] Other technologies, such as the use of supercritical fluids, subcritical water extraction, enzyme-assisted extraction, ultrasound-assisted extraction, and microwave-assisted extraction, allow the compounds to be sustainable, allowing significant management of horticultural waste for the generation of new food products.[48]

A trend in the innovation of raw materials as functional food is the combination of probiotics and prebiotics from different sources; for example, vegetable fibers from husk residues are composed of cellulose, hemicellulose, lignin, pectins, and waxes, which first become integral flours that are complemented with traditional multigrains such as rye, sunflower seeds, and sesame obtained from the oil industry. Other similar raw materials are apple fiber, a residue generated in the production of beverages such as cider, cocoa epicarp, which is separated in the first stages of processing, as well as coffee husks.[100]

Bio-residues generated from oil extraction processes from oilseed oil are recognized for their active phytoconstituents, which are mainly present in the extraction cakes. In soybean (*Glycine max*) isoflavones, aglycones, glycosides, and phenolic acids have been extracted and in sunflower cake (*Helianthus annuus*) mainly polyphenols, in rapeseed cake (*Brassica napus*) synapic acid and sinapin are obtained, An innovative trend is also to take advantage of the wastewater from these processes, which has been characterized as a high source of bioactive phytochemicals with biological activities.[113]

The food industry in general has managed to innovate to meet commercial demand, and recently conventional foods have been generated which offer the consumer high nutritional quality; organic eggs enriched with functional molecules such as omega-3 fatty acids by improving the diet of poultry have had great reception, since the main long-chain omega-3 (PUFA) contained in the yolk are represented by eicosapentaenoic acid (EPA, C20: 5 n-3) and docosahexaenoic (DHA, C22:6 n-3) which are formed through elongation and desaturation reactions of α-linolenic acid (ALA, C18-3 n-3) and are considered highly healthful and functional.[106]

Another important development has been the biotechnological processes where the fermentation of food wastes from fruits and vegetables from various agroindustries forges a valuable generation of secondary nutrients, one of the most valued compounds is the production of organic acids, which can be produced through fermentation using molds, yeasts, and bacteria; citric acid and lactic acid can be produced.[48] Fermentation substrates can range from cassava bran or cassava fiber to apple pomace; pineapple, mandarin, and mixed fruit residues have also been used, which can produce up to 50% of organic acids.[67]

All molecules of interest in the development of functional foods and nutraceuticals have characteristics due to their susceptibility to biological reactions such as oxidation and production variables that facilitate their degradation during processing, particularly temperature. For this reason, the encapsulation of bioactive compounds is one of the most widely used alternatives to optimize and preserve functional properties; it is also considered a valuable technology to add desirable sensory attributes such as aroma, texture, color, and flavor, due to this is a continuous development due to its unique potential, its high versatility, and its wide range of applications.[74]

It is possible to encapsulate bioactive compounds derived from food waste at the micro- and nano-encapsulation scale, preserving them under favorable environmental conditions and optimizing the controlled release of the compound once the food is consumed.[6] Compounds such as lutein, for example, are of great interest because the human body cannot synthesize it; its bioavailability is important after ingestion from external sources.[54] Therefore, gelatin, protein, and maltodextrin microcapsules obtained by spray-dried dispersions (SDD) can be used to effectively protect lutein from degradation.[110]

The main purpose of functional foods and nutraceuticals is to generate both physiological and medicinal effects on human health, which is why the trend is projected toward a variety of products such as prebiotics, probiotics, dietary fibers, fatty acids (polyunsaturated), antioxidants, spices, herbs, nutrients, and dietary supplements, which are feasible to obtain from by-products of food and agricultural sources. The compounds finally obtained are substances that interact effectively with proteins, DNA, and other biological components to achieve the desired results, some used by the pharmaceutical industry are exploited to generate natural medicines.

3.4.2 DEVELOPMENT OF FUNCTIONAL FOODS

The production of functional foods and nutraceuticals demands the use of ingredients that promote the health of consumers; however, their incorporation is a challenge for the industrial production of such foods, since they are usually bio-compounds such as antioxidants, vitamins, and minerals, extracted from other sources that lose stability and consequently generate short bioavailability.[1] Using colloidal delivery systems has been shown to improve the stability, dispersibility, and bioaccessibility of encapsulated compounds.[17]

In this sense, one of the current trends in the production of functional foods and nutraceuticals is the encapsulation of bioactive compounds (Fig. 3.1), so that their ability to be captured by cells is improved. In encapsulation it is possible to enclose materials such as cells or bioactive molecules, like antioxidants, enzymes, polyphenols, and micronutrients for later delivery into a system; encapsulation can be called microencapsulation for particle sizes from 3 to 800 μm or nanoencapsulation for particle sizes from 10 to 1000 nm.[89]

According to Sarao and Arora,[98] the encapsulation process has three steps: the first step is the incorporation of the bioactive component in a solid by agglomeration or adsorption or in a liquid by dissolution or dispersion matrix, in the second step, the liquid matrix is dispersed or sprayed in the

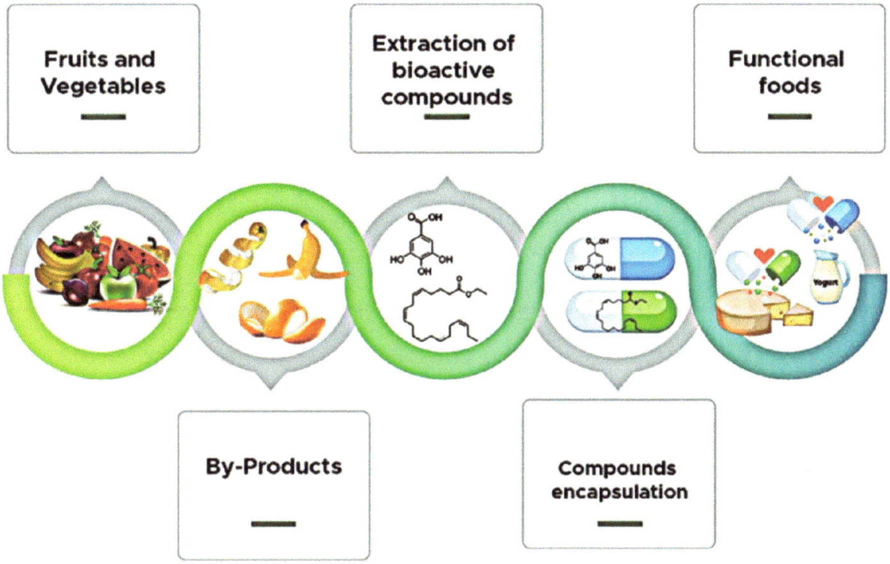

FIGURE 3.1 Schematic representation of the production of functional foods.

case of a solid matrix solution, the third step corresponds to the stabilization of the system through a physical treatment like evaporation, solidification, and coalescence, or by chemical treatment such as polymerization or physical-chemical process like gelation.

The encapsulation of bioactive compounds in nanoparticle-based delivery systems has been used in functional foods and nutraceuticals formulations since by decreasing the size of the particles a faster absorption can be given through the mucus layer that covers the cells of the gastrointestinal epithelium and also could prevent the adverse interactions with other food ingredients and decrease undesirable flavor profiles.[24,46,112]

Some of the methods of nutraceutical nanoparticle production include lipid formulations, natural nanocarriers, and biopolymer nanoparticles. The encapsulation of substances in lipid-based nanoparticles is used especially in bioactive compounds that are hydrophobic such as triglycerides, carotenoids, tocopherols, flavonoids, polyphenols, phytosterols, oil-soluble vitamins, flavors, and preservatives, which allows it to be possible to incorporate such substances in foods and water-based beverages.[46] The lipid-based systems formulated from digestible lipids, release free fatty acids and monoacylglycerols when they are digested by lipases in the gastrointestinal tract and then interact synergistically with bile salts and phospholipids in the small intestine to increase the bio-accessibility of encapsulated bioactives by forming mixed micelles.[17]

In emulsion-based systems are used oil, water, and emulsifiers, although other components may also be added to enhance the stability or to modify the physicochemical properties, such as thickening agents, gelling agents, or weighting agents.[69] Nanoemulsions can be formed by high- or low-energy methods, high-energy systems generate intense disruptive forces to break up the oil and water phases, some devices are used such as high-shear mixers, colloid mills, high-pressure valve homogenizers, microfluidizers, and sonicators, whereas low-energy methods are based on the spontaneous formation of emulsions when the composition or temperature of particular kinds of surfactant, oil, and water is changed in a specific form, and includes techniques such as phase inversion temperature or spontaneous emulsification.[64,71] The incorporation of nanoparticles into food can be done in liquid or powder form using techniques such as spray–drying or freeze–drying, which generates good dispersibility in water, high physical stability, high optical clarity, and good oral bioavailability.[46] Nanoliposomes are another kind of structure with a bilayer membrane formed by dissolving phospholipids especially lecithin and cholesterol in aqueous solutions, which favors the encapsulating of both hydrophilic and hydrophobic bioactives at the same time.[77]

Another technique commonly used in the encapsulation of bioactive compounds is the biopolymer-based colloidal delivery systems of hydrophilic bioactives within a biopolymer particle, which is usually made from proteins and/or polysaccharides; in this system, the composition, internal structure, size, and charge of biopolymer particles can be controlled, which allows diverse applications.[70] Fabrication methods of biopolymer-based colloidal delivery systems are extrusion, phase separation, antisolvent precipitation, and templating methods, each one of them with advantages and disadvantages related to production costs, type of operation, scale-up, and the capacity to manufacture microgels with various properties.[71]

In the biopolymer-based colloidal delivery systems, diverse food-grade host molecules have been utilized to entrap a bioactive molecule through noncovalent interactions,[17] the most common host molecules are:

- Cyclodextrins are nontoxic oligopolymers of glucose that can house a lot of small hydrophobic guest molecules, and has a hydrophilic exterior surface, which improves their water-solubility in food matrices, also can mask odors, and have been widely studied in the area of drug delivery for their ability to encapsulate, either partially or fully other molecules.[83]

- Micelles of casein are food-grade nanocarriers for enclosing and delivering unstable hydrophobic nutraceuticals that can be incorporated into food. These proteins have the ability to easily bind with hydrophobic molecules; they can perform interactions with other proteins/polymers such as covalent conjugations and noncovalent interactions; they also have reassembly and self-assembly properties, gelation properties, and surface activity. Therefore, caseins have unique structural and functional properties that make them suitable nanocarriers for bioactive and nutraceutical components.[24,45]
- Amylose is a linear polymer of D-glucopyranose joined by α-1-4 glucosidic bond and amylopectin, a branched polymer of glucose joined by α-1-4 glucosidic bond and α-1-6 glycosidic bond for ramification,[98] such has the ability to build a wide variety of nano and microstructures based on crystalline, glassy, gel types, and due to their characteristics of the physical adsorption of aroma compounds, the starch is able to form inclusion complexes with volatile compounds, especially small nonpolar molecules.[9]

According to Jafari,[45] and Torres-Maravilla et al.[105] the polymer wall materials commonly used in the nanoencapsulation of bioactives are:

a. Polysaccharides, such as gum Arabic, modified starches, maltodextrins, alginates, pectin, carrageenan, cellulose derivates, chitosan, cyclodextrins, starch particles.
b. Fast and waxes, such as hydrogenated vegetable oils, bees wax, lecithin, medium chain triglycerides, glyceryl behenate.
c. Proteins such as gelatin, whey proteins, sodium caseinate, soy proteins, gluten caseins, zein, silk fibroin.
d. Synthetic compounds such as paclitaxel, mPEG5000-b-p, polyacrylonitrile, polycaprolactone, polylactic acid.

Some important considerations in selecting a suitable host material for bioactive compound encapsulation are economic viability, ease of large-scale production, physicochemical properties of the material such as polarity, solubility, transition temperature, loading capacity, safety, interference in the sensory characteristics of the food, and the good protection characteristics of the bioactive for its biostability.[24]

3.5 PERSPECTIVES

In recent years, the inclusion of bioactive compounds obtained from fruit residues for the elaboration and development of functional foods has been increasing. These compounds, such as polyphenols, carotenoids, probiotics, among others, have beneficial properties for health, such as preventing chronic diseases, strengthening the immune system, and improving mood. It is expected that, in the coming years, with innovation in the food industry, new extraction and preservation technologies will be implemented to obtain biocompounds more efficiently and with better bioavailability. This would allow manufacturers to develop functional foods with a higher concentration of bioactive compounds, which would result in greater health benefits for the consumer.

In addition to research into new technologies, new trends are also developing in the field of functional foods. One of these is the trend toward personalized functional foods, i.e., foods designed to provide specific health benefits for everyone. This trend is driven by the development of nutrigenomics and metabolomics, disciplines that study the relationship between genetics, diet, and health. New trends and perspectives in this field point to a future in which consumers will have access to a wide range of functional foods that meet their individual health needs.

KEYWORDS

- **agro-industrial byproducts**
- **agribusiness**
- **bioactive compounds**
- **valorization**

REFERENCES

1. Abuhassira-Cohen, Y.; Livney, Y. D. Enhancing the Bioavailability of Encapsulated Hydrophobic Nutraceuticals: Insights From In Vitro, In Vivo, and Clinical Studies. *Curr. Opin. Food Sci.* **2022,** *45,* 100832.
2. Afshin, A.; Sur, P. J.; Fay, K. A.; Cornaby, L.; Ferrara, G.; Salama, J. S.; Mullany, E. C.; Abate, K. H.; Abbafati, C.; Abebe, Z.; Afarideh, M.; Aggarwal, A.; Agrawal,

S.; Akinyemiju, T.; Alahdab, F.; Bacha, U.; Bachman, V. F.; Badali, H.; Badawi, A.; Bensenor, I. M.; et al. Health Effects Of Dietary Risks in 195 Countries, 1990–2017: A Systematic Analysis for the Global Burden of Disease Study 2017. *The Lancet* **2019,** *393* (10184), 1958–1972.

3. Aguilar-Hernández, G.; De Lourdes García-Magaña, M.; De los Ángeles Vivar-Vera, M.; Sáyago-Ayerdi, S. G.; Sánchez-Burgos, J. A.; Morales-Castro, J.; Anaya-Esparza, L. M.; González, E. M. Optimization of Ultrasound-Assisted Extraction Of Phenolic Compounds From *Annona muricata* By-Products And Pulp. *Molecules* **2019,** *24* (5).
4. Ajila, C. M.; Brar, S. K.; Verma, M.; Tyagi, R. D.; Valéro, J. R. Solid-State Fermentation Of Apple Pomace Using *Phanerocheate chrysosporium* - Liberation and Extraction of Phenolic Antioxidants. *Food Chem.* **2011,** *126* (3), 1071–1080.
5. Alongi, M.; Anese, M. Re-Thinking Functional Food Development Through a Holistic Approach. *J. Funct. Foods*; Elsevier Ltd., 2021.
6. Alu'datt, M. H.; Alrosan, M.; Gammoh, S.; Tranchant, C. C.; Alhamad, M. N.; Rababah, T.; Zghoul, R.; Alzoubi, H.; Ghatasheh, S.; Ghozlan, K.; Tan, T. C. Encapsulation-Based Technologies For Bioactive Compounds and Their Application In The Food Industry: A Roadmap For Food-Derived Functional And Health-Promoting Ingredients. *Food Biosci.*, **2022.**
7. An, Y.; Lu, W.; Li, W.; Pan, L.; Lu, M.; Cesarino, I.; Li, Z.; Zeng, W. Dietary Fiber in Plant Cell Walls-The Healthy Carbohydrates. *Food Qual. Saf.* **2022,** *6* (June 2021), 1–17.
8. Araújo, J. M. S.; de Siqueira, A. C. P.; Blank, A. F.; Narain, N.; de Aquino Santana, L. C. L. A Cassava Starch–Chitosan Edible Coating Enriched with *Lippia sidoides* Cham. Essential Oil and Pomegranate Peel Extract for Preservation of Italian Tomatoes (*Lycopersicon esculentum* Mill.) Stored at Room Temperature. *Food Bioproc. Tech.;* **2018,** *11* (9), 1750–1760.
9. Assadpour, E.; Mahdi Jafari, S. A Systematic Review on Nanoencapsulation of Food Bioactive Ingredients and Nutraceuticals by Various Nanocarriers. *Crit. Rev. Food Sci. Nutr.;* **2019,** *59* (19), 3129–3151.
10. Azizi, M.; Seidavi, A. R.; Ragni, M.; Laudadio, V.; Tufarelli, V. Practical Applications of agricultural Wastes in Poultry Feeding in Mediterranean and Middle East regions. Part 1: Citrus, grape, Pomegranate and Apple Wastes. *Worlds Poult. Sci. J.;* **2018,** *74* (3), 489–498.
11. Ballesteros-Vivas, D.; Álvarez-Rivera, G.; Morantes, S. J.; Sánchez-Camargo, A. del P.; Ibáñez, E.; Parada-Alfonso, F.; Cifuentes, A. An Integrated Approach For The Valorization Of Mango Seed Kernel: Efficient Extraction Solvent Selection, Phytochemical Profiling And Antiproliferative Activity Assessment. *Food Res. Int.* **2019,** *126* (May), 108616.
12. Banožić, M.; Babić, J.; Jokić, S. Recent Advances in Extraction of Bioactive Compounds From Tobacco Industrial Waste-A Review. *Ind. Crops. Prod.;* **2020,** *144* (November 2019).
13. Barba, F. J.; Zhu, Z.; Koubaa, M.; Sant'Ana, A. S.; Orlien, V. Green Alternative Methods for the Extraction of Antioxidant Bioactive Compounds From Winery Wastes and By-Products: A Review. *Trends Food Sci. Technol.;* **2016,** *49,* 96–109.
14. Bar-Ya'akov, I.; Tian, L.; Amir, R.; Holland, D. Primary Metabolites, Anthocyanins, and Hydrolyzable Tannins in The Pomegranate Fruit. *Front Plant. Sci.;* **2019,** *10* (May), 1–19.

15. Baysal, S. S.; Ülkü, M. A. Food Loss and Waste: A Sustainable Supply Chain Perspective. In *Disruptive Technologies and Eco-Innovation For Sustainable Development*; 2022; pp 90–108.
16. Bruno, M.; Cittadini, E.; Grenoville, S. Dinámica de la Generación de Residuos Sólidos y Desperdicio de Alimentos en los Mercados Concentradores de Frutas y Verduras del Área Metropolitana de Buenos Aires (AMBA). *Siembra* **2023,** *10* (1).
17. Caballero, S.; Li, Y. O.; McClements, D. J.; Davidov-Pardo, G. Encapsulation and Delivery Of Bioactive Citrus Pomace Polyphenols: A Review. *Crit. Rev. Food. Sci. Nutr.;* **2022,** *62* (29), 8028–8044.
18. Cerón-Martínez, L. J.; Hurtado-Benavides, A. M.; Ayala-Aponte, A.; Serna-Cock, L.; Tirado, D. F. A Pilot-Scale Supercritical Carbon Dioxide Extraction To Valorize Colombian Mango Seed Kernel. *Molecules* **2021,** *26* (8).
19. Chamorro, F.; Carpena, M.; Fraga-Corral, M.; Echave, J.; Riaz Rajoka, M. S.; Barba, F. J.; Cao, H.; Xiao, J.; Prieto, M. A.; Simal-Gandara, J. Valorization of Kiwi Agricultural Waste And Industry By-Products By Recovering Bioactive Compounds And Applications As Food Additives: A Circular Economy Model. *Food Chem.*; **2022,** *370* (September 2021), 131315.
20. Chang, Y.; Harmon, P. F.; Treadwell, D. D.; Carrillo, D.; Sarkhosh, A.; Brecht, J. K. Biocontrol Potential of Essential Oils in Organic Horticulture Systems: From Farm to Fork. *Front Nutr.*; **2022,** *8* (January), 1–26.
21. Choo, W. S.; Saik, A. Y. H. Valorization of Fruit and Vegetable Waste for Bioactive Pigments: Extraction and Utilization. In *Valorization of Agri-Food Wastes and By-Products: Recent Trends, Innovations and Sustainability Challenges*, INC, 2021.
22. Da Silva, L. M. R.; De Figueiredo, E. A. T.; Ricardo, N. M. P. S.; Vieira, I. G. P.; De Figueiredo, R. W.; Brasil, I. M.; Gomes, C. L. Quantification of Bioactive Compounds in Pulps and by-Products of Tropical Fruits From Brazil. *Food Chem.* **2014,** *143,* 398–404.
23. Dable-Tupas, G.; Otero, M. C. B.; Bernolo, L. Functional Foods and Health Benefits. In *Functional Foods and Nutraceuticals: Bioactive Components, Formulations and Innovations*; 2020; pp 1–11.
24. Davidov-Pardo, G.; Joye, I. J.; Mcclements, D. J. Food-Grade Protein-Based Nanoparticles and Microparticles for Bioactive Delivery: Fabrication, Characterization, and Utilization. *Adv. Protein. Chem. Struct. Biol.* 1st ed.;Elsevier Inc., 2015; Vol. 98.
25. de Andrade, J. C.; Galvan, D.; Kato, L. S.; Conte-Junior, C. A. Consumption of Fruits and Vegetables Contaminated with Pesticide Residues in Brazil: A Systematic Review With Health Risk Assessment. *Chemosphere* **2023,** 322.
26. de la Rosa, L. A.; Moreno-Escamilla, J. O.; Rodrigo-García, J.; Alvarez-Parrilla, E. Phenolic Compounds. *Postharvest Physiology and Biochemistry of Fruits and Vegetables*,Elsevier Inc., 2018.
27. De Laurentiis, V.; Corrado, S.; Sala, S. Quantifying Household Waste of Fresh Fruit and Vegetables in the EU. *Waste Manag.* **2018,** *77,* 238–251.
28. Divekar, P. A.; Narayana, S.; Divekar, B. A.; Kumar, R.; Gadratagi, B. G.; Ray, A.; Singh, A. K.; Rani, V.; Singh, V.; Singh, A. K.; Kumar, A.; Singh, R. P.; Meena, R. S.; Behera, T. K. Plant Secondary Metabolites as Defense Tools against Herbivores for Sustainable Crop Protection. *Int. J. Mol. Sci.* **2022,** *23* (5).
29. Durán, R.; Valenzuela, A. Artículos De Actualización La Experiencia Japonesa Con Los Alimentos Foshu ¿Los Verdaderos Alimentos Funcionales? the Japanese Experience with Foshu Foods the True Functional Foods? *Rev. Chil. Nutr.* **2010,** *37.*

30. El-Beltagi, H. S.; Parmar, A.; Chen, T.; El-Mogy, M. M. Editorial: Advances in Pre- and Postharvest Applications to Reduce Qualitative and Quantitative Food Loss and Waste. *Front. Plant Sci.* 2023.
31. FAO. *Frutas y Verduras – Esenciales en tu dieta:Año Internacional de las Frutas y Verduras, 2021*. Frutas y verduras – esenciales en tu dieta, 2020.
32. Feng, X.; Zhang, W.; Wu, W.; Bai, R.; Kuang, S.; Shi, B.; Li, D. Chemical Composition and Diversity of the Essential Oils of Juniperus Rigida Along the Elevations in Helan and Changbai Mountains and Correlation with the Soil Characteristics. *Ind. Crops. Prod.* **2021,** *159* (June 2020), 113032.
33. Ferreira, S. M.; Santos, L. Incorporation of Phenolic Extracts from Different by-products in Yoghurts to Create Fortified and Sustainable foods. *Food Biosci.* **2023,** *51* (September 2022), 102293.
34. Fonteles, T. V.; Rodrigues, S. Prebiotic in Fruit Juice: Processing Challenges, Advances, and Perspectives. *Curr. Opin. Food. Sci* . **2018,** *22,* 55–61.
35. Fonteles, T. V.; Rodrigues, S.; Patra, J. K.; Das, G.; Lee, S.; Kang, S. S.; Shin, H. S.; Niego, A. G.; Rapior, S.; Thongklang, N.; Raspé, O.; Jaidee, W.; Lumyong, S.; Hyde, K. D.; FAO, Agricultura, O. de las N. para la alimentación, Liu, Z.; de Souza, T. S. P.; Holland, B.; Dunshea, F.; Barrow, C.; et al. Molecular Mechanisms That Link Oxidative Stress, Inflammation, and Fibrosis in the Liver. *Antioxidants* **2020,** *11* (4), 103272.
36. Fuentes Berrio, L.; Acevedo Correa, D.; Chantré, C. A.; Gelvez Ordoñez, V. M. Alimentos Funcionales: Impacto y retos Para El Desarrollo Y Bienestar De La Sociedad Colombiana. *Biotecnología en el Sector Agropecuario y Agroindustrial* **2015,** *13* (2), 140.
37. Gaharwar, S. S.; Mohakar, V. N.; Kumar, A. Value Addition of Fruit and Vegetable Waste: a Nutraceutical Perspective. *Fruit and Vegetable Waste Utilization and Sustainability*, INC, 2023.
38. Galanakis, C. M. Functionality of Food Components and Emerging Technologies. *Foods*, MDPI AG, 2021.
39. Garg, A.; Sharma, R.; Dey, P.; Kumar, A. Antioxidants Effects in Health Chapter1. *3 - Food Auto-Oxidation : An Overview* **2022,** 2022–2023.
40. Giacosa, S.; Ferrero, L.; Paissoni, M. A.; Río Segade, S.; Gerbi, V.; Rolle, L. Grape Skin Anthocyanin Extraction from Red Varieties During Simulated Maceration: Influence of Grape Seeds and Pigments Adsorption on Their Surface. *Food Chem.* **2023,** 424.
41. González G.; C. G. Frutas y Verduras Perdidas y Desperdiciadas, una Oportunidad Para Mejorar el Consumo. *Revista Chilena de Nutricion* 2018.
42. Gullón, P.; Gullón, B.; Astray, G.; Carpena, M.; Fraga-Corral, M.; Prieto, M. A.; Simal-Gandara, J. Valorization of by-products from Olive Oil Industry and Added-value Applications for Innovative Functional Foods. *Food Res. Int.* **2020,** *137* (August).
43. Hussain, S.; Jõudu, I.; Bhat, R. Dietary fiber from Underutilized Plant Resources-A positive Approach for Valorization of Fruit and Vegetable Wastes. *Sustain. (Switzerland)* **2020,** *12* (13).
44. Innovation in Fruit and Vegetable Supply Chains. *Nat. Food* 2022.
45. Jafari, S. M. An Overview of Nanoencapsulation Techniques and Their Classification. *Nanoencapsulation Technologies for the Food and Nutraceutical Industries*,Elsevier Inc., 2017.
46. Jafari, S. M.; McClements, D. J. Nanotechnology Approaches for Increasing Nutrient Bioavailability. *Adv. Food Nutr. Res.*, 1st ed.;Elsevier Inc., 2017; Vol. 81.

47. Järvinen, R.; Erkkilä, A. T. Tocopherols: Physiology and Health Effects. In *Encyclopedia of Food and Health* 2015; vol. 3; pp 303–308.
48. Kainat, S.; Arshad, M. S.; Khalid, W.; Zubair Khalid, M.; Koraqi, H.; Afzal, M. F.; Noreen, S.; Aziz, Z.; Al-Farga, A. Sustainable Novel Extraction of Bioactive Compounds from Fruits and Vegetables Waste for Functional Foods: A Review. *Int. J. Food Prop.* 2022.
49. Kamath, V.; Rajini, P. S. The Efficacy of Cashew Nut (*Anacardium occidentale* L.) Skin Extract as a Free Radical Scavenger. *Food Chem.* **2007,** *103* (2), 428–433.
50. Kehili, M.; Kammlott, M.; Choura, S.; Zammel, A.; Zetzl, C.; Smirnova, I.; Allouche, N.; Sayadi, S. Supercritical CO2 Extraction and Antioxidant Activity of Lycopene and β-carotene-enriched Oleoresin from Tomato (*Lycopersicum esculentum* L.) peels by-product of a Tunisian industry. *Food Bioprod. Process.* **2017,** 102.
51. Kumar, K.; Srivastav, S.; Sharanagat, V. S. Ultrasound Assisted Extraction (UAE) of Bioactive Compounds from Fruit and Vegetable Processing by-Products: A Review. *Ultrason. Sonochem.* **2021,** *70* (July 2020), 105325.
52. Kumari, B.; Tiwari, B. K.; Hossain, M. B.; Brunton, N. P.; Rai, D. K. Recent Advances on Application of Ultrasound and Pulsed Electric Field Technologies in the Extraction of Bioactives from Agro-Industrial By-products. *Food Bioproc. Tech.* 2018.
53. Lee, S. Y.; Liang, Y. N.; Stuckey, D. C.; Hu, X. Single-step Extraction of Bioactive Compounds from Cruciferous Vegetable (kale) Waste using Natural Deep Eutectic Solvents. *Sep. Purif. Technol.* **2023,** *317,* 123677.
54. Lei, Z.; Langrish, T. A Review of the Extraction and Closed-loop Spray Drying-assisted Micro-Encapsulation of Algal Lutein for Functional Food Delivery. *Processes* 2021.
55. Leong, Y. K.; Chang, J. S. Valorization of Fruit Wastes for Circular Bioeconomy: Current Advances, Challenges, and Opportunities. *Bioresour. Technol.* 2022.
56. Lin, D.; Xiao, M.; Zhao, J.; Li, Z.; Xing, B.; Li, X.; Kong, M.; Li, L.; Zhang, Q.; Liu, Y.; Chen, H.; Qin, W.; Wu, H.; Chen, S. An overview of Plant Phenolic Compounds and Their Importance in Human Nutrition and Management of type 2 Diabetes. *Molecules* **2016,** *21* (10).
57. Linares-Luna, R. G.; Castro, F. I. G.; González-Guerra, G. M.; del Pilar Restrepo-Elorza, M.; Montiel-Carrillo, A. P.; Álvarez-Rivera, K. Y.; Hernández, S. Biocombustibles A Partir DE Residuos DE frutas Y Vegetales: Procesos DE Transformación Y Áreas DE Oportunidad. *J. Energy, Eng. Optim. Sustain.* **2022,** *6* (2), 57–76.
58. Liu, X.; Le Bourvellec, C.; Yu, J.; Zhao, L.; Wang, K.; Tao, Y.; Renard, C. M. G. C.; Hu, Z. Trends and Challenges on Fruit and Vegetable Processing: Insights into Sustainable, Traceable, Precise, Healthy, Intelligent, Personalized and Local Innovative Food Products. *Trends Food. Sci. Technol.* 2022.
59. Liu, Y.; Benohoud, M.; Galani Yamdeu, J. H.; Gong, Y. Y.; Orfila, C. Green Extraction of Polyphenols From Citrus Peel By-products and Their Antifungal Activity Against *Aspergillus flavus*. *Food Chem. X*. **2021,** *12,* 100144.
60. Liu, Z.; de Souza, T. S. P.; Holland, B.; Dunshea, F.; Barrow, C.; Suleria, H. A. R. Valorization of Food Waste to Produce Value-Added Products Based on Its Bioactive Compounds. *Processes* **2023,** *11* (3).
61. Lu, J.; He, R.; Sun, P.; Zhang, F.; Linhardt, R. J.; Zhang, A. Molecular Mechanisms of Bioactive Polysaccharides from *Ganoderma lucidum* (Lingzhi), a Review. *Int. J. Biol. Macromol.* 2020.

62. Ma, J.; Qiu, D.; Pang, Y.; Gao, H.; Wang, X.; Qin, Y. Diverse roles of Tocopherols in Response to Abiotic and Biotic Stresses and Strategies for Genetic Biofortification in Plants. *Mol. Breed.* **2020,** *40* (2), 1–27.
63. Macagnan, F. T.; da Silva, L. P.; Hecktheuer, L. H. Dietary fibre: The Scientific Search for an Ideal Definition and Methodology of Analysis, and Its Physiological Importance as a Carrier of Bioactive Compounds. *Food Res. Int.* **2016,** *85,* 144–154.
64. Manzoor, M.; Sharma, P.; Murtaza, M.; Jaiswal, A. K.; Jaglan, S. Fabrication, Characterization, and Interventions of Protein, Polysaccharide and Lipid-based Nanoemulsions in Food and Nutraceutical Delivery Applications: A Review. *Int. J. Biol. Macromol.* **2023,** *241* (April), 124485.
65. Maoka, T. Carotenoids as natural functional pigments. *J. Nat. Med.* **2020,** *74* (1), 1–16.
66. Mason-D'Croz, D.; Bogard, J. R.; Sulser, T. B.; Cenacchi, N.; Dunston, S.; Herrero, M.; Wiebe, K. Gaps Between Fruit and Vegetable Production, Demand, and Recommended Consumption at Global and National Levels: An Integrated Modelling Study. *Lancet Planet Health* **2019,** *3* (7).
67. Matheus, J. R. V.; Andrade, C. J. de, Miyahira, R. F.; Fai, A. E. C. Persimmon (*Diospyros kaki* L.): Chemical Properties, Bioactive Compounds and Potential Use in the Development of New Products–A Review. *Food Rev. Int.* 2022.
68. Mattsson, L.; Williams, H.; Berghel, J. Waste of Fresh Fruit and Vegetables at Retailers in Sweden – Measuring and Calculation of Mass, Economic Cost and Climate Impact. *Resour. Conserv. Recycl.* **2018,** 130.
69. McClements, D. J. *Food Emulsions*, 3rd Ed.; CRC Press, 2015a.
70. McClements, D. J. Encapsulation, Protection, and Release of Hydrophilic Active Components: Potential and Limitations of Colloidal Delivery Systems. *Adv. Colloid. Interface Sci.* **2015b,** *219,* 27–53.
71. McClements, D. J. Advances in Nanoparticle and Microparticle Delivery Systems for Increasing the Dispersibility, Stability, and Bioactivity of Phytochemicals. *Biotechnol. Adv.* **2020,** *38,* #pagerange#.
72. Meng, X.; Zhou, J.; Zhao, C. N.; Gan, R. Y.; Li, H. Bin Health Benefits and Molecular Mechanisms of Resveratrol: A Narrative Review. *Foods*,MDPI Multidisciplinary Digital Publishing Institute, 2020.
73. Mir-Cerdà, A.; Nuñez, O.; Granados, M.; Sentellas, S.; Saurina, J. An Overview of the Extraction and Characterization of Bioactive Phenolic Compounds from Agri-food Waste within the Framework of Circular Bioeconomy. *TrAC - Trends Anal. Chem.* **2023,** 161.
74. Mohammed, N. K.; Tan, C. P.; Manap, Y. A.; Muhialdin, B. J.; Hussin, A. S. M. Spray Drying for the Encapsulation of Oils—A Review. *Molecules* 2020.
75. Mor, R. S.; Panghal, A.; Kumar, V. Challenges and Opportunities of Circular Economy in Agri-Food Sector. In *Environmental Footprints and Eco-design of Products and Processes*; Mor, R. S., Panghal, A., Kumar, V., Eds.;Springer Singapore: Singapore, 2021.
76. More, P. R.; Jambrak, A. R.; Arya, S. S. Green, Environment-friendly and Sustainable Techniques for Extraction of Food Bioactive Compounds and Waste Valorization. *Trends Food Sci. Technol.* **2022,** *128* (May), 296–315.
77. Nejatian, M.; Darabzadeh, N.; Bodbodak, S.; Saberian, H.; Rafiee, Z.; Kharazmi, M. S.; Jafari, S. M. Practical Application of Nanoencapsulated Nutraceuticals in Real Food Products; A Systematic Review. *Adv. Colloid. Interface Sci.* **2022,** *305* (May), 102690.

78. Nirmal, N. P.; Khanashyam, A. C.; Mundanat, A. S.; Shah, K.; Babu, K. S.; Thorakkattu, P.; Al-Asmari, F.; Pandiselvam, R. Valorization of Fruit Waste for Bioactive Compounds and Their Applications in the Food Industry. *Foods* 2023.
79. Otálora-Orrego, D.; Martin G.; D. A. Técnicas Emergentes de Extracción de β-caroteno para la Valorización de Subproductos Agroindustriales de la Zanahoria (*Daucus carota* L.): Una revisión. *Informador. Técnico.* **2020,** *85* (1), 83–106.
80. Parniakov, O.; Barba, F. J.; Grimi, N.; Lebovka, N.; Vorobiev, E. Extraction Assisted by Pulsed Electric Energy as a Potential Tool for Green and Sustainable Recovery of Nutritionally Valuable Compounds From Mango Peels. *Food Chem.* **2016,** *192,* 842–848.
81. Pathania, S.; Kaur, N. Utilization of Fruits and Vegetable By-products for Isolation of Dietary Fibres and its Potential Application as Functional Ingredients. *Bioact. Carbohydr. Diet. Fibre* **2022,** *27* (February 2021), 100295.
82. Picazo-Aragonés, J.; Terrab, A.; Balao, F. Plant Volatile Organic Compounds Evolution: Transcriptional Regulation, Epigenetics and Polyploidy. *Int. J. Mol. Sci.* **2020,** *21* (23), 1–18.
83. Poulson, B. G.; Alsulami, Q. A.; Sharfalddin, A.; El Agammy, Emam. F.; Mouffouk, F.; Emwas, A.-H.; Jaremko, L.; Jaremko, M. Cyclodextrins: Structural, Chemical, and Physical Properties, and Applications. *Polysaccharides* **2021,** *3* (1), 1–31.
84. Pradhan, D.; McCarthy, B.; Stapleton, K.; Jaiswal, S.; Jaiswal, A. K. Fruit and Vegetable Wastes for Nutraceuticals, Functional Foods, and Speciality Chemicals. *Fruit and Vegetable Waste Utilization and Sustainability* INC, 2023.
85. Ramírez-Osorio, L. J.; Villaruel-Lopez, A.; Villagrán, Z.; Anaya-Esparza, L. M. Residuos Alimenticios: Fuente de Componentes Bioactivos Para la Elaboración de Alimentos Funcionales. *Acta de Ciencia en Salud* **2022,** *6* (16), 17–26.
86. Ramos-Tovar, E.; Muriel, P. Molecular Mechanisms That Link Oxidative Stress, Inflammation, and Fibrosis in the Liver. *Antioxidants* MDPI, 2020.
87. Ray, A.; Dubey, K. K.; Marathe, S. J.; Singhal, R. Supercritical fluid extraction of bioactives from Fruit Waste and its Therapeutic Potential. *Food Biosci.* **2023,** *52* (January), 102418.
88. Reis, P. M. C. L.; Dariva, C.; Vieira, G. Â. B.; Hense, H. Extraction and Evaluation of Antioxidant Potential of the Extracts Obtained from Tamarind Seeds (*Tamarindus indica*), Sweet Variety. *J. Food Eng.* **2016,** *173,* 116–123.
89. Reque, P. M.; Brandelli, A. Encapsulation of Probiotics and Nutraceuticals: Applications in Functional food Industry. *Trends Food Sci. Technol.* **2021,** *114* (April), 1–10.
90. Rifna, E. J.; Misra, N. N.; Dwivedi, M. Recent Advances in Extraction Technologies for Recovery of Bioactive Compounds Derived from Fruit and Vegetable Waste Peels: A Review. *Crit. Rev. Food. Sci. Nutr.* 2023.
91. Rodrigo Durán, C.; Alfonso Valenzuela, B. La Experiencia Japonesa con Los Alimentos Foshu ¿Los Verdaderos Alimentos Funcionales? *Revista Chilena de Nutricion* **2010,** *37* (2), 224–233.
92. Rondon, Estefany.; Szantó, M.; Pacheco, J.; Contreras, E.; Gálvez, A. Guía general para la gestión de residuos sólidos domiciliarios. *Manuales de la CEPAL* 2016.
93. Ruesgas-Ramón, M.; Suárez-Quiroz, M. L.; González-Ríos, O.; Baréa, B.; Cazals, G.; Figueroa-Espinoza, M. C.; Durand, E. Biomolecules Extraction from Coffee and Cocoa by- and co-Products using Deep Eutectic Solvents. *J. Sci. Food Agric.* **2020,** *100* (1).

94. Safdar, M. N.; Kausar, T.; Nadeem, M. Comparison of Ultrasound and Maceration Techniques for the Extraction of Polyphenols From the Mango Peel. *J. Food Process Preserv.* **2017,** *41* (4), e13028.
95. Sagar, N. A.; Pareek, S.; Sharma, S.; Yahia, E. M.; Lobo, M. G. Fruit and Vegetable Waste: Bioactive Compounds, Their Extraction, and Possible Utilization. *Compr. Rev. Food Sci. Food Saf.* **2018,** *17* (3), 512–531.
96. Sánchez-Camargo, A. del P.; Gutiérrez, L. F.; Vargas, S. M.; Martinez-Correa, H. A.; Parada-Alfonso, F.; Narváez-Cuenca, C. E. Valorisation of Mango Peel: Proximate Composition, Supercritical Fluid Extraction of Carotenoids, and Application as An Antioxidant Additive for an Edible Oil. *J. Supercrit. Fluid.* **2019,** *152,* 104574.
97. Santana de Oliveira, M.; Vostinaru, O.; Rigano, D.; de Aguiar Andrade, E. H. Editorial: Bioactive Compounds Present in Essential Oils: Advances and Pharmacological Applications. *Front. Pharmacol.* **2023,** *14* (January), 11–12.
98. Sarao, L. K.; Arora, M. Probiotics, Prebiotics, and Microencapsulation: A Review. *Crit. Rev. Food. Sci. Nutr.* **2017,** *57* (2), 344–371.
99. Sartori, A. G. de O.; da Silva, M. V. Main Food Sources of Carotenoids, According to the Purpose and Degree Of Processing, for Beneficiaries of the "Bolsa Família" in Brazil. *Food Sci. Technol.* **2014,** *34* (2), 408–415.
100. Sasae, Y. Y. A.; Londok, J. J. M. R.; Tulung, B.; Rahasia, C. A. Pengaruh Pemberian Sumber Serat Berbeda Dalam Pakan Terhadap Kecernaan Semu Serat Kasar Dan Hemiselulosa Pada Ayam Pedaging Strain Cobb. *Zootec* **2020,** *40* (1).
101. Shirahigue, L. D.; Ceccato-Antonini, S. R. Agro-industrial Wastes as Sources of Bioactive Compounds for Food and Fermentation Industries. *Ciencia Rural* **2020,** *50* (4).
102. Siciliano, R. A.; Reale, A.; Mazzeo, M. F.; Morandi, S.; Silvetti, T.; Brasca, M. Paraprobiotics: A new Perspective for Functional Foods and Nutraceuticals. Nutrients, 2021April 1. MDPI AG.
103. Sorrenti, V.; Burò, I.; Consoli, V.; Vanella, L. Recent Advances in Health Benefits of Bioactive Compounds from Food Wastes and By-Products: Biochemical Aspects. *Int. J. Mol. Sci.* **2023,** *24* (3).
104. Sruthi, P.; Naidu, M. M. Cashew nut (*Anacardium occidentale* L.) testa as a Potential Source of Bioactive Compounds: A Review on its Functional Properties and Valorization. *Food Chem. Advanc.* **2023,** *3* (March), 100390.
105. Tahir, A.; Shabir Ahmad, R.; Imran, M.; Ahmad, M. H.; Kamran Khan, M.; Muhammad, N.; Nisa, M. U.; Tahir Nadeem, M.; Yasmin, A.; Tahir, H. S.; Zulifqar, A.; Javed, M. Recent approaches for Utilization of Food Components as Nano-encapsulation: A Review. *Int. J. Food Prop.* **2021,** *24* (1), 1074–1096.
106. Toomer, O. T.; Hulse-Kemp, A. M.; Dean, L. L.; Boykin, D. L.; Malheiros, R.; Anderson, K. E. Feeding High-oleic Peanuts to Layer Hens Enhances Egg Yolk Color and Oleic Fatty Acid Content in Shell Eggs. *Poult. Sci.* **2019,** *98* (4).
107. Torres-Maravilla, E.; Méndez-Trujillo, V.; Hernández-Delgado, N. C.; Bermúdez-Humarán, L. G.; Reyes-Pavón, D. Looking inside Mexican Traditional Food as Sources of Synbiotics for Developing Novel Functional Products. *Fermentation* **2022,** *8* (3), 1–22.
108. Viana, R. B.; Cardoso, E. D. C.; Gouveia, I. M. De, Rezende, M. L. G.; Monteiro, B. M.; Araújo, C. V. De Avaliação da eficiência do Somaticell® para o diagnóstico da

contagem indireta de células somáticas no leite de búfalas. *Revista de Ciências Agrárias* **2010,** *53* (1), 24–30.
109. Vodnar, D. C.; Călinoiu, L. F.; Mitrea, L.; Precup, G.; Bindea, M.; Păcurar, A. M.; Szabo, K.; Ştefănescu, B. E. A new Generation of Probiotic Functional Beverages Using Bioactive Compounds from Agro-Industrial Waste. *Funct. Med. Beverages* **2019,** 483–528. Elsevier.
110. Wang, Y.; Ye, H.; Zhou, C.; Lv, F.; Bie, X.; Lu, Z. Study on the Spray-drying Encapsulation of Lutein in the Porous Starch and Gelatin Mixture. *Eur. Food Res. Technol.* **2012,** *234* (1).
111. Wong-Paz, J. E.; Aguilar-Zárate, P.; Veana, F.; Muñiz-Márquez, D. B. Impacto de las Tecnologías de Extracción Verdes Para la Obtención de Compuestos Bioactivos de los Residuos de Frutos Cítricos. *TIP Revista Especializada en Ciencias Químico-Biológicas* **2020,** *23* (1), 1–11.
112. Yao, M.; McClements, D. J.; Xiao, H. Improving oral Bioavailability of Nutraceuticals by Engineered Nanoparticle-Based Delivery Systems. *Curr. Opin. Food Sci.* **2015,** *2,* 14–19.
113. Zafar, S.; Ashraf, M. Y.; Ali, Q.; Ashraf, A.; Anwar, S.; Iqbal, N.; Kausar, A.; Nouman, A.; Ali, M.; Zafar, M. A.; Feroz, K. Antioxidant Activity and Secondary Metabolites in Selected Vegetables Irrigated with Sewage Water. *Appl. Ecol. Environ. Res.* **2016,** *14* (5).
114. Zhang, H. F.; Yang, X. H.; Wang, Y. Microwave Assisted Extraction of Secondary Metabolites from Plants: Current Status and Future Directions. *Trends Food Sci. Technol.* 2011.
115. Zhu, Y.; Luan, Y.; Zhao, Y.; Liu, J.; Duan, Z.; Ruan, R. Current Technologies and Uses for Fruit and Vegetable Wastes in a Sustainable System: A Review. *Foods* **2023,** *12* (10), 1949.
116. Zia, A.; Farkhondeh, T.; Pourbagher-Shahri, A. M.; Samarghandian, S. The Role of Curcumin in Aging and Senescence: Molecular Mechanisms. In *Biomedicine and Pharmacotherapy*; Elsevier Masson s.r.l, 2021.

CHAPTER 4

Extraction of Compounds of Interest Present in Fruit Waste Using High Hydrostatic Pressures

EBBER ADDÍ QUINTANA-OBREGÓN[1,2],
LETICIA XÓCHITL LÓPEZ-MARTÍNEZ[1,2], JESÚS AYALA-ZAVALA[3], and
MANUEL VARGAS-ORTÍZ[4*]

[1]*CONAHCYT-Centro de Investigación en Alimentación y Desarrollo A.C. Sonora, México*

[2]*Coordinación de Alimentos de Origen Vegetal, Hermosillo, Sonora, México*

[3]*Centro de Investigación en Alimentación y Desarrollo), Laboratorio de calidad, autenticidad y trazabilidad de los alimentos, Hermosillo, Sonora, México*

[4]*CONAHCYT-Centro de Investigación en Alimentación y Desarrollo, Laboratorio de Calidad, Autenticidad y Trazabilidad de los Alimentos, Coordinación de Tecnología de Alimentos de Origen Animal, Hermosillo, Sonora, México*

*Corresponding author.

ABSTRACT

The increase in population necessitates an increase in food production; however, an increase in production inevitably leads to the generation of waste, by-products, and residues. In particular, the production and processing of fruits and vegetables generate large quantities of waste that serve as a significant source of bioactive compounds, minerals, fiber, and vitamins. These compounds are of interest to the pharmaceutical, energy, chemical, and food industries. Many of these compounds are thermolabile,

Agroindustrial Waste Management and Natural Resources Conservation: Environmental Challenges and Transformation into Value-Added Products. José Juan Buenrostro-Figueroa, Juan Manuel Tirado-Gallegos, Mónica L. Chávez-González, A. K. Haghi, and Shaheemath Suhara K. K. (Eds.)

thus requiring nonthermal processes for their extraction. High hydrostatic pressures (HHPs) are a nonthermal technology primarily used to ensure the safety of food. Due to their effects on cellular integrity, the potential of this technology has been evaluated as a method for extracting compounds of interest present in plant tissues.

4.1 INTRODUCTION

The production of fruits and vegetables has steadily increased each year, reaching 900 million tons of fruit and 1128 million tons of vegetables. The fruits with the highest production and commercialization volumes include bananas, watermelons, apples, oranges, grapes, mangoes, pineapples, avocados, and papayas. On the contrary, the vegetables with the highest production and commercialization volumes are tomatoes, cucumbers, cabbages, eggplants, and onions.[1–6] The increase in production is a response to the growing global population; however, per capita consumption of fruits and vegetables is below the recommended levels by the World Health Organization (WHO). To meet these recommendations, measures must be taken to improve distribution and access to these foods, but most importantly, to reduce losses, waste, and inefficiencies throughout the entire value chain. It is estimated that between 5 and 38% of the total production of fruits and vegetables is lost, with most of these losses being unavoidable due to various factors (mechanical damage, pests, diseases, postharvest physiology, economic factors, processing).[5,7–11] Fruits and vegetables are the primary sources of waste and by-products, as they generate some form of waste at each stage of marketing or processing, often unavoidably. In addition to losses due to human mishandling, which should be considered waste.[12–16] All waste, by-products, and wastage of plant origin are a source of substances of interest to the pharmaceutical, cosmetic, food, chemical, and energy industries, and different technologies are being evaluated to recover them.[17–19] Most compounds of interest in waste, by-products, and plant wastage are thermolabile, so nonthermal technologies are necessary to harness them. High hydrostatic pressures (HHPs) meet this requirement and involve subjecting the biomass to pressurization conditions (usually between 100 and 400 megapascals (MPa) for periods of 1–10 min) using a fluid (typically water) as a pressure-transmitting medium, which is confined in a chamber where pressure is uniformly generated simultaneously across the entire biomass surface.[20–24] It has been proven that HHPs cause damage to the structure and composition of the cell membrane of plant tissues, leading to an imbalance in the entry and exit of cellular components. This promotes greater extraction of compounds of interest; furthermore, to make the extraction more efficient, the use of combinations of water and organic solvents has been evaluated as pressure-transmitting media.[24–29] Currently, HHPs are widely applied to inactivate pathogenic microorganisms and those responsible for food deterioration. However, this chapter aims to demonstrate the potential of HHPs to promote the extraction of compounds of interest from plant tissues, thus becoming a strategy for utilizing waste, by-products, or wastage generated in the value chain of fruits and vegetables.

4.2 GENERATION OF WASTE AND BY-PRODUCTS IN FRUIT PRODUCTION AND PROCESSING

With the advancement of agronomic science applied to cultivation, fruit production has increased every year. In 2019, notably, 883 million tons of fruits were produced, while in 2020, the production reached 887 million tons, and in 2021, it reached a production of 900 million tons of fruits. The Asian continent is where the regions with the highest production are located.[1–3] The most in-demand fruits account for 57% of the total fruit production, with bananas contributing 18% of the total fruit volume, watermelons at 11%, apples at 10%, and oranges and grapes each contributing 9%. India is the largest producer of bananas, China is the largest producer of watermelons, apples, and grapes, and Brazil is the largest producer of oranges.[1,2] On the contrary, vegetable production has also seen an increase; in 2020, 446 million tons were produced, and by 2021, the figure reached 1128 million tons. Tomatoes are the highest-producing vegetable, accounting for 16% of total global production, followed by onions at 9% of the worldwide total. Cucumbers represent 8%, cabbages contribute 6%, and eggplants account for 5%. Among the top five demanded vegetable species, China is the largest producer of tomatoes, cucumbers, cabbages, and eggplants, while India is the largest producer of onions.[1] The increase in fruit and vegetable production is in response to the growing demand for food due to global population growth and increased life expectancy. This necessitates a higher production volume for raw materials and processed foods. However, the increase in fruit and vegetable production also leads to a rise in fruit waste.[4,5] Due to various factors, such as demand, but primarily climatic and geographical factors, the fruits with the highest production are not necessarily the most traded. For this reason, mangoes, pineapples, avocados, and papayas are the fruits with the highest export volumes.[6]

Despite the increase in fruit and vegetable production, it is estimated that, on average, a person consumes only 390 g out of the 400 g of fruits and vegetables recommended by the WHO.[7] One of the main reasons for not meeting the WHO recommendation is the scarcity of fruits and vegetables due to losses throughout the value chain, whether they are intended for fresh consumption or processing. In both cases, these losses originate in the cultivation field and result in waste.[8] Due to their physiology, high water content, enzymes, and the fragility of their tissues, fruits, and vegetables are highly perishable, leading to waste and wastage from the moment they are separated from the plant, and even before in the case of those that need to be harvested when fully ripe. It is estimated that between 14 and 50% of fruits and vegetables produced are lost during the harvesting process or immediately afterward.[5,6,9,10]

In general, food loss refers to a reduction in the quantity of edible biomass, and this reduction can occur at any stage of the production chain, from the field to processing, marketing, or in the home of the final consumer.[11] On the contrary, in the production of fruits and vegetables, food waste and food by-products refer to the waste (parts of the product that were not intended to be food) and by-products (which were not the main part of the food) generated during the production, processing, marketing, or consumption of food. However, this definition also includes as waste any biomass that, although intended to be food, is too deteriorated to be processed or marketed.[11,12] In addition, the definition of waste and by-product is also subject to the stage at which the waste is generated. For example, at the field level, we can consider plant stems, leaves, roots, immature fruits, or those unfit for harvesting as waste.[13] In the subsequent links of the value chain, waste can be generated due to activities such as selection, storage, transportation, processing, and consumption of fruit or vegetables, as well as deterioration caused by pests or microorganisms.[11] The reduction of food specific to each stage of the value chain is called waste; however, when food loss or waste is due to human mishandling, it should be considered food wastage.[14]

Considering the proportion and volumes of production, the FAO estimates that, in general terms, the greatest food loss and waste occur in the form of fruits and vegetables.[10,11] One of the major ecological concerns is the large volumes of waste generated from processing fruits and vegetables. Although fruit and vegetable waste is biodegradable, it has a negative impact on the environment due to the sheer volume, making it difficult to manage. Furthermore, it serves as a breeding ground for crop pests and pathogens.[15,16] Up to 25% of the total annual production is estimated to be lost or wasted globally, which amounts to approximately 225 million tons. Tubers, roots, and oilseeds report the highest losses, followed by fruits and vegetables.[8] The percentage of loss or waste varies depending on the stage of the production chain. Specifically for fruits and vegetables, between 5 and 35% are lost or wasted during harvesting, between 5 and 32% during storage, between 9 and 15% during transportation, between 2 and 38% during processing, and between 7 and 24% during marketing.[8]

The largest amount of waste is generated during the processing of fruits and vegetables. This is because industrialization involves sorting, washing, cleaning, peeling, cooking, packaging, and storage, generating waste and by-products.[9,17] The percentage of waste and by-products generated in the industrialization of fruits and vegetables depends on the technological level of the agroindustry, as well as the organizational and cultural level of the company.[18] The ecological issue posed by the accumulation of large volumes of waste and by-products from fruit and vegetable processing has led to the search for an alternative for their utilization. It has been proven that these waste and by-products are still a good source of vitamins, organic acids, dietary fiber, pigments, minerals, antimicrobials, antioxidants, anti-inflammatory agents, aromatic compounds, proteins, oils, and sugars, all of which are of interest to the food, pharmaceutical, cosmetic, chemical, and energy industries.[8,10,16,18,19]

4.3 HIGH HYDROSTATIC PRESSURES (HHPS)

Various technologies have been employed to recover all the elements and compounds of interest present in the waste and by-products of fruits and vegetables. For example, technologies such as microwave-assisted extraction, supercritical fluid extraction, ultrasonic-assisted extraction, and HHP have been used. These technologies are designed to facilitate the extraction and maintain the stability of the compounds of interest.[10]

The HHP process is based on isostatic compression (constant pressure) and Le Chatelier's principle. The first principle refers to the fact that the transmission of pressure occurs uniformly and instantaneously regardless of the geometry and size of the food. The second principle refers to the fact that any phenomenon in equilibrium, whether it is a chemical reaction, phase transition, or change in molecular configuration, is accompanied by a decrease in volume, which can be enhanced by hydrostatic pressure.[20,21]

HHP processing systems (Fig. 4.1) consist of chambers (refrigerated or not) where all types of foods can be processed, regardless of their size,

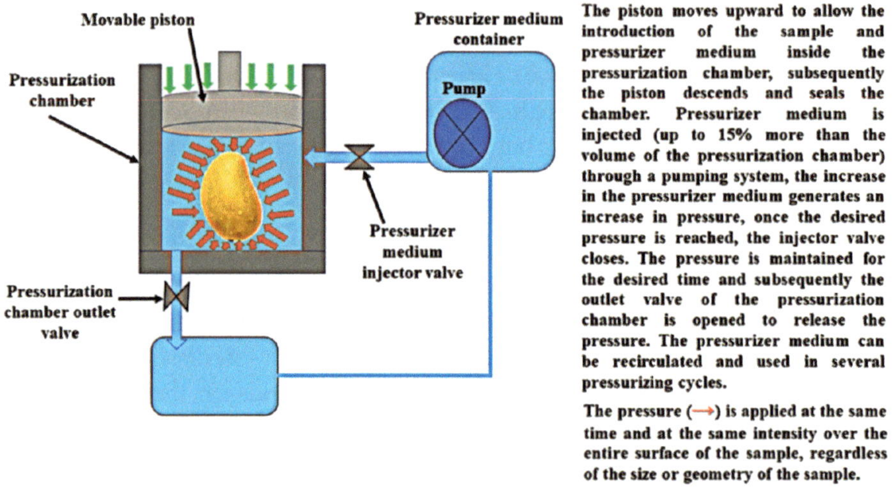

FIGURE 4.1 General diagram of the operation of high hydrostatic pressure equipment.

composition, aggregation state, and geometry, thanks to the principle of isostatic compression.[22] Typically, there is no heat exchange between the system and its surroundings, making the process adiabatic. Inside the chamber, the food is surrounded by a pressure-transmitting fluid (such as water, oil, or propylene glycol), which can be reused for multiple processes, making it an environmentally friendly technology. Once the process begins, the fluid is compressed, either using a manual piston system to increase pressure or by pumps or intensifiers. The pressure is increased approximately 100 MPa/min until it reaches the isostatic state at the working pressure, which can be maintained indefinitely. Residence times generally range from 1 to 15 min. The pressure is instantly released at the end of the process, causing a temperature reduction in the system. Therefore, the final temperature of the food can be lower than its initial temperature if the process is carried out under adiabatic conditions.[23]

This processing technology offers several advantages compared to conventional thermal treatments. Among the most notable is the retention of the nutritional, nutraceutical, and sensory value of foods.[24] At the molecular level, HHP (provided that the pressure level is below 1 GPa) does not cause drastic changes in the chemical structures of food molecules, as it only affects noncovalent bonds (van der Waals forces, hydrogen bonds, and electrostatic interactions). Therefore, many biomacromolecules and bioactive compounds are preserved almost entirely after processing.[25–27]

Currently, HHP is primarily used to inactivate microorganisms, and only a few studies have explored its application for other purposes. The primary effect of pressurization on microorganisms is damage to the integrity and permeability of the cell membrane, resulting in effects such as loss of turgor pressure and imbalance in the flow of materials both into and out of the cell. These damages are reversible at pressures below 150 MPa, but cell damage becomes permanent when pressures of 150 MPa are applied, regardless of the duration or temperature during pressurization.[28] The damage to the membrane is mainly caused by the effect of pressurization on the phospholipids and proteins that constitute it, leading to the leakage of electrolytes and other components.[28,29] High-pressure treatment affects different components of plant tissues. For instance, at pressures exceeding 200 MPa, starch gelatinization, protein denaturation, changes in the phase transition of lipids, and enzyme inactivation are induced. These effects can be reversible depending on the time and temperature of pressurization, but they become irreversible when the pressure exceeds 300 MPa.[30]

4.4 HIGH HYDROSTATIC PRESSURE EXTRACTION (HHPE)

HHPE is an emerging extraction method shown to be faster and more efficient than other extraction methods. Diverse studies have stated that HHPE can reduce processing time and obtain higher extraction yields than other methods and has no detrimental effects on the activity or structure of bioactive components.[31–33] However, there is limited information on applying high hydrostatic pressure to extract phenolic compounds and other bioactive such as sulforaphane.

4.4.1 EXTRACTION OF PHENOLIC COMPOUNDS WITH HIGH HYDROSTATIC PRESSURES (HHPS)

Phenolic compounds are among the most extensively distributed naturally bioactive substances. More than 8000 phenolic compounds have been identified from plants.[34] These compounds are responsible for the color and taste properties of fruits, vegetables, and their products, and it has been demonstrated that they have beneficial biological activities for health,[35] so it is important to preserve them and avoid their reduction during the different treatments to which fruits and/or vegetables are subjected for processing and preservation, it is equally important to apply extraction methods that can preserve their beneficial characteristics. HHPE has not been extensively applied to extract phenolics from fruit and vegetable wastes.

In this context, Briones-Labarca et al.[31] investigated the influence of different treatment conditions (400 MPa/5, 10, and 15 min) using water as a pressure-transmitting medium on the changes in TPC from Chilean papaya seeds. Results indicated that different times used during HHP result in significantly different content of TPC. The extraction values increased from 467, 515, and 647-mg GAE/100 g seeds for 5, 10, and 15 min, respectively. This result showed that the extraction for HHP increased with increasing extraction time. This behavior could be explained by the fact that the equilibrium of pressure between the inside and outside of the cells is rapid during a short time (until 15 min) under the high-pressure condition or due to the diffusion speed of the solvent being very high.[36] In the same study, the effect of HHPE on flavonoids was evaluated. The flavonoid content of papaya seed extract increased from 114, 159 to 268 mg QE/100 g seeds for HHPE at 500 MPa for 5, 10, and 15 min, respectively. These results confirmed that HHP effectively increases the phenolic compound yield from Chilean papaya seeds.

Similarly, Okur et al.[32] studied the effect of HHP at 400 and 500 MPa for 1, 5, and 10 min at 20°C on sour cherry derived from mashed cherries for the obtention of phenolic compounds. In this study, the increment in treatment time from 1 to 10 min enhances the extraction yield of TPC from 145 to 210-mg GAE/100 g at 400 MPa and from 170 to 227-mg GAE/100 g at 500 MPa. The increment in pressure from 400 to 500 MPa improved the extracted TPC from 210- to 227-mg GAE/100 g at 10 min. Under the process of HHPE, the differential pressure between the cell interior and the exterior of cell membranes is so large that it will lead to rapid permeation. Consequently, the concentration between the cell interior and the exterior of cell membranes can reach equilibrium in a short time. Then, the higher the hydrostatic pressure, the more solvent can enter into the cell, and more compounds can permeate the cell membrane.[31]

Teles et al.[33] evaluated the effect of HHPE from grape pomace at different pressures (50, 100, and 200 MPa) and times (0–30 min) (Table 4.1). Grape pomace presented a 5-min maximum release of TPC at 200 MPa (870-mg GAE/100 g), with a marked reduction afterward. The process at 100 MPa had a steady increase up to 15 min (750-mg GAE/100 g) and a decrease after this period. For 50 MPa, the highest TPC release occurred at 10 min (840-mg GAE/100 g), and after that time, there was a gradual decrease. The results showed that TPC shows a sharp increase in the first 10 min of the

TABLE 4.1 Extraction of Total Phenolic Compounds (mg GAE/100 g) from Grape Pomace by HHP.

Pressure/Time (MPa)	0 min	5 min	10 min	15 min	30 min
50	0	820	840	800	760
100	0	626	680	750	610
200	0	870	720	630	700

process, with a stability period between 15 and 30 min in each treatment; this behavior could be attributed to the depolymerization of phenolic compounds in lower molecular weight compounds not quantified by the Folin-Ciocalteu assay.[37] The authors stated that the differences in susceptibility to time and HHP could be associated with the phenolic profile, considering that each compound has a specific behavior during the process. As the differences in the susceptibility of phenolic compounds to time and HHP may be related to the phenolic profile of the extracts, considering that each compound has a specific performance during different pressures and times.

On the contrary, in a study conducted by Inada et al.[38] they evaluated the effect of the condition of HHPE of pressure (200, 350, and 500 MPa) and time (1, 5.5, and 10 min) on jaboticaba peels and seed. The authors concluded that processing was ineffective in increasing TPC in comparison to the unprocessed sample (Fig. 4.2). Higher pressure and longer time decreased from 17% (350 MPa for 5.5 min) to 43% (200 MPa for 10 min) in TPC.

According to the authors, the decrease in phenolic compounds could be attributed to a rise in the rates of nonenzymatic oxidative and hydrolytic reactions caused by the increases in pressure and time of processing; on the contrary, it is known that pressurization leads to adiabatic heating, which may have contributed to the degradation of phenolic compounds.[39]

The impact of time (5 and 10 min), temperature (25, 35, 45, and 55°C), and solvents (water, 1% HCl, 50 and 70% methanol with and without addition of HCl, and 50 and 70% ethanol), at a constant pressure of 600 MPa on tomato peel, has been evaluated in by Grassino et al.[40] With respect to the yield of TPC. The average TPC in tomato peel obtained by HHPE is 29.80 and 29.51-mg GAE/g for 5 and 10 min of extraction, respectively. Time did not affect TPC yields, which might be related to the retention of phenolic by the wall cell components.[41] In contrast to time, a rise of temperature from 25 to 45°C significantly enhanced the solubility of phenols and the highest values are obtained at 45°C (40.44 and 39.47-mg GAE/g, for 5 and 10 min). The high extraction efficiency is also obtained using ethanol; although 1% HCl provided a great recovery of phenols, its addition to methanol decreased TPC yields.

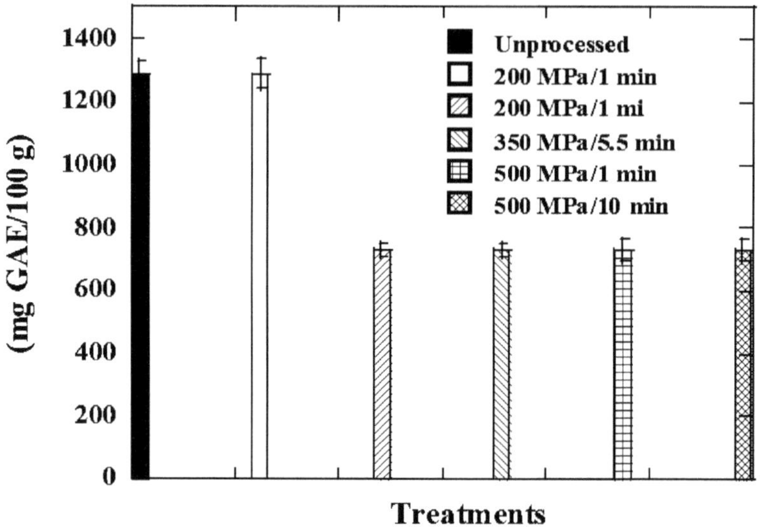

FIGURE 4.2 Effect of treatment on phenolic compound contents on jaboticaba peel and seed.

4.4.2 OTHER BIOACTIVES

Sulforaphane (4-methylsulfinybutyl isothiocyanate) is an organosulfur compound belonging to a family of isothiocyanates[42] and is present in cruciferous vegetables.

The sulforaphane content of papaya seed extract increased from 4144, 5289, and 5014 mg/100 g seed for HHPE at 500 MPa for 5, 10, and 15 min, respectively, relative to conventional extraction (915-mg/100-g seed). The authors concluded that although the extraction yields significantly increased when applying HHPE, the maximum extraction yield could be reached within a much shorter extraction time.[31]

4.4.3 DIETARY FIBER EXTRACTION WITH HIGH HYDROSTATIC PRESSURES (HHPS)

Dietary fiber is made up of various carbohydrate polymers found in plant cell walls, such as cellulose, hemicellulose, phenolic compounds, gums, pectin, and others. Based on their solubility, fiber can be classified into two categories, namely soluble dietary fiber (SDF) and insoluble dietary fiber (IDF).[43] Fiber is an essential part of our diet that offers many health benefits.

It promotes regular bowel movements, helps to lower blood glucose and cholesterol levels, and increases fermentation in the colon. Fiber also positively influences the microflora in our gut, which can lead to improved overall health. Additionally, it can help increase satiety, reduce adiposity, and prevent the development of colon cancer.[44,45] Also, the fiber confers combined technological and biological functional qualities in food systems.[43,44] When it comes to analyzing the fiber content of food products, it is for obtaining quantitative data. The extraction methods used in this process have been extensively researched and studied. However, when it comes to agrifood waste, the extraction methods used have a particular significance, as they are crucial in obtaining valuable by-products.

A method for fiber extraction is HHP; this technology has applications for maximization of obtention dietary fiber, assisted and nonassisted[46] compared to other methods. In apple by-product, the process with HHP increased the yield of total dietary fiber to 13% at 200 mPa and 600 MPa (15 min); in contrast, there is a decreasing effect on soluble fiber to 400 and 500 MPa.[47] In the other matrix, pressure and time of residence, the trend is a conversion or IDF to SFD; some examples are mature pear pomace (300 MPa, 15 min)[47] peel mango (600 MPa, 10 min),[48] peel orange (600 MPa, 10 min),[48] cabbage powder (*Brassica oleracea* var. capitata) (600 MPa, 10 min)[49] and others. The conversion of IDF to SFD is caused by partial solubilization of IDF for effects on unions molecular modified the structure of macromolecules and generation of smaller molecular fragments.[48,50,51] Of course, the yield and SFD/IFD ratio will depend on the source of the fiber, pressure, residence time, and if there is an assisted method, opening the possibility to many combinations of variables; but in general terms, if want to obtain a recovery higher soluble fiber content, HHP is a practical method to revalorized waste from agrifood.

4.4.4 PECTIN EXTRACTION WITH HIGH HYDROSTATIC PRESSURES (HHPS)

Pectin is a complex structural polysaccharide found in plant tissue in the middle lamella and primary and secondary cell walls that, depending on the type of pectin obtained, has physicochemical features useful in the food and pharmaceutical industries.[52,53] This complex polysaccharide is part of soluble fiber and is one of the main reasons to extract fiber from agrifood waste. The structural variability of pectin depends on the source, extraction method, and treatments during the processing extraction; for this reason, broad extraction processes have been evaluated as additional alternatives to conventional procedures, including the use of organic acid, enzyme-assisted, microwave, ultrasound, ohmic, pulsed electric field, subcritical water, and high-pressure processing, although the majority of these proposals are still in the research phase.[52] However, HHP is an increasingly utilized technique in the food industry and is a technique assayed and used to extract pectin from fruits and vegetables.[54]

Nonthermal, near-instantaneous, and shape- and size-independent use are among the benefits of HHP for pectin extraction.[55] Also, this technology can increase pectin recovery.[40,56] In some structures where pectin is most readily available as the pulp, the molecular weight of pectin extraction is increased; for example, in mango pulp slices at 300, 400, and 500 MPa (5 min) increase compared to a traditional method,[55] even when pectin it is accessible in solution, the main chain (galacturonic acid) did not change[57] and can increase their viscosity.[58] But in other sources or conditions of extraction (pressure), where pectin is part of a more complex matrix or has been exposed to previous physicochemical or mechanical processes, the HHP can be chemical structure modified, and the molecular weight of pectin[58] by depolymerization, degradation or modification of the chains original.[59–61] Both conditions can be advantageous depending on the application of the extracted pectin.

The esterification grade is a crucial determinant of pectin application in the food sector. The esterification grade shows the presence of methoxyl groups, and pectin can be classified into high-methoxyl pectin and low-methoxyl pectin, both of which can be used as ingredients in the food industry.[62] HHP allows to obtain pectin of commercial food quality containing greater than 65% galacturonic acid, and the esterification grade is unaffected by this process.[60,63] On the contrary, the depolymerization of pectin for the extraction process can favor the pharmaceutical industry in the obtention monosaccharides or oligosaccharides pectin of interest pharmaceutical or nutritional. The degradation of pectin can increase monosaccharides content such as mannose and arabinose.[47,61] Mannose can be applied to food, pharmaceutical, and poultry industries,[64] while arabinose can be used in the food and flavors industry and has properties for the treatment of human diabetes, precursor to produce amino acids, bacteriological diagnostic, and others.[65] Also, oligosaccharides derived from pectin have antioxidant activity; prebiotic properties; prevention and treatment of infectious diseases and cancer; lipid reduction properties; potential of absorption of minerals; and application in agriculture as a growth stimulator, ripening agent, and others.[66] HHP is a method that can be used to obtain pectin or pectin derivatives with applications in the pharmaceutical and even agricultural food industry.

4.5 FUTURE PERSPECTIVES

As mentioned earlier, the waste, by-products, and wastage from fruits and vegetables serve as reservoirs for various valuable components for different industries, so research aimed at their utilization will continue to evolve. Despite the well-known effects of high pressures on microorganisms and plant tissues, few studies still apply high pressures as promoters of the extraction of valuable components.[67] HHPs appear to be a suitable method for the extraction of thermolabile compounds because they do not require an increase in temperature to promote cell rupture. Therefore, they could be one of the best technologies for extracting antioxidant compounds, vitamins, and pigments.[68] One opportunity for applying HHPs to increase the content of phenolic compounds lies in the fact that many waste, by-products, and wastage of fruits and vegetables consist of live tissues, meaning they have not undergone pasteurization or sterilization processes that cause immediate cell death. It has been reported that the application of moderate pressures (between 15 and 100 MPa) can stress plant cells and activate the synthesis mechanism of phenolic compounds, increasing the phenol content inside the cell. However, this depends on many factors that make it difficult to obtain conclusive results. Nonetheless, encouraging results of this phenomenon have been found when mangoes, strawberries, and carrots are pressurized.[69–71] On the contrary, promising results have been found in sour cherry, grape, and apricot nectar subjected to pressures ranging from 400 to 550 MPa, increasing the extraction and bioaccessibility of phenolic compounds. In this case, the increase was caused by greater damage to the cell structure and the breaking of bonds between phenolic compounds and fiber.[32,72,73] This suggests that in the future, high pressures may become the primary method for extracting valuable components from food waste, by-products, and wastage. Additionally, it is an environmentally friendly technology because the water used for pressurization can be reused for many pressurization cycles, and the energy cost is relatively low compared to thermal technologies. Another potential advantage is that extraction can be promoted in aqueous media or with combinations of water and solvents,[74,75] providing great versatility to this technology and making it more feasible to be applied.

4.6 CONCLUSIONS

HHPs meet the condition of preserving compounds of interest because they are a nonthermal technology. In addition, they primarily generate a mechanical effect on tissues, reducing the chemical modification of bioactive compounds, thus ensuring that their potential benefits are preserved. Furthermore, high pressures are versatile and not limited by the sample's shape, size, or texture, making them suitable for all variants (solids, semi-solids, or liquids) of waste, by-products, or residues from vegetables. Perhaps the only limitation is the design of equipment to process the large volumes generated, but this can be compensated for by the speed at which pressurization treatments can be applied (5–10 min per pressurization cycle). Another limitation is the cost of the equipment, as being a nontraditional technology, there are limited companies distributing such equipment. One advantage of HHPs treatments is that the pressurizing medium can be reused in multiple cycles, reducing costs and being environmentally friendly. Despite all the advantages they may present, there are currently limited studies aimed at evaluating HHPs as promoters of compound extraction. So far, promising results have been found, especially in reducing extraction times and increasing extraction yields. The possibility of using extraction solvents as part of the pressurizing medium enhances the potential and versatility of HHPs as an extraction technology. HHPs have the potential to become one of the primary strategies for extracting bioactive compounds; however, there may still be biomass that needs to be directed to a secondary process.

KEYWORDS

- **pressure**
- **extraction**
- **bioactive compounds**
- **fruit waste**
- **by-products**

REFERENCES

1. FAO. Agricultural Production Statistics 2000–2020. FAOSTAT Analytical Brief Series No. 41, Rome, 2022.
2. FAO. In *World Food and Agriculture – Statistical Yearbook 2021*; Rome, 2021.
3. STATISTA. [Online] 2023. https://es.statista.com/estadisticas/635296/fruta-produccion-global-de-fruta-por-region/. (Accessed Aug 1, 2023).
4. Salim, N.; Singh, A.; Raghavan, V. Potential Utilization of Fruit and Vegetable Wastes for Food through Drying or Extraction Techniques. *Nov. Tech. Nutr. Food Sci.* **2017,** *1*, 1–12.
5. FAO. In *The State of Food and Agriculture 2019: Moving forward on Food Loss and Waste Reduction*; FAO: Rome, Italy, 2019.
6. FAO. Major Tropical Fruits: Market Review 2020, Rome, 2021.
7. FAO. International Year of Fruits and Vegetables 2021, Rome, 2021.
8. FAO and CIRAD. In *Fruit and Vegetables – Opportunities and Challenges for Small-Scale Sustainable Farming*; Rome, 2021.
9. Moon, K. M.; Kwon, E. B.; Lee, B.; Kim, C. Y. Recent Trends in Controlling the Enzymatic Browning of Fruit and Vegetable Products. *Molecules* **2020,** *25*, 2754.
10. Zhu, Y.; Luan, Y.; Zhao, Y.; Liu, J.; Duan, Z.; Ruan, R. Current Technologies and Uses for Fruit and Vegetable Wastes in a Sustainable System: A Review. *Foods* **2023,** *12*, 1949.
11. FAO. Global Food Losses and Food Waste – Extent, Causes and Prevention, Rome, 2011.
12. OECD (Organisation for Economic Co-operation and Development). Glossary of Environment Statistics, Studies in Methods, Series F, No. 67, United Nations, New York, 1997.
13. Obi, F. O.; Ugwuishiwu, B. O.; Nwakaire, J. N. Agricultural Waste Concept, Generation, Utilization and Management. *Nigerian J. Technol.* **2016,** *35* (4), 957–964.
14. FAO. Food Wastage Footprint, Impacts on Natural Resources, Rome, 2013.
15. Sagar, N. A.; Pareek, S.; Sharma, S.; Yahia, E. M.; Lobo, M. G. Fruit and Vegetable Waste: Bioactive Compounds, their Extraction, and Possible Utilization. *Compr. Rev. Food Sci. Food Saf.* **2018,** *17*, 512–531.
16. Lucarini, M.; Durazzo, A.; Bernini, R.; Campo, M.; Vita, C.; Souto, E. B.; Lombardi-Boccia, G.; Ramadan, M. F.; Santini, A.; Romani, A. Fruit Wastes as a Valuable Source of Value-Added Compounds: A Collaborative Perspective. *Molecules* **2021,** *26*, 6338.
17. Schieber, A. Side Streams of Plant Food Processing as a Source of Valuable Compounds: Selected Examples. *Ann. Rev. Food Sci. Technol.* **2017,** *8*, 97–112.
18. Donner, M.; Verniquet, A.; Broeze, J.; Kayser, K.; De Vries, H. Critical Success and Risk Factors for Circular Business Models Valorising Agricultural Waste and By-Products. *Resour. Conserv. Recycl.* **2021,** *165*, 105236.
19. Torres-León, C.; Ramírez-Guzman, N.; Londoño-Hernandez, L.; Martinez-Medina, G. A.; Díaz-Herrera, R.; Navarro-Macias, V.; Alvarez-Pérez, O. B.; Picazo, B.; Villarreal-Vázquez, M.; Ascacio-Valdes, J.; Aguilar, C. N. Food Waste and Byproducts: An Opportunity to Minimize Malnutrition and Hunger in Developing Countries. *Front. Sustain. Food Syst.* **2018,** *2*, 52.
20. Augusto, P. E. D.; Tribst, A. A. L.; Cristianini, M. High Hydrostatic Pressure and High-Pressure Homogenization Processing of Fruit Juices. In *Fruit Juices: Extraction,*

Composition, Quality and Analysis; Rajauria, G., Tiwari, B. K., Eds.; Academic Press: Massachusetts, 2017; pp 393–421.
21. Picart-Palmade, L.; Cunault, C.; Chevalier-Lucia, D.; Belleville, M. P.; Marchesseau, S. Potentialities and Limits of Some Non-thermal Technologies to Improve Sustainability of Food Processing. *Front. Nutr.* **2019,** *5*, 130.
22. Chawla, R.; Patil, G. R.; Singh, A. K. High Hydrostatic Pressure Technology in Dairy Processing: A Review. *J. Food Sci. Technol.* **2011,** *48*, 260–268.
23. Balasubramaniam, V. M. B.; Martínez-Monteagudo, S. I.; Gupta, R. Principles and Application of High Pressure-Based Technologies in the Food Industry. *Annu. Rev. Food Sci.Technol.* **2015,** *6*, 435–462.
24. Rawson, A.; Patras, A.; Tiwari, B. K.; Noci, F.; Koutchma, T.; Brunton, N. Effect of Thermal and Non Thermal Processing Technologies on the Bioactive Content of Exotic Fruits and their Products: Review of Recent Advances. *Food Res. Int.* **2011,** *44,* 1875–1887.
25. Martín-Belloso, O.; Soliva-Fortuny, R.; Elez-Martínez, P.; Marsellés-Fontanet, A. R.; Vega-Mercado, H. Non-thermal Processing Technologies. In *Food Safety Management: A Practical Guide for the Food Industry*; Motarjemi, Y., Lelieveld, H., Eds.; Academic Press: Massachusetts, 2014; pp 443–465.
26. Mújica-Paz, H.; Valdez-Fragoso, A.; Samson, C. T.; Welti-Chanes, J.; Torres, A. High-Pressure Processing Technologies for the Pasteurization and Sterilization of Foods. *Food Bioproc. Tech.* **2011,** *4*, 969–985.
27. Rojas, M. L.; Miano, A. C.; Aguilar, K.; Augusto, P. E. D. Emerging Technologies for Noncarbonated Beverages Processing. In *Trends in Non-alcoholic Beverages*; Galanakis C. M., Ed.; Academic Press: Massachusetts, 2020; pp 233–261.
28. Rux, G.; Gelewsky, R.; Schlüter, O.; Herppich, W. B. High Hydrostatic Pressure Treatment Effects on Selected Tissue Properties of Fresh Horticultural Products. *Innov. Food Sci. Emerg. Technol.* **2020,** *61*, 102326.
29. Rux, G.; Gelewsky, R.; Schlüter, O.; Herppich, W. B. High Hydrostatic Pressure Effects on Membrane-Related Quality Parameters of Fresh Radish Tubers. *Postharvest Biol. Technol.* **2019,** *151*, 1–9.
30. San Martin, M. F.; Barbosa-Cánovas, G. V.; Swanson, B. G. Food Processing by High Hydrostatic Pressure. *Crit. Rev. Food Sci. Nutr.* **2002,** *42*, 627–645.
31. Briones-Labarca, V.; Plaza-Morales, M.; Giovagnoli-Vicuña, C.; Jamett, F. High Hydrostatic Pressure and Ultrasound Extractions of Antioxidant Compounds, Sulforaphane and Fatty Acids from Chilean Papaya (*Vasconcellea pubescens*) Seeds: Effects of Extraction Conditions and Methods. *LWT - Food Sci. Technol.* **2015,** *60* (1), 525–534.
32. Okur, İ.; Baltacıoğlu, C.; Ağçam, E.; Baltacıoğlu, H.; Alpas, H. Evaluation of the Effect of Different Extraction Techniques on Sour Cherry Pomace Phenolic Content and Antioxidant Activity and Determination of Phenolic Compounds by FTIR and HPLC. *Waste Biomass Valori.* **2019,** *10*, 3545–3555.
33. Teles, A. S. C.; Chávez, D. W. H.; Coelho, M. A. Z.; Rosenthal, A.; Gottschalk, L. M. F.; Tonon, R. V. Combination of Enzyme-Assisted Extraction and High Hydrostatic Pressure for Phenolic Compounds Recovery from Grape Pomace. *J. Food Eng.* **2021,** *288*, 110128.
34. Alara, O. R.; Abdurahman, N. H.; Ukaegbu, C. I. Extraction of Phenolic Compounds: A Review. *Curr. Res. Food Sci.* **2021,** *4,* 200–214.

35. Altemimi, A.; Lakhssassi, N.; Baharlouei, A.; Watson, D. G.; Lightfoot, D. A. Phytochemicals: Extraction, Isolation, and Identification of Bioactive Compounds from Plant Extracts. *Plants* **2017,** *6* (4), 42.
36. Prasad, N. K.; Yang, B.; Zhao, M.; Wang, B. S.; Chen, F.; Jiang, Y. Effects of High-Pressure Treatment on the Extraction Yield, Phenolic Content and Antioxidant Activity of Litchi (*Litchi chinensis* Sonn.) Fruit Pericarp. *Int. J. Food Sci. Technol.* **2009,** *44*, 960–966.
37. Chen, X.; Li, L.; You, Y.; Mao, B.; Zhao, W.; Zhan, J. The Effects of Ultra-High Pressure Treatment on the Phenolic Composition of Red Wine. *South African J. Enol. Vitic.* **2012,** *33*, 203–213.
38. Inada, K. O. P.; Nunes, S.; Martinez-Blazquez, J. A.; Tomás-Barberán, F. A.; Perrone, D.; Monteiro, M. Effect of High Hydrostatic Pressure and Drying Methods on Phenolic Compounds Profile of Jaboticaba (*Myrciaria jaboticaba*) Peel and Seed. *Food Chem.* **2020,** *309*, 125794.
39. Tauscher, B. Pasteurization of Food by Hydrostatic High Pressure: Chemical Aspects. *Z. Lebensm. Unters. Forsch.* **1995,** *200*, 3–13.
40. Grassino, A. N.; Pedisić, S.; Dragović-Uzelac, V.; Karlović, S.; Ježek, D.; Bosiljkov, T. Insight into High-Hydrostatic Pressure Extraction of Polyphenols from Tomato Peel Waste. *Plant Foods Hum. Nutr.* **2020,** *75*, 427–433.
41. Liu, S.; Xu, Q.; Li, X.; Wang, Y.; Zhu, J.; Ning, C.; Chang, X.; Meng, X. Effects of High Hydrostatic Pressure on Physicochemical Properties, Enzymes Activity, and Antioxidant Capacities of Anthocyanins Extracts of Wild *Lonicera caerulea berry. Innov. Food Sci. Emerg. Technol.* **2016,** *36*, 48–58.
42. Panjwani, A. A.; Liu, H.; Fahey, J. W. Crucifers and Related Vegetables and Supplements for Neurologic Disorders: What is the Evidence?. *Curr. Opin. Clin. Nutr. Metab. Care.* **2018,** *21*, 451–457.
43. Maphosa, Y.; Jideani, V. A. Dietary Fiber Extraction for Human Nutrition—A Review. *Food Rev. Int.* **2016,** *32*, 98–115.
44. Hussain, S.; Jõudu, I.; Bhat, R. Dietary Fiber from Underutilized Plant Resources—A Positive Approach for Valorization of Fruit and Vegetable Wastes. *Sustainability* **2020,** *12* (13), 5401.
45. Fuller, S.; Beck, E.; Salman, H.; Tapsell, L. New Horizons for the Study of Dietary Fiber and Health: A Review. *Plant Foods Hum. Nutr.* **2016,** *71*, 1–12.
46. De la Peña Armada, R.; Villanueva-Suárez, M. J.; Mateos-Aparicio, I. High Hydrostatic Pressure Processing Enhances Pectin Solubilisation on Apple By-Product Improving Techno-Functional Properties. *Eur. Food Res. Technol.* **2020,** *246*, 1691–1702.
47. Yan, L.; Li, T.; Liu, C.; Zheng, L. Effects of High Hydrostatic Pressure and Superfine Grinding Treatment on Physicochemical/ Functional Properties of Pear Pomace and Chemical Composition of its Soluble Dietary Fiber. *LWT - Food Sci. Technol.* **2019,** *107*, 171–177.
48. Tejada-Ortigoza, V; García-Amezquita, L. E.; Serna-Saldívar, S. O.; Welti-Chanes, J. The Dietary Fiber Profile of Fruit Peels and Functionality Modifications Induced by High Hydrostatic Pressure Treatments. *Food Sci. Technol. Int.* **2017,** *23* (5), 396–402.
49. Chen, Y. R.; Wu, S. J. Effects of High-Hydrostatic Pressure and High-Pressure Homogenization on the Biological Activity of Cabbage Dietary Fiber. *J. Sci. Food Agric.* **2022,** *102* (14), 6299–6308.

50. Yang, B.; Jiang, Y.; Wang, R.; Zhao, M.; Sun, J. Ultra-High Pressure Treatment Effects on Polysaccharides and Lignins of Longan Fruit Pericarp. *Food Chem.* **2009,** *112* (2), 428–431.
51. Larrea, M. A.; Chang, Y. K.; Martínez-Bustos, F. Effect of Some Operational Extrusion Parameters on the Constituents of Orange Pulp. *Food Chem.* **2005,** *89* (2), 301–308.
52. Quintana-Obregón, E. A.; Martínez-Téllez, M. A.; Aispuro-Hernández, E.; Vargas-Ortiz, M. A. Pectin from Byproducts in the Food Industry Trends and Opportunities. In *Food Byproducts Managements and their Utilization*; Gómez-García, R., Vilas-Boas, A. A., Campos, D. A., Pintao, M. M., Aguilar, C. N., Eds.; Apple Academic Press: Palm Bay, FL, 2024 [IN PRESS].
53. Roy, S.; Priyadarshi, R.; Łopusiewicz, Ł.; Biswas, D.; Chandel, V.; Rhim, J. W. Recent Progress in Pectin Extraction, Characterization, and Pectin-Based Films for Active Food Packaging Applications: A Review. *Int. J. Biol. Macromol.* **2023,** 124248.
54. Khubber, S.; Kazemi, M.; Amiri Samani, S.; Lorenzo, J. M.; Simal-Gandara, J.; Barba, F. J. Structural-Functional Variability in Pectin and Effect of Innovative Extraction Methods: An Integrated Analysis for Tailored Applications. *Food Rev. Int.* **2023,** *39* (4), 2352–2377.
55. Hu, K.; Chen, D.; Chen, M.; Xiang, A.; Xie, B.; Sun, Z. Mechanistic Insights into Changes in Endogenous Water Soluble Pectin and Carotenoid Bioaccessibility in Mango Beverage Upon High Pressure Processing. *Food Hydrocoll.* **2023,** *140*, 108623.
56. Naghshineh, M.; Olsen, K.; Georgiou, C. A. Sustainable Production of Pectin from Lime Peel by High Hydrostatic Pressure Treatment. *Food Chem.* **2013,** *136* (2), 472–478.
57. Zhong, L.; Li, X.; Duan, M.; Song, Y.; He, N.; Che, L. Impacts of High Hydrostatic Pressure Processing on the Structure and Properties of Pectin. *LWT - Food Sci. Technol.* **2021,** *148*, 111793.
58. Peng, X.; Mu, T.; Zhang, M.; Sun, H.; Chen, J.; Yu, M. Effects of pH and High Hydrostatic Pressure on the Structural and Rheological Properties of Sugar Beet Pectin. *Food Hydrocoll.* **2016,** *60*, 161–169.
59. Tian, X.; Yang, Y.; Zhang, W.; Wang, Y. Effect of High Hydrostatic Pressure and Thermal Processing on the Shelf Life and Quality Attributes of Apple-Kiwi-Carrot Puree Blend. *J. Food Qual.* **2023,** 1631285.
60. Duan, X.; Zhu, Y.; Shu, C.; Gao, J.; Liu, F.; Pan, S. Extraction of Pectin from Satsuma mandarin Peel: A Comparison of High Hydrostatic Pressure and Conventional Extractions in Different Acids. *Molecules* **2022,** *27* (12), 3747.
61. Xie, F.; Zhang, W.; Lan, X.; Gong, S.; Wu, J.; Wang, Z. Effects of High Hydrostatic Pressure and High Pressure Homogenization Processing on Characteristics of Potato Peel Waste Pectin. *Carbohydr. Polym.* **2018,** *196*, 474–482.
62. Mungure, T. E.; Roohinejad, S.; Bekhit, A. E.; Greiner, R.; Mallikarjunan, K. Potential Application of Pectin for the Stabilization of Nanoemulsions. *Curr. Opin. Food Sci.* **2018,** *19*, 72–76.
63. Huang, B.; Zhao, K.; Zhang, Z.; Liu, F.; Hu, H.; Pan, S. Changes on the Rheological Properties of Pectin-Enriched Mango Nectar by High Intensity Ultrasound. *LWT - Food Sci. Technol.* **2018,** *91*, 414–422.
64. Hu, X.; Shi, Y.; Zhang, P.; Miao, M.; Zhang, T.; Jiang, B. D-Mannose: Properties, Production, and Applications: An Overview. *Compr. Rev. Food Sci. Food Saf.* **2016,** *15* (4), 773–785.

65. Fehér, C. Novel Approaches for Biotechnological Production and Application of L-arabinose. *J. Carbohydr. Chem.* **2018,** *37* (5), 251–284.
66. Babbar, N.; Dejonghe, W.; Gatti, M.; Sforza, S.; Elst, K. Pectic Oligosaccharides from Agricultural by-Products: Production, Characterization and Health Benefits. *Crit. Rev. Biotechnol.* **2016,** *36* (4), 594–606.
67. Navarro-Baez, J. E.; Martínez, L. M.; Welti-Chanes, J.; Buitimea-Cantúa, G. V.; Escobedo-Avellaneda, Z. High Hydrostatic Pressure to Increase the Biosynthesis and Extraction of Phenolic Compounds in Food: A Review. *Molecules* **2022,** *27*, 1502.
68. Scepankova, H.; Martins, M.; Estevinho, L.; Delgadillo, I.; Saraiva, J. A. Enhancement of Bioactivity of Natural Extracts by Non-Thermal High Hydrostatic Pressure Extraction. *Plant Foods Hum. Nutr.* **2018,** *73*, 253–267.
69. Hu, K.; Peng, D.; Wang, L.; Liu, H.; Xie, B.; Sun, Z. Effect of Mild High Hydrostatic Pressure Treatments on Physiological and Physicochemical Characteristics and Carotenoid Biosynthesis in Postharvest Mango. *Postharvest Biol. Technol.* **2020,** *172*(10), 47.
70. Viacava, F.; Ortega-Hernández, E.; Welti-Chanes, J.; Cisneros-Zevallos, L.; Jacobo-Velázquez, D. A. Using High Hydrostatic Pressure Processing Come-Up Time as an Innovative Tool to Induce the Biosynthesis of Free and Bound Phenolics in Whole Carrots. *Food Bioprocess Technol.* **2020,** *13*, 1717–1727.
71. Kim, T. E.; Gil, B.; Kim, C. T.; Cho, Y. J. Enrichment of Phenolics in Harvested Strawberries by High-Pressure Treatment. *Food Bioprocess Technol.* **2017,** *10*, 222–227.
72. Morata, A.; Loira, I.; Vejarano, R.; Bañuelos, M. A.; Sanz, P. D.; Otero, L.; Suárez-Lepe, J. A. Grape Processing by High Hydrostatic Pressure: Effect on Microbial Populations, Phenol Extraction and Wine Quality. *Food Bioprocess Technol.* **2014,** *10*, 3545–3555.
73. Huang, W.; Bi, X.; Zhang, X.; Liao, X.; Hu, X.; Wu, J. Comparative Study of Enzymes, Phenolics, Carotenoids and Color of Apricot Nectars Treated by High Hydrostatic Pressure and High Temperature Short Time. *Innov. Food Sci. Emerg. Technol.* **2013,** *18*, 74–82.
74. Jun, X. High-Pressure Processing as Emergent Technology for the Extraction of Bioactive Ingredients from Plant Materials. *Crit. Rev Food Sci Nutr.* **2013,** *53*, 837–852.
75. Pinela, J.; Prieto, M. A.; Barros, L.; Carvalho, A. M.; Oliveira, M. B. P. P.; Saraiva, J. A.; Ferreira, I. C. F. R. Cold Extraction of Phenolic Compounds from Watercress by High Hydrostatic Pressure: Process Modelling and Optimization. *Sep. Purif. Technol.* **2017,** *192*, 501–512.

CHAPTER 5

Fruit Waste Seeds as Natural Sources of Food Proteins

JOSÉ ARMANDO ULLOA* and NITZIA THALÍA FLORES JIMÉNEZ

Centro de Tecnología de Alimentos, Universidad Autónoma de Nayarit, Ciudad de la Cultura Amado Nervo, Tepic, Nayarit, México

*Corresponding author.

ABSTRACT

The continuous growth of the world population has increased the demand for food, which also expands the generation of by-products. To date, most of the food by-products are discarded as waste, exacerbating the environmental situation of the places where they are dumped. In the case of fruits, the proportion of by-products can be up to 80% based on the weight of the food, which includes pulp, peel/skin, seeds, and stones. Of particular interest are the seeds as fruit residues for protein recovery. In addition, the use of high-intensity ultrasound, an emerging technology considered friendly to the environment, has been beneficial both in the recovery of vegetable proteins and in modifying their physicochemical and structural properties. Generally, the modification of the physicochemical and structural properties of proteins influences the improvement of their functional properties, thus expanding their possibilities of use for food purposes. This chapter summarizes the relevance of fruit seeds as a source of protein, the typical process for the recovery of the said component, and the use of ultrasound as a beneficial treatment in preparing protein concentrates or isolates that can be used as food ingredients.

5.1 INTRODUCTION

The world population has experienced a dramatic growth that has resulted in an increase in its basic needs. According to the United Nations Population

Agroindustrial Waste Management and Natural Resources Conservation: Environmental Challenges and Transformation into Value-Added Products. José Juan Buenrostro-Figueroa, Juan Manuel Tirado-Gallegos, Mónica L. Chávez-González, A. K. Haghi, and Shaheemath Suhara K. K. (Eds.)

Fund,[139] the world population of 7.25 billion in 2013 will increase to 8.05 billion this year, a number projected to 9.5 billion by 2050. As a consequence of this, the world food demand also will be tripled.[45] For these reasons, strategies have been outlined to face the challenges that these problems imply. In the year 2015, the United Nations[137] approved the 2030 Agenda for Sustainable Development, as an opportunity to embark on a new path to improve the life of world society, eradicating poverty, protecting the planet, and ensuring prosperity for all. Of the 17 goals defined by the UN for Sustainable Development, which are interrelated, two of them are closely related to food: Goal 2—Zero Hunger and Goal 12—Responsible Consumption and Production.

To contribute to the zero hunger and responsible production and consumption goals, actions related to the chains that involve the production, conservation, and transformation of foods, as well as food accessibility to the population have been implemented. However, the Food and Agriculture Organization of the United Nations[41] estimates that in the world, approximately 14% of food is lost annually between harvest and market retail, while 17% is wasted at the retail and consumer level.

At present, livestock products, such as meat, dairy products and eggs, fish and seafood, and foods rich in plant-based proteins, such as cereals, legumes, nuts, and seeds, continue to be the main dietary sources of protein for human consumption. Nonetheless, animal proteins are widely recognized as having the best nutritional quality, due to their amino acid composition, digestibility, and the content of other nutrients such as iron and calcium.[29] Unfortunately, the production of proteins of animal origin, especially dairy and red meats, causes greenhouse gas emissions from livestock and loss of terrestrial biodiversity, due to human intervention with the consequent negative impact on the environment, being essential a more sustainable production and consumption of food of animal origin.[140]

In addition to sustainability issues, excessive meat consumption is linked to the development of cardiovascular disease, type 2 diabetes, and some types of cancer.[54] Meat consumption also has an impact on public health, due to the risk of outbreaks of foodborne diseases caused by pathogens such as bacteria, viruses, and parasites. Therefore, a change in the pattern of protein intake seems necessary, including the partial replacement of those of animal origin by vegetables. However, the substitution of animal proteins for those of plant origin could decrease the quality of the human diet, due to its lower digestibility, bioavailability, and presence of antinutritional factors.[6]

Among the sustainable alternatives to meet the global requirements of vegetable proteins, cakes and meals from the oilseed industry represent the most important agro-industrial by-products, which are used as raw material in the production of protein concentrates or isolates through various technologies.[12,82] For many years and to date, the most important vegetable protein in the form of isolate or protein concentrate worldwide is soy.[31]

On the other hand, another important source to recover nutrients as proteins could be the wastes from fruit processing, which if used will diminish environmental contamination because they are generally disposed of by the sanitary landfill procedures.[9,37] Some estimations indicate that the global fruit processing industry generates more than 0.5 billion tons of waste,[15] including peel, core, crown, seeds, and pomace, that represent between 10% and 80% of the fruit weight.[84]Recent studies have demonstrated that seeds from fruit processing are a good source to produce protein concentrates or isolates for food use.[38,43,109,116,120]

Unfortunately, plant proteins in their native form and those recovered by the various isolation or concentration methods generally have limited applications due to their poor functionality.[99] Nevertheless, there are many physical, chemical, or biological technologies available to modify the protein's physicochemical properties, with which their functional properties and food applications can be improved and diversified.[3] Among them, high-intensity ultrasound, which is a physical technology, stands out for offering better results and efficiency, as well as for being an economical and environmentally friendly treatment.[25]

This chapter presents a review of the most recent information about fruit seeds as a source of protein, including its functional properties and its enhancement by high-intensity ultrasound, and its application as ingredients or for the development of new foods.

5.2 SUSTAINABLE USE OF FRUIT WASTE FOR FOOD APPLICATIONS

Faced with the current climate crisis and its worsening, the food industry has been committed to accelerating the transition from its traditional approaches to more sustainable ones.[141] Approximately one-third of human-caused greenhouse gas emissions are related to food. This is derived from an extensive chain that involves its production, processing, transportation, distribution, preparation, and consumption, as well as its disposal. Each of these stages generates greenhouse gases that trap the sun's heat and contribute to climate change.[138]

Most food-related greenhouse gases come from agriculture and land use, including methane from livestock,[121] nitrous oxide from fertilizers,[8] carbon dioxide from clearing forests for farm expansion,[75] agricultural emissions from manure management,[117] burning crop residues,[36] and fuel use on farms.[86] A smaller proportion of greenhouse gas emissions from food are caused by preservation methods and industrial transformation processes, production of packaging materials, transportation, and waste management.[85,130]

Currently, worldwide, 88% of waste is disposed of in landfills or incinerated, which is linked to the linear economic model characterized by "produce, use and dispose of." In contrast, an alternative model called circular economy has been developed, where consumption of materials and the production of waste is minimized, while the recycling and reuse of waste to elaborate new materials is favored.[95] Therefore, this model could contribute to sustainable development, better environmental quality, and economic improvement.[22]

According to the most recent report of the FAO (2022),[40] fruit production in the world for the 2020 year was 887 million ton, which represented 9.5% of the production of primary crops. Besides, fruits and vegetables are one of the most consumed food groups worldwide, either as raw, processed, or as a complement to other foods, representing more than 42% of total food waste.[48] Fruit processing industries generate waste in the form of pods, leaves rind, seeds, kernel, peels, pulp, or pomace.[115] The contribution of fruit waste depends on the characteristics of each one of them and the processing to which they are subjected.

The olive,[23] grape,[95] orange,[104] and apple[20] processing industries generate waste that represents around 75% (leaves, pomace, and skins), 25% (pomace, seeds, and skins), 30–40% (pulp, seeds, and skins), and 30% (skin, seeds, stems, and pulp) of the weight of the fruit, respectively. From the use of the edible part of other fruits such as pomegranate, banana, dragon fruit, guava, durian, mango, papaya, pineapple, and jackfruit residues of up to 50% (peel and marc), 35% (peel), 30–45% (peel and seed), 10% (seed and skin), 60–70% (skin and seed), 45% (kernel and peel), 45–50% (skin and seed), 10–20% (skin and seed), 33% (skin and core), and 60–70% (skin and core) are produced based on the weight of the fruit, respectively.[97]

Many studies have confirmed that fruit processing wastes are source of primary metabolites such as proteins, carbohydrates, and lipids,[10,64,88,106,133] as well as secondary metabolites or biologically active components such as phenolic compounds, terpenes, terpenoids, and alkaloids.[72,98,129,142,143]

The main applications of bioactive compounds recovered from fruit waste is as functional food ingredients. Processed functional foods are those that have been carefully designed by adding bioactive substances, which, beyond satisfying nutritional needs, also present health benefits, although in their general characteristics they resemble conventional foods.[128] The health benefits associated with the consumption of functional foods are the protection of nervous and immune systems, oxidative damage, and cardiovascular diseases, as well as control of blood glucose level, health skin and bones, general enhancement of physical performances, and some forms of cancer.[7,34] Some examples of functional foods are ice cream, crackers, cookies, cheese, yogurt, bread, flour, and pasta.[107]

In addition to bioactive compounds, the fruit waste have been used natural ingredients of proteins, amino acids, carbohydrates, dietetic fiber, pectin, organic acids, oil, essential oils, minerals, vitamins, pigments, and others. The main functions of these molecules are to improve the nutritional and sensory quality of food products, as well as for preservation purposes.[101,108] Current technologies for the processing of fruit wastes in a sustainable system include microwave-assisted extraction, supercritical fluid extraction, ultrasonic-assisted extraction, pressurized liquid extraction, high hydrostatic pressure technique, pulsed electric field, and enzyme-assisted extraction.[155]

5.3 PROTEINS IN FRUIT SEEDS

Fruits are highly appreciated for their sensory and nutritional attributes. They are consumed fresh in their due state of maturity, also in processed products that contain them as ingredients. Fruit processing makes it possible to use them to overcome excess production during the harvest season. Md Sani et al.[88] reported that during the fruits processing is generated as waste corresponding to seeds about 25–40% of its total weight, which contain nutrients such as vitamins, minerals, carbohydrates, fats, lipids, and proteins, as well as bioactive compounds.

Although the protein content of fresh fruit seeds is generally low, peeling and drying significantly raise its concentration. After the husk or covering and the water from the fruit seeds have been removed, milling is applied to obtain the corresponding flour, which is typically used as raw material for protein recovery.[81] The defatting of the flour from fruit seeds also increase its protein concentration. The protein content in a defatted flour from fruit seeds depends on fruit type, chemical composition of the seeds, remotion or not the seed shells, fat extraction, and the used solvent.[50]

The protein contents of raw and defatted flours from fruit seeds are shown in Table 5.1. As can be observed, the range in the protein content of such flours is very wide. For example, raw mango seed flour has a protein content of 4%, while that the protein content of raw melon seed flour is 28.6%. On the other hand, the increase of the protein content in defatted flours by remotion of oil in raw flours also is very variable. The augment in the protein concentration by defatting of raw flour from jackfruit seed is only 16.1% in the respective defatted flour, while in the case of the raw flour from fruit passion seed is of 416%. The variability in these results could be due to the contents of oil and protein in the raw flours, as well as of the oil quantity removed during the defatting stage for the preparation of the defatted flours.

TABLE 5.1 Protein Content from Fruit Seeds as Raw or Defatted Flours.

Seed source	Protein content (%)		References
	Raw flour	Defatted flour	
Orange	17.8–17.9	45.5	[13,120]
Fruit passion	12.5	64.5	[38]
Guamuchil	11.6	33.2	[43]
Mango	4.4	8.2	[73,109]
Jackfruit	13.7–14.0	16.0	[91,116,136]
Soursop	15.3	27.8	[24]
Guava	7.2	22.7	[18,44]
Melon	18.8–28.6	37.3–57.5	[33,127]
Watermelon	17.1–27.6	54.5–61.3	[46,66,69]
Persian lime	20.6	34.4	[42]
Papaya	28.0	32.4	[1,67]
Pumpkin	19.6–36.5	52.4–54.0	[30,35]
Neem	12.0	46.3	[2]
Tamarind	13.8	26.9	[19]

The protein content of raw mango (4.4%) and guava (7.2%) seed flours is below the typical protein content of cereal grains (10–15%), while those from seeds passion fruit (12.5%), guamuchil (11.6%), Jackfruit seeds (13.7–14.0%), soursop (15.3%), neem (12.0%), and tamarind (13.8%) seeds (Table 5.1), precisely contain the level of protein corresponding to the food group of cereals (Shewry, 2023). Nevertheless, the protein content of the raw flours from orange (17.8%), melon (18.8–28.6%), watermelon (17.1–27.6%), Persian lime (20.6%), papaya (28.9%), and pumpkin (19.6–36.5%) seeds, fall within (17–28%) or exceeds the protein contribution range of legumes.[122]

Raw flours from fruit seeds have been proposed as food ingredients not only as a source of protein, but also of other nutrients such as carbohydrates and lipids, as well as bioactive compounds.[4] However, the presence of lipids in some fruit seed flour could shorten its useful life by oxidation reactions that cause undesirable flavors.[101]

On the other hand, the protein content of the defatted flours obtained of seeds from orange (45.5%), fruit passion (64.5%), guamuchil (33.2%), soursop (27.8%), guava (22.7%), melon (37.3–57.5%), watermelon (54.5–61.3%), Persian lime (34.4%), papaya (32.4%), pumpkin (52.4–54.0%), neem (46.3%), and tamarind (26.9%), fall in the range of the protein values for the most important oilseed press cakes (soybean, sunflower seed, sesame seed, cottonseed), which do not surpass the 55% of protein. A high protein content of the defatted flours from fruit seeds is a determinant factor for its consideration as raw material in the production of protein isolates or concentrates.[33]

5.4 METHODS FOR RECOVERY OF PROTEINS

The methods for the recovery of protein from vegetable sources have as purpose the obtention of products with a higher content of such polymers with respect to the raw matter, using several procedures of extraction, purification, and/or concentration. These products are named protein concentrates or isolates if its protein contents are ≤80 or >80%, respectively.[3] According to Hadidi et al.,[58] the three most common methods to produce protein concentrates or isolate protein from plant sources are alkaline extraction-isoelectric precipitation (AE-IP), micellar precipitation (MP), and salt extraction-dialysis (SE).

The AE-IP method is currently the most popular protein extraction method due to its rapidity, simplicity, and affordability.[94] This method adjusts a suspension of plant material to strongly alkaline conditions (pH 8–12) to solubilize the proteins, removing the insoluble matter by sedimentation, centrifugation, or filtration. The dispersion containing the soluble protein fraction is then adjusted to the isoelectric point of the proteins and an aggregate is formed, which is collected by remotion of the aqueous phase by sedimentation, centrifugation, or filtration. Generally, the protein precipitate is re-solubilized at pH 7 and then converted into a powder using spray or freezing drying technologies. This process is highly influenced by many parameters such as pH, temperature, time, ionic strength, and the ratio of solvent-flour.[144]

The MP method for the preparation of plant protein isolates involves three main stages. First, most of the salt-soluble proteins (globulin proteins) are solubilized in common food-grade salt solutions (sodium, calcium, and potassium chloride) at nearly neutral pH (~7, 10–60 min) to salt-in the proteins at moderate temperature (15–35 °C). Then, soluble proteins in the salt extract are recovered by reducing the ionic strength via membrane separation (dialysis) or precipitation by dilution (water) at a lower temperature (1–4°C). Finally, the recovered proteins are spray-dried or lyophilized.[134]

Unlike the MP method, the procedure for the preparation of protein concentrates or isolates by SE, after the proteins from plant sources are solubilized under alkaline conditions in a buffer containing salts as NaCl or KCl, the recovered supernatant is directly lyophilized or spray-dried.[39] Combining different extraction methods can also be used to reduce extraction time, increase extraction yield, and overcome the limitations of individual methods,[14] which can be chemical, enzymatic, and physical.[78] Enzymatic hydrolysis has been used to enhance protein extraction efficiency by cell wall polysaccharide degrading enzymes such as alcalase, cellulase, xylanase, and protease, which makes easier toward the recovery of proteins.[146]

Physical methods such as microwave, pulsed electric field, gamma radiation, pressurized liquid, and ultrasound extraction (USon-E), also have been used to reduce operating costs and volumes of the toxic organic solvents, as well as to increase extraction yields and improve functional properties of proteins.[43] The principal mechanism on which the assisted extraction methods are based is their ability to cause physical damage of the cell walls of the materials, facilitating mass transfer and the release of cell content into the medium.[71]

The content of protein of the isolates from fruit seeds in Table 5.2 is in the range from 68.3% (tamarind seed) to 97.9% (plum seed), while for the concentrates is from 20.8% (pomegranate seed) to 74.9% (white melon seed). Protein content of the concentrates and isolates varies widely depending on several factors, among which the protein source and the extraction method stand out.[123]

Under the same conditions and the AE-IP method, different protein contents were obtained for the orange (86.4%) and passion fruit (96.2%) seed isolates. A similar situation occurred with the protein isolates of guamuchil and passion fruit seeds. For protein isolates from jackfruit seeds by AE-IP, variations in the pH values of protein extraction and isoelectric precipitation also signified differences in protein content.

TABLE 5.2 Protein Concentrates and Isolates from Fruit Seeds.

Fruit seed	Protein content (%)[a]	Method of obtention	References
Protein isolates			
Orange	86.4	AE-IP (pH 12.0, 4.5)	[120]
Fruit passion	96.2	AE-IP (pH 12.0, 4.5)	[38]
Guamuchil	75.2	AE-IP (pH 12.0, 4.0)	[43]
Jackfruit	69.6	AE-IP (pH 12.0, 4.0)	116
	76.9	AE-IP (pH 9.0, 4.5)	[60]
Persian lime	81.4	AE-IP (pH 9.0, 4.0)	[42]
Papaya	88.1	AE-IP (pH 10.0, 4.0)	[67]
Plum	97.9	AE-IP (pH 10.0, 5.0)	[21]
Neem	86.3	SE (0.5 M NaCl)	[2]
	81.0	AE-IP (pH 10.7, 4.5)	[112]
Pumpkin	92.6	AE (pH 9.0) - SE (1.0 M NaCl)	[145]
	86.1	USon-E (195 W, 19 min and 32°C) - AE-IP (pH 9.5, 5.0)	[30]
	89.0	AE-IP (pH 11.0, 4.0)	[35]
Watermelon	88.0	AE-IP (pH 11.0, 4.5)	[66]
Tamarind	68.3	AE-IP (pH 12.0, 4.0)	[19]
Protein concentrate			
White melon	74.9	SE (0.5 M NaCl)	[103]
Pomegranate	20.8	SE (6.4 % KCl)	[27]
Grape	37.1	SE (6.4 % KCl)	[27]
Mango	70.0	AE-IP (pH 11.0, 2.0)	[109]
Wood apple	76.5	AE-IP (pH 11.0, 4.5)	[16]

[a]The protein content values and their classification into protein isolates or concentrates are based on what is cited in the works of each author.

AE-IP, alkaline extraction and isoelectric precipitation; the first pH corresponds to AE and the second to IP; SE, salt extraction; USon-E, ultrasound extraction.

A higher protein content was obtained for the neem seed isolate with SE (86.3%) than with AE-IP (81.0%). The protein content values of the pumpkin seed protein isolates were 86.1, 89.0, and 92.6% when obtained by USon-E, AE-IP, and SE, respectively. In the case of protein concentrates, the same response could be expected from protein isolates, in terms of the variability of the protein content because of the protein source and the conditions and methods of preparation, as well as the combination between them.

Research to improve the protein recovery of agro-industrial by-products or not-edible sources, such as some fruit seeds, to reintroduce them into the food chain will continue to be a challenge. In the same sense, the use of environmentally friendly technologies, the reduction of operating costs and the minimization of the impact of processing on the functionality of proteins are some relevant aspects of study.[14]

5.5 FUNCTIONALITY OF PROTEINS

Protein functionality is the set of non-nutritive functions that affect the behavioral properties of said polymers in food products during processing, preparation, and storage.[28] The functional properties of proteins in food depend on its intrinsic properties and environmental conditions. The intrinsic properties of proteins include their structures, amino acid composition, shape, size, charge, hydrophobicity, hydrophilicity, and interaction with other food components. Temperature, pH, ionic strength, preservatives and additives, and processing conditions are the main environmental conditions that affect the functional properties of proteins.[83] The basic functional properties of proteins in food are solubility, water holding capacity, oil holding capacity, and those corresponding to emulsification, foaming, and gelation.[118]

Values of protein solubility (PSo), water holding capacity (WHC), oil holding capacity (OHC), emulsifying activity (EmA), emulsification stability (EmS), foaming capacity (FoC), foaming stability (FoS), and least gelation concentration (LGC) of some the protein concentrates or isolates obtained from fruit seeds are shown in Table 5.3. As can be seen, protein concentrates or isolates from fruit seeds show great variability in functional properties. The minimum (7% at pH 8) and maximum (93% at pH 7) PSo values are for mango and passion fruit seed proteins, and grape seeds, respectively. The WHC and OHC values in the range of 1.3–11.5% and 1.0–8.4% correspond to proteins obtained from guamuchil and mango seeds and guamuchil and soursop seeds, respectively. At pH 8, jackfruit seed proteins have the highest values of EmA (118%), EmS (130%), FoC (222%), and FoS (65%). However, at pH 7, passion fruit seed proteins show the highest FoC (215%) and the neem seed proteins the highest FoS (94%). Mango seed protein meet the best gelling properties with the lowest value of 3% (LGC).

Functionality determines the suitability of proteins to be used as food ingredients, depending on their properties such as solubility, fluid binding, emulsifying, foaming, and gelling.[59] The PSo is the most important of the

TABLE 5.3 Functional Properties of Protein Concentrates and Isolates Obtained from Fruit Seeds.

Source	Functional properties								
	PSo (pH)	WHC	OHC	EmA (pH)	EmS (pH)	FoC (pH)	FoS (pH)	LGC (pH)	Ref
Orange	30 (8)	3.0	3.6	57.5 (7)	55.5 (7)	1.3 (7)	27.7 (7)	14	[120]
Jackfruit	66 (8)	6.5	6.1	118 (8)	130 (8)	222 (8)	65 (8)	9 (6)	[136]
Passion fruit	17 (8)	N/A	N/A	45.7 (7)	47.6 (7)	215 (7)	63.6 (7)	4 (7)	[38]
Guamuchil	N/A	1.3	1.0	3.3[a]	94.8[a]	116	113	20	[43]
Mango	17 (8)	11.5	3.2	70 (6)	68 (6)	17 (8)	76 (8)	3 (8)	[109]
Papaya	88 (8)	N/A	N/A	68 (8)[a]	55 (8)[a]	31 (8)	19.5 (8)	N/A	[67]
Lime	37 (8)	3.5	4.7	23 (8)[a]	10 (8)[a]	25 (8)	33 (8)	N/A	[42]
Pomegranate	41 (7)	1.8	3.4	15.3	N/A	N/A	N/A	N/A	[27]
Grape	93 (7)	2.6	4.1	22.3	N/A	N/A	N/A	N/A	[27]
Melon	45 (8)	2.1	1.2	N/A	N/A	30	N/A	10	[103]
Soursop	38 (7)	N/A	8.4	18.8	N/A	57	81	N/A	[144]
Watermelon	68 (7)	3.4	4.2	31[a]	51[a]	N/A	N/A	4	[46]
Neem	91 (8)	1.7	3.2	37 (7)	74 (7)	42 (7)	94 (7)	N/A	[2]

[a]Emulsifying activity index (m^2/g) and emulsion stability index (%) instead of EmA and EmS, respectively.

EmA, emulsifying activity (%); EmS, emulsion stability (%); FoC, foaming capacity (%); FoS, foam stability (%); LGC, least gelation concentration; OHC, oil holding capacity (g oil/g protein); N/A, not available; PSo, protein solubility (%);Ref., references; WHC, water holding capacity (g water/g protein).

functional properties since many of the others of this type are affected by such a characteristic. As a particular property, the PSo is important for the successful addition of plant proteins into many aqueous-based food systems.[149] The fluid-binding properties (WHC and OHC) in foods are crucial for the control of fluid holding as oil or water, which influences flavor retention and texture such as tenderness, juiciness, and mouthfeel, as well as in the performance of processes, including aspects as cook yield, water loss, and so on.[144] The emulsifying and foaming properties of the plant proteins represented by EmA and EmS and FoC and FoS, respectively, are critical for applications where stable emulsions or foams are needed. The gelling properties typically measured as LGC are of interest because they determine the formation of food gels, contributing to the water retention, flexibility, and desirable textural attributes of semisolid foods.[59] The functional properties of the proteins recovered as concentrates and isolates are highly dependent on the process used.[61]

5.6 ULTRASOUND FOR PROTEIN FUNCTIONALITY IMPROVEMENT

The production processes of protein concentrates and isolates often affect their physicochemical and functional properties, which limits their use. However, various physical, enzymatic, or chemical treatments, or combinations of them, have been applied to modify the physicochemical and functional properties of proteins and improve their applications.[99] Within the emerging physical technologies, ultrasound represents a fast, efficient, and reliable alternative to improve the quality and functionality of proteins.[70]

Ultrasound is mechanical waves whose frequency is above the hearing capacity of the human ear (20 kHz) and is considered a green, safe, fast, nontoxic, and environmentally friendly technology.[131] In general, the application of ultrasound in the field of food depends on the frequency range (20 kHz–10 MHz) that is used, so it can be classified into two categories: (1) low-intensity ultrasound (<1 W cm^{-2}) and high-frequency (1–10 MHz) and (2) high-intensity ultrasound (>5 W cm^{-2} or 10–1000 W cm^{-2}) and low-frequency (20–100 kHz).[5,153] High-intensity (low-frequency) ultrasound is used to alter the physical or chemical properties of food materials as is the case of protein isolates or concentrates.[135]

The application of high-intensity ultrasound in proteins has been carried out using devices such as ultrasonic baths or ultrasound probes. In addition to the used equipment, another factor that influences the impact of ultrasound on the protein functionality is the sonication conditions such as intensity, frequency, amplitude, time, density, and pulsations, as well as conditions of pretreatments as pH, temperature, and ionic strength.[62,132] The sound waves emitted by these devices cause a phenomenon called cavitation, which modifies the native properties of proteins and with it also their functional properties.[77,93]The reasons for the changes in the functional properties of the proteins induced by high-intensity ultrasound are based on three aspects: (1) the structural modification, mainly due to the modification in the size of the particles and the rise in hydrophobicity and sulfhydryl groups, (2) the denaturation and/or aggregation (unfolding and/or folding) of protein molecules, with significant impacts on protein-water interactions and protein-protein interactions, and (3) the alteration of the ultrasound microenvironment (pH, temperature, ionic strength) that cause shifts in protein behavior.[49,148,151]

Table 5.4 shows the effect of the high-intensity ultrasound on the functional properties of the protein isolates from fruit seeds. In general, the functional properties of the protein isolates or concentrates were improved by the effect of ultrasound, showing increases in PSo, OHC, FoC, FoS, EmA, and EmS, as well as a reduction in LGC, independently of the ultrasound system and conditions used. The proteins of the jackfruit, guamuchil, orange, and apple seeds showed a decrease in WHC due to ultrasound, while those of tamarind and watermelon showed an increase. In some proteins, high-intensity ultrasound does not modify certain functional properties. For example, no changes were observed by ultrasound in PSo, EmA, and EmS for jackfruit seed proteins, in FoC, FoS, EmA, EmS for orange seeds, and in FoS for pumpkin seed protein.

The extent of improvement or deterioration, as well as no change in the functional properties of proteins by ultrasound depends on the sonication systems and conditions. Flores-Jiménez et al.[43] and Biswas and Sit[19] reported an improvement in PSo up to of 115.5% for guamuchil proteins and at least 17.3% for tamarind proteins, respectively, compared to native proteins. Augments in OHC of 3.61, 22.1, 25.5, 39.8, 50.0, and 108.8% by ultrasound were observed for seed protein isolates of orange,[120] tamarind,[19] watermelon,[46] guamuchil,[43] apple,[47] and jackfruit,[116] respectively. On the other hand, reductions due to the effect of ultrasound of 15.2 and 49.3% in WHC were determined for the protein isolates of orange[120] and jackfruit.[116]

Regarding the foaming properties, increases by ultrasound effect of 110.2% were observed in FoC for the tamarind seed proteins,[19] while for FoS the increases were 58.1, 87.1 and 18.9% for those obtained from the guamuchil,[43] tamarind,[19] and apple[47] seeds, respectively. Nevertheless, ultrasound did not modify the FoS of protein isolates from jackfruit,[116] orange,[120] and pumpkin[35] seeds, but in the case of proteins from fruit passion seeds,[38] a reduction of 18.5% for FoS was observed.

In the case of emulsifying properties, ultrasound caused an improvement in fruit seed proteins, except for orange and passion fruit seed proteins. Rises of 15.6 and 87.4% in EmA were reported for the proteins of apple[47] and guamuchil[43] seeds, respectively. A 30.8% increase in EmS was shown by ultrasound-treated watermelon proteins, in contrast to non-sonicated proteins.[46] An enhancement of the gelling properties by ultrasound was found for proteins from orange[120] and fruit passion[38] seeds, reducing their LGC by 500 and 100%, respectively.

Ultrasound treatment is a valuable tool to modify the characteristics of proteins in by-products (seeds) for use in food. Therefore, studies must be carried out to investigate the impact of these changes with ultrasound on the properties of each protein concentrate and isolate to be obtained.

5.7 USES OF PROTEINS AS FOOD INGREDIENTS

Proteins recovered from plant sources in the form of protein concentrates or isolates are widely used as ingredients in the food industry, due to their nutritional benefits and their functionality in the formulation or development

TABLE 5.4 Effect of Ultrasound on Functional Properties of Protein Isolates or Concentrates from Fruit Seeds.

Protein source	Ultrasound system	Conditions	Functional properties								References
			PSo	WHC	OHC	FoC	FoS	EmA	EmS	LGC	
Jackfruit	Probe	f = 20 kHz; P = 200, 400, and 600 W; t = 15 min	↑	↓	↑	↑	N/C	↑	↑	↓	[116]
Passion fruit	Bath	f = 40 kHz; P = 130 W; t = 15 and 30 min	↑	N/A	N/A	↑	↓	N/C	N/C	↓	[38]
Orange	Bath	f = 42 kHz; P = 130 W; t = 15 and 30 min	↑	↓	↑	N/C	N/C	N/C	N/C	↓	[120]
Guamuchil	Probe	f = 20 kHz; P = 200, 400, and 600 W; t = 15 y 30 min	↑	↓	↑	↑	↑	↑	↓	↓	[43]
Tamarind	Probe	f = 25 kHz; P = 100 and 200 W; t = 15 and 30 min	↑	↑	↑	↑	↑	↑	↑	N/A	[19]
Pumpkin	Probe	f = 20 kHz; P = 300 and 500 W; t = 30 min	N/A	N/A	N/A	↑	N/C	N/A	N/A	N/A	[35]
Apple	Probe	f = 40 kHz; P = 200 W; t = 30 min	N/C	↓	↑	↑	↑	↑	↑	N/A	[47]
Watermelon	Bath	f = 25 and 40 kHz; P = 200 W; t = 5, 10, 20, and 30 min	N/A	↑	↑	N/A	N/A	↑	↑	↓	[46]

A, amplitude; I, intensity; EmA, emulsifying activity; EmS, emulsifying stability; f, frequency; FoC, foaming capacity; FoS, foaming stability; LGC, least gelation concentration; OHC, oil absorption capacity; P, power; PSo, protein solubility; t, time; WHC, water absorption capacity; ↑, increase; ↓, decrease; N/C, not change; N/A, not available.

of food products.[89] Although the nutritional value of vegetable proteins is an important quality factor, the functionality of these polymers is equally important when used as a food ingredients, since they affect the appearance, color, flavor, odor, and texture of the products.[3]

The plant-based foods market is one of the fastest growing sectors of the modern food industry. In the United States the sale of plant-based food in all categories grew by nearly 43% from 2019 to 2022, reaching a market value of around $ 8 billion. According to the Good Food Institute,[52] the main categories where the proteins are the most important functional ingredient to mimic animal-base foods are milk, meat, creamer, yogurt, butter, cheese, drinks, dressings, spreads, and eggs. Consumers demand that each of these products meet the characteristic quality attributes, which implies great challenges for formulation and production in the plant-based food industry. Predictions the global plant-based protein market indicates that it will reach a value of more than $35 billion by 2024.[87]

In the world, the main group of foods in which plant proteins are applied as ingredients is that of analogues of animal products, however, they are also very important in bakery products, pastas, emulsions, dressings, hydrogels, desserts, films, and coating materials (Table 5.5).

TABLE 5.5 Main Applications of Plant-Proteins in Foods.

Applications	Examples of foods	References
Analogues of:		
Meat	Sausages, burgers, nuggets	[11,17,63,74,150,]
Fish	Fillets, slices, burgers, sticks	[102,126]
Eggs	Fluid eggs, eggs slices	[80,154]
Milk and fermented milk	Beverages from several sources and yogurt	[65,96,100,105]
Cheese	Mozzarella, cheddar, processed and pasteurized cheese, tofu, cream cheese	[53,68,92]
Other:		
Bakery goods	Breads, cookies, crackers	[51,111,118]
Pasta	Noodles and pasta	[90,119,111]
Emulsions or dressings	Mayonnaise and salad dressings	[26,124,152]
Hidrogels and desserts	Jellies, jams, ice cream, and other confectionery products	[55,76,110]
Films	Edible packaging and coatings	[56,79]
Coating material	Encapsulated products: nutrients, bioactive compounds, essential oils	[32,57]

5.8 CONCLUSIONS

Given the growing demand for protein due to the increase in population, alternative sources to those typically used should be considered. A potential source for the recovery of proteins are the seeds of fruits derived from their processing, which are generally discarded as waste. Protein concentrates or isolates can be obtained from fruit seeds by various methods, with acceptable functional properties for use as food ingredients. In addition, fruit seed protein concentrates or isolates treated with ultrasound substantially improve their functional properties for better use in food. The protein concentrates or isolates obtained from fruit seeds and treated with ultrasound could be used as a source of protein in the preparation of analogs of animal foods such as meat, fish, dairy products, and eggs. Other important applications of these protein isolates or concentrates could be as ingredients in bakery products, pastas, emulsions, hydrogels, desserts, or used as the preparation of food films or support encapsulation material.

ACKNOWLEDGMENT

The authors thank the National Council of Humanities, Science and Technology of Mexico (CONAHCYT) for the postdoctoral fellowship awarded (601948) to Ph.D. Nitzia Thalía Flores Jiménez.

KEYWORDS

- **fruit seeds**
- **protein isolates**
- **ultrasound**
- **functional properties**
- **wastes**

REFERENCES

1. Abdel-Hameed, S. M.; Abd Allah, N. A. R.; Hamed, M. M.; Soltan, O. I. A. Papaya Fruit By-Products as Novel Food Ingredients in Cupcakes. *Ann. Agric. Sci.* **2023,** *68*, 60–74.

2. Acho, M. A.; Osemwegie, O. O.; Amonsou, E. O.; Arise, R. O. Physico-Chemical and Functional Characteristics of Neem Seed Protein Isolate. *Sci. Afri.* **2023,** *21*, e01774.
3. Akharume, F. U.; Aluko, R. E.; Adedeji, A. A. Modification of Plant Proteins for Improved Functionality: A Review. *Compr. Rev. Food Sci. Food Saf.* **2021,** *20*, 198–224.
4. Allaqaband, S.; Dar, A. H.; Patel, U.; Kumar, N.; Nayik, G. A.; Khan, S. A.; Ansari, M. J.; Alabdallah, N. M.; Kumar, P.; Pandey, V. K.; Kovács, B.; Shaikh, A. M. Utilization of Fruit Seed-Based Bioactive Compounds for Formulating the Nutraceuticals and Functional Food: A Review. *Front. Nutr.* **2022,** *9*, 902554.
5. Alarcon-Rojo, A. D.; Carrillo-Lopez, L. M.; Reyes-Villagrana, R.; Huerta-Jiménez, M.; García-Galicia, I. A. Ultrasound and Meat Quality: A Review. *Ultrason. Sonochem.* **2018,** *55*, 369–382.
6. Almeida Sá, A. G.; Franco Moreno, Y. M.; Mattar Carciofi, B. A. Plant Proteins as High-Quality Nutritional Source for Human Diet. *Trends Food Sci. Technol.* **2020**, *97*, 170–184.
7. Alongi, A.; Anese, M. Re-thinking Functional Food Development through a Holistic Approach. *J. Funct. Foods* **2021,** *81*, 104466.
8. Aneja, V. P.; Schlesinger, W. H.; Li, Q.; Battye, W. H. Characterization of Atmospheric Nitrous Oxide Emissions from Global Agricultural Soils. *SN Appl. Sci.* **2019,** *1*, 1662.
9. Antonisamy, A. J.; Marimuthu, S.; Malayandi, M.; Rajendran, K.; Lin, Y. C.; Andaluri, G.; Lee, S. L.; Ponnusamy, V. K. Sustainable Approaches on Industrial Food Wastes to Value-Added Products- A Review on Extraction Methods, Characterizations, and its Biomedical Applications. *Environ. Res.* **2022,** *217*, 114758.
10. Araújo, R. G.; Rodriguez-Jasso, R. M.; Ruiz, H. A.; Pintado, M. M. E.; Aguilar, C. N. Avocado by-Products: Nutritional and Functional Properties. *Trends Food Sci. Technol.* **2018,** *80*, 51–60.
11. Arora, S.; Kataria, P.; Nautiyal, N.; Tuteja, T.; Sharma, S.; Ahmad, F.; Haque, S.; Shahwan, M., Capanoglu, E.; Vashishth, R.; Gupta, A. K. Comprehensive Review on the Role of Plant Protein as a Possible Meat Analogue: Framing the Future of Meat. *ACS Omega* **2023,** *8*, 23305–23319.
12. Arrutia, F.; Binner, E.; Williams, P.; Waldron, K. W. Oilseeds Beyond Oil: Press Cakes and Meals Supplying Global Protein Requirements. *Trends Food Sci. Technol.* **2020,** *100*, 88–102.
13. Atta, M. B.; El-Shenawi, G. M. Functional Properties and In-Vitro Digestibility of Bitter Orange (*Citrus aurantium*) Seed Flour. *Merit Res. J. Agric. Sci. Soil Sci.* **2013,** *1*, 42–47.
14. Avelar, Z.; Rodrigues, R. M.; Pereira, R. N.; Vicente, A. A. Chapter 15; Future Food Proteins-Trends and Perspectives. In *Future Foods*; Bhat, R., Ed.; Academic Press, 2022; pp 267–285. https://doi.org/10.1016/B978-0-323-91001-9.00007-4
15. Banerjee, J.; Singh, R.; Vijayaraghavan, R.; MacFarlane, D.; Patti, A. F.; Arora, A. Bioactives from Fruit Processing Wastes: Green Approaches to Valuable Chemicals. *Food Chem.* **2017,** *225*, 10–22.
16. Banupriya, L.; Kumar, T. P. V. Centesimal Composition and Amino Acid Profiling of Wood Apple (*Limonia acidissima*) Seed: An Underutilized Protein Source. *Int. J. of Adv. Res.* **2016,** 4, 1288–1296.
17. Beniwal, A. S.; Singh, J.; Kaur, L.; Hardacre, A.; Singh, H. Meat Analogs: Protein Restructuring during Thermomechanical Processing. *Compr. Rev. Food Sci. Food Saf.* **2021,** *20*, 1221–1249.

18. Bernardino-Nicanor, A.; Scilingo, A. A.; Añón, M. C.; Dávila-Ortíz, G. Guava Seed Storage Protein: Fractionation and Characterization. *LWT - Food Sci. Technol.* **2006,** *39*, 902–910.
19. Biswas, B.; Sit, N. Effect of Ultrasonication on Functional Properties of Tamarind Seed Protein Isolates. *J. Food Sci. Technol.* **2020,** *57*, 2070–2078.
20. Caliceti, C.; Malaguti, M.; Marracino, L.; Barbalace, M. C.; Rizzo, P.; Hrelia, S. Agri-Food Waste from Apple, Pear, and Sugar Beet as a Source of Protective Bioactive Molecules for Endothelial Dysfunction and its Major Complications. *Antioxidants* **2022,** *11*, 1786.
21. Čakarević, J. C.; Vidović, S. S.; Vladić, J. Z.; Jokić, S. D.; Pavlović, N. S.; Popović, L. M. Plum Oil Cake Protein Isolate: A Potential Source of Bioactive Peptides. *Food Feed Res.* **2019,** *46*, 171–178.
22. Chamorro, F.; Carpena, M.; Fraga-Corral, M.; Echave, J.; Riaz Rajoka, M. S.; Barba, F. J.; Cao, H.; Xiao, J.; Prieto, M. A.; Simal-Gandara, J. Valorization of Kiwi Agricultural Waste and Industry by-Products by Recovering Bioactive Compounds and Applications as Food Additives: A Circular Economy Model. *Food Chem.* **2022,** *370*, 131315.
23. Chanioti, S.; Katsouli, M.; Tzia, C. Novel Processes for the Extraction of Phenoliccompounds from Olive Pomace and their Protection by Encapsulation. *Molecules* **2021,** *26*, 1781.
24. Chaparro, S. P.; Tavera, M. L.; Martinez, J. J.; Gil, J. H. Propiedades funcionales de la harina y de los aislados proteicos de la semilla de guanábana (*Annona muricata*). *Rev. U D C Actual. Divulg. Cient.* **2014,** *17*, 151–159.
25. Chen, W.; Ma, H.; Wang, Y. Y. Recent Advances in Modified Food Proteins by High Intensity Ultrasound for Enhancing Functionality: Potential Mechanisms, Combination with Other Methods, Equipment Innovations and Future Directions. *Ultrason. Sonochem.* **2022,** *85*, 105993.
26. Choi, H. W.; Ham, S. H.; Hahn, J.; Choi, Y. J. Developing Plant-Based Mayonnaise using Pea Protein-Xanthan Gum Conjugates: A Maillard Reaction Approach. *LWT-Food Sci. Technol.* **2023,** *185*, 115137.
27. Coşkun, Ö.; Gülseren, İ. Aqueous Extraction and Functionality of Protein Concentrates Manufactured from Cold Press Meals of Pumpkin, Pomegranate, and Grape Seeds. *Nutrire* **2020,** *45*, 11.
28. Dash, D. R.; Singh S. K.; Singha, P. Recent Advances on the Impact of Novel Non-thermal Technologies on Structure and Functionality of Plant Proteins: A Comprehensive Review. *Crit. Rev. Food Sci. Nutr.* [Online] Oct 11, 2022. DOI: 10.1080/10408398.2022.2130161. https://www.tandfonline.com/ (accessed Aug 3, 2023).
29. Day, L.; Cakebread, J. A.; Loveday; S. M. Food Proteins from Animals and Plants: Differences in the Nutritional and Functional Properties. *Trends in Food Sci. Technol.* **2022,** *119*, 428–442.
30. Das, M.; Devi, L. M.; Badwaik, L. S. Ultrasound-Assisted Extraction of Pumpkin Seeds Protein and its Physicochemical and Functional Characterization. *Appl. Food Res.* **2022,** *2*, 100121.
31. Deak, N. A.; Johnson, L. A.; Lusas, E. W.; Rhee, K. C. Chapter19; Soy Protein Products, Processing, and Utilization. In *Soybeans*; Johnson, L. A., White, P. J., Galloway, R., Ed.; AOCS Press, 2008; pp 661–724. https://doi.org/10.1016/B978-1-893997-64-6.50022-6.

32. Dehnad, D.; Emadzadeh, B.; Ghorani, B.; Rajabzadeh, G.; Tucker, N.; Jafari, S. M. Bioactive-Loaded Nanovesicles Embedded within Electrospun Plant Protein Nanofibers; A Double Encapsulation Technique. *Food Hydrocoll.* **2023,** *141*, 108683.
33. Devi, L. M.; Badwaik, L. S. Influence of Temperature, Time and Alkali Concentration on Protein Extraction from Muskmelon Seed Meal. *Indian Chem. Eng.* **2021,** *64*, 219–226.
34. Domínguez Díaz, L.; Fernández-Ruiz, V.; Cámara, M. An International Regulatory Review of Food Health-Related Claims in Functional Food Products Labeling. *J. Funct. Foods* **2020,** *68*, 103896.
35. Du, H.; Zhang, J.; Wang, S.; Manyande, A.; Wang, J. Effect of High-Intensity Ultrasonic Treatment on the Physicochemical, Structural, Rheological, Behavioral, and Foaming Properties of Pumpkin (*Cucurbita moschata* Duch.) Seed Protein Isolates. *LWT-Food Sci. Technol.* **2022,** *155*, 112952.
36. Dutta, A.; Patra, A.; Hazra, K. K.; Nath, C. P.; Kumar, K.; Rakshit, A. A State of the Art Review in Crop Residue Burning in India: Previous Knowledge, Present Circumstances and Future Strategies. *Environ. Chall.* **2022,** *8*, 100581.
37. Esparza, I.; Jiménez-Moreno, N.; Bimbela, F.; Ancín-Azpilicueta, C.; Gandía, L. M. Fruit and Vegetable Waste Management: Conventional and Emerging Approaches. *J. Environ. Manage.* **2020,** *265*, 110510.
38. Espinosa-Murillo, N. C.; Ulloa, J. A.; Urías-Silvas, J. E.; Rosas-Ulloa, P.; Ramírez-Ramírez, J. C.; Gutiérrez-Leyva, R.; Ulloa-Rangel, B. E. Impact of High-Intensity Ultrasound on the Physicochemical and Functional Properties of a Protein Isolate from Passion Fruit (*Passiflora edulis*) Seeds. *Int. J. Food Eng.* **2021,** *17*, 609–618.
39. Fang, B.; Chang, L.; Ohm, J. B.; Chen, B.; Rao, J. Structural, Functional Properties, and Volatile Profile of Hemp Protein Isolate as Affected by Extraction Method: Alkaline Extraction–Isoelectric Precipitation vs Salt Extraction. *Food Chem.* **2023,** *405* (Part B), 135001.
40. FAO. Food and Agriculture Organization of the United Nations. In *World Food and Agriculture-Statistical Yearbook 2022*; Food and Agriculture Organization of the United Nations: Rome, 2022.
41. FAO. Food and Agriculture Organization of the United Nations. Food Loss and Waste, 2023. https://www.fao.org/nutrition/capacity-development/food-loss-and-waste (accessed Aug 8, 2023).
42. Fathollahy, I.; Farmani, J.; Kasaai, M. R.; Hamishehkar, H. Characteristics and Functional Properties of Persian Lime *(Citrus latifolia)* Seed Protein Isolate and Enzymatic Hydrolysates. *LWT-Food Sci. Technol.* **2021,** *140*, 110765.
43. Flores-Jiménez, N. T.; Ulloa, J. A.; Urías-Silvas, J. E.; Ramírez-Ramírez, J. C.; Bautista-Rosales, P. U.; Gutiérrez-Leyva, R. Influence of High-Intensity Ultrasound on Physicochemical and Functional Properties of a Guamuchil *Pithecellobium dulce* (Roxb.) Seed Protein Isolate. *Ultrason. Sonochem.* **2022,** *84*, 105976.
44. Fontanari, G. G.; Batistuti, J. P.; Bannach, G.; Pastre, I. A.; Ionashiro, E. Y.; Fertonani, F. L. Thermal Study and Physico-Chemical Characterization of Some Functional Properties of Guava Seeds Protein Isolate (*Psidium guajava*). *J. Therm. Anal. Calorim.* **2006,** *83*, 709–713.
45. Fukase, E.; Martin. W. Economic Growth, Convergence, and World Food Demand and Supply. *World Dev.* **2020,** *132*, 104954.
46. Gadalkar, S. M.; Rathod, V. K. Extraction of Watermelon Seed Proteins with Enhanced Functional Properties using Ultrasound. *Prep. Biochem. Biotechnol.* **2020,** *50*, 133–140.

47. Gani, A.; Ashraf, Z. U.; Noor, N.; Wani, I. A. Ultrasonication as an Innovative Approach to Tailor the Apple Seed Proteins into Nanosize: Effect on Protein Structural and Functional Properties. *Ultrason. Sonochem.* **2022,** *86*, 106010.
48. Ganesh, S. K.; Sridhar, A.; Vishali, S. Utilization of Fruit and Vegetable Waste to Produce Value-Added Products: Conventional Utilization and Emerging Opportunities-A Review. *Chemosphere* **2022,** *287* (Pt 3), 132221.
49. Gharibzahedi, S. M. T.; Smith, B. The Functional Modification of Legume Proteins by Ultrasonication: A Review. *Trends Food Sci. Technol.* **2020,** *98*, 107–116.
50. Giwa, S. O.; Akanbi, T. O. Mechanization of Melon Processing and Novel Extraction Technologies: A Short Review. *Sci. Afr.* **2020,** *9*, e00478.
51. Goldstein, N.; Reifen, R. The Potential of Legume-Derived Proteins in the Food Industry. *Grain Oil Sci. Technol.* **2022,** *5*, 167–178.
52. GFI. Good Food Institute. U.S. Retail Market Insights for the Plant-Based Industry [Online] 2023. https://gfi.org/marketresearch/#overall (accessed Aug 8, 2023).
53. Grossmann, L.; McClements, D. J. The Science of Plant-Based Foods: Approaches to Create Nutritious and Sustainable Plant-Based Cheese Analogs. *Trends Food Sci. Technol.* **2021,** *118* (Part A), 207–229.
54. Grummon, A. H.; Musicus, A. A.; Salvia, M. G.; Thorndike, A. N.; Rimm, E. B. Impact of Health, Environmental, and Animal Welfare Messages Discouraging Red Meat Consumption: An Online Randomized Experiment. *J. Acad. Nutr. Diet.* **2023,** *123*, 466–476.
55. Guo, Y.; Zhang, Z. G.; Cai, J.; Li, W. R.; Chen, L. Y.; Wu, W. C. Co-Folding of Soy Protein Isolates and Shellac by Structural Interplays to Induce Hydrogels. *Food Hydrocoll.* **2023,** *139*, 108527.
56. Hadidi, M.; Jafarzadeh, S.; Forough, M.; Garavand, F.; Alizadeh, S.; Salehabadi, A.; Khaneghah, A. M.; Jafari, S. M. Plant Protein-Based Food Packaging Films; Recent Advances in Fabrication, Characterization, and Applications. *Trends Food Sci. Technol.* **2022a,** *120*, 154–173.
57. Hadidi, M.; Boostani, S.; Jafari, S. M. Pea Proteins as Emerging Biopolymers for the Emulsification and Encapsulation of Food Bioactives. *Food Hydrocoll.* **2022b,** *126*, 107474.
58. Hadidi, H.; Aghababaei, F.; McClements, D. J. Enhanced Alkaline Extraction Techniques for Isolating and Modifying Plant-Based Proteins. *Food Hydrocoll.* **2023a,** *145*, 109132.
59. Hadidi, H.; Hossienpour, Y.; Nooshkam, M.; Mahfouzi, M.; Gharagozlou, M.; Aliakbari, F. S.; Aghababaei, F.; McClement D. J. Green Leaf Proteins: A Sustainable Source of Edible Plant-Based Proteins. *Crit. Rev. Food Sci. Nutr.* **2023b,** *3*, 1–18.
60. Haque, M. A.; Akter, F.; Rahman, H.; Baqui, M. A. Jackfruit Seeds Protein Isolate by Spray Drying Method: The Functional and Physicochemical Characteristics. *Food Nutr. Sci.* **2020,** *11*, 355–374.
61. Hewage, A.; Olatunde, O. O.; Nimalaratne, C.; Malalgoda, M.; Aluko, R. E.; Bandara, N. Novel Extraction Technologies for Developing Plant Protein Ingredients with Improved Functionality. *Trends Food Sci. Technol.* **2022,** *129*, 492–511.
62. Higuera-Barraza, O. A.; Del Toro-Sanchez, C. L.; Ruiz-Cruz, S.; Márquez-Ríos, E. Effects of High-Energy Ultrasound on the Functional Properties of Proteins. *Ultrason. Sonochem.* **2016,** *31*, 558–562.
63. Ismail, I.; Hwang, Y. H.; Joo, S. T. Meat Analog as Future Food: A Review. *J. Animal Sci. Technol.* **2020,** *62*, 111–120.

64. Iqbal, A.; Schulz, P.; Syed S. H.; Rizvi, S. H. Valorization of Bioactive Compounds in Fruit Pomace from Agro-Fruit Industries: Present Insights and Future Challenges. *Food Biosci.* **2021,** *44* (Part A), 101384.
65. Jeske, S.; Zannini, E.; Arendt, E. K. Evaluation of Physicochemical and Glycaemic Properties of Commercial Plant-Based Milk Substitutes. *Plant Foods Hum. Nutr.* **2017,** *72*, 26–33.
66. Jyothi lakshmi, A.; Kaul, P. Nutritional Potential, Bioaccessibility of Minerals and Functionality of Watermelon *(Citrullus vulgaris)* Seeds. *LWT - Food Sci. Technol.* **2011,** *44*, 1821–1826.
67. Kadiri, O.; Akanbi, C. T.; Saka, O. Gbadamosi, S. O. Isolation of Protein Isolated from Defatted *Carica papaya* linn. Seeds: Influence of pH and NaCl on its Solubility and Functional Properties. *Proceed. Niger. Acad. Sci.* **2020,** *13*, 86–96.
68. Kamath, R.; Basak, S.; Gokhale, J. Recent Trends in the Development of Healthy and Functional Cheese Analogues-A Review. *LWT-Food Sci. Technol.* **2022,** *155*, 112991.
69. Khalid, I. I. Functional Properties of Watermelon (*Citrullus lanatus*) and Pumpkin Seed Flours and Protein Isolate. *Int. J. Nutr.* **2023,** *7*, 4–15.
70. Kim, Y. J.; Lee, M. H.; Kim, S. M.; Kim, B. K.; Yong, H. I.; Choi, Y. S. Improvement of Structural, Physicochemical, and Rheological Properties of Porcine Myofibrillar Proteins by High-Intensity Ultrasound Treatment for Application as Pickering Stabilizers. *Ultrason. Sonochem.* **2023,** *92*, 106263.
71. Kumar, K.; Srivastav, S.; Sharanagat, V. S. Ultrasound Assisted Extraction (UAE) of Bioactive Compounds from Fruit and Vegetable Processing by-Products: A Review. *Ultrason. Sonochem.* **2021,** *70*, 105325.
72. Kumari, N.; Radha; Kumar, M.; Puri, S.; Zhang, B.; Rais, N.; Pundir, A.; Chandran, D.; Raman, P.; Dhumal, S.; Dey, A.; Senapathy, M.; Kumar, S.; Pokharel, B. R.; Deshmukh, V.; Damale, R. D.; Thiyagarajan, A.; Balamurgan, V.; Sathish, G.; Singh, S.; Lorenzo, J. M. Peach (*Prunus persica* (L.) Batsch) Seeds and Kernels as Potential Plant-Based Functional Food Ingredients: A Review of Bioactive Compounds and Health-Promoting Activities. *Food Biosci.* **2023,** *54*, 102914.
73. Kumoro, A. C.; Alhanif, M.; Wardhani, D. H. A Critical Review on Tropical Fruits Seeds as Prospective Sources of Nutritional and Bioactive Compounds for Functional Foods Development: A Case of Indonesian Exotic Fruits. *Int. J. Food Sci.* **2020,** *2*, 1–15.
74. Kurek, M. A.; Onopiuk, A.; Pogorzelska-Nowicka, E.; Szpicer, A.; Zalewska, M.; Płtorak, A. Novel Protein Sources for Applications in Meat-Alternative Products-Insights and Challenges. *Foods* **2022,** *11*, 957.
75. Lawrence, D.; Coe, M.; Walker, W.; Verchot, L.; Vandecar, K. The Unseen Effects of Deforestation: Biophysical Effects on Climate. *Front. For. Glob. Change* **2022,** *5*, 756115.
76. Li, J.; Wu, M.; Wang, Y.; Li, K.; Du, J.; Bai, Y. Effect of pH-Shifting Treatment on Structural and Heat Induced Gel Properties of Peanut Protein Isolate. *Food Chem.* **2020,** *325*, 126921.
77. Liang, Y.; Teng, F.; He, M.; Jiang, L.; Yu, J.; Wang, X.; Li, Y.; Wang, Z. Effects of Ultrasonic Treatment on the Structure and Rehydration Peculiarity of Freeze-Dried Soy Protein Isolate Gel. *Food Struct.* **2021,** *28*, 100169.
78. Likittrakulwong, W.; Poolprasert, P.; Srikaeo, K. Effects of Extraction Methods on Protein Properties Obtained from Paddy Rice and Germinated Paddy Rice. *Peer J.* **2021,** *9*, e11365.

79. Liu, L.; Yang, S.; Chen, C.; Fang, Y.; Li, L.; Ban, Z. High-Performance Films Fabricated by Food Protein Nanofibrils Loaded with Vanillin: Mechanism, Characterization and Bacteriostatic Effect. *Food Packag. Shelf Life.* **2023,** *37*, 101080.
80. Lu, Z.; Lee, P. R.; Yang, H. Synergistic Adsorption of Surface-Active Components at the Air-Liquid Interface Improves Foaming Properties of Plant-Based Egg Analogues. *Food Hydrocoll.* **2023,** *137*, 108414.
81. Lu Martínez, A. A.; Báez González, J. G.; Bautista Villarreal, M.; García Alanis, K. G.; Galindo Rodríguez, S. A.; García Márquez, E. Studied of Defatted Flour and Protein Concentrate of *Prunus serotine* and Applications. *Foods* **2019,** *9*, 29.
82. Ma, C. Y. Soybean: Soy Concentrates and Isolates. In *Encyclopedia of Food Grains*, Second Edition; Wrigley, C., Corke, H., Seetharaman, K., Jon Faubion. J., Eds.; Academic Press: New York, 2016; pp 482–488.
83. Małecki, J.; Muszynski, S.; Sołowiej, B. G. Proteins in Food Systems-Bionanomaterials, Conventional and Unconventional Sources, Functional Properties, and Development Opportunities. *Polymers* **2021,** *13*, 2506.
84. Manhongo, T. T.; Chimphango, A. F. A.; Thornley, P.; Röder, M. Current Status and Opportunities for Fruit Processing Waste Biorefineries. *Renew. Sust. Energ. Rev.* **2022,** *155*, 111823.
85. Martindale, W.; Hollands, T. Æ.; Jagtap, S.; Hebishy, E.; Duong, L. Turn-Key Research in Food Processing and Manufacturing for Reducing the Impact of Climate Change. *Int. J. Food Sci.* [Online] June 12, 2023. https://www.wiley.com (accessed Aug 2, 2023).
86. Mathur, S.; Waswani, H.; Singh, D.; Ranjan, R. Alternative Fuels for Agriculture Sustainability: Carbon Footprint and Economic Feasibility. *AgriEng.* **2022,** *4*, 993–1015.
87. McClements, D. J.; Grossmann, L. The Science of Plant-Based Foods: Constructing Next-Generation Meat, Fish, Milk, and Egg Analogs. *Compr. Rev. Food Sci. Food Saf.* **2021,** *20*, 4049–4100.
88. Md Sani, A. N.; Mohd Adzahan, N.; Ismail-Fitry, M. R. Valorization of Malaysian Tropical Fruit Seeds: A Review of their Nutrition, Bioactivity, Processing and Food Application. *Food Biosci.* **2022,** *50* (Part B), 102156.
89. Meganaharshini, M.; Sudhakar, V.; Dhivyabharathi, N.; Deepak, S. Review on Recent Trends in the Application of Protein Concentrates and Isolates-A Food Industry Perspective. *Food Human.* **2023,** *1*, 308–325.
90. Messia, M. C.; Cuomo, F.; Falasca, L.; Trivisonno, M. C.; De Arcangelis, E.; Marconi, E. Nutritional and Technological Quality of High Protein Pasta. *Foods* **2021,** *10*, 589.
91. Miah, M. Y.; Bhattacharjee, S.; Sultana, A.; Bhowmik, S.; Sarker, A. K.; Paul, S. C.; Islam, M. S.; Zaman, A. Evaluation of Amino Acid Profile of Jackfruit (*Artocarpus heterophyllus*) Seed and its Utilization for Development of Protein Enriched Supplementary Food. *J. Noakhali Sci. Technol. Uni.* **2017,** *1*, 77–84.
92. Michel, S. E. S.; Scheermeijer, R.; Ambühl, M.; Farrés, I. F. Novel Plant-Based Cream Cheese: A Tribology Perspective. *J. Food Eng.* **2022,** *335*, 111172.
93. Mir, N. A.; Riar, C. S.; Singh, S. Structural Modification of Quinoa Seed Protein Isolates (QPIs) by Variable Time Sonification for Improving its Physicochemical and Functional Characteristics. *Ultrason. Sonochem.* **2019,** *58*, 104700.
94. Miranda, C. G.; Speranza, P.; Kurozawa, L. E., Kawazoe Sato, A. C. Lentil Protein: Impact of Different Extraction Methods on Structural and Functional Properties. *Heliyon* **2022,** *8*, e11775.

95. Mir-Cerdà, A.; Nuñez, O.; Granados, M.; Sentellas, S.; Saurina, J. An Overview of the Extraction and Characterization of Bioactive Phenolic Compounds from Agri-Food Waste within the Framework of Circular Bioeconomy. *Trends Anal. Chem.* **2023,** *161*, 116994.
96. Montemurro, M.; Pontonio, E.; Coda, R.; Rizzello, C. G. Plant-Based Alternatives to Yogurt: State-of-the-Art and Perspectives of New Biotechnological Challenges. *Foods* **2021,** *10*, 316.
97. More, P. R.; Jambrak, A. R.; Arya, S. S. Green, Environment-Friendly and Sustainable Techniques for Extraction of Food Bioactive Compounds and Waste Valorization. *Trends Food Sci. Technol.* **2022,** *128*, 296–315.
98. Mwaurah, P. W.; Kumar, S.; Kumar, N.; Panghal, A.; Attkan, A. K.; Singh, V. K.; Garg, M. K. Physicochemical Characteristics, Bioactive Compounds and Industrial Application of Mango Kernel and its Products: A Review. *Compr. Rev. Food Sci. Food Saf.* **2020,** *19*, 2421–2446.
99. Nasrabadi, N. N.; Sedaghat Doost, A.; Mezzenga, R. Modification Approaches of Plant-Based Proteins to Improve their Techno-Functionality and use in Food Products. *Food Hydrocoll.* **2021,** *118*, 106789.
100. Nawaz, M. A.; Tan, M.; Oiseth, S.; Buckow, R. An Emerging Segment of Functional Legume-Based Beverages: A Review. *Food Rev. Int.* **2022,** *38*, 1064–1102.
101. Nirmal, N. P.; Khanashyam, A. C.; Mundanat, A. S.; Shah, K.; Babu, K. S.; Thorakkattu, P.; Al-Asmari, F.; Pandiselvam, R. Valorization of Fruit Waste for Bioactive Compounds and their Applications in the Food Industry. *Foods* **2023,** *12*, 556.
102. Nowacka, M.; Trusinska, M.; Chraniuk, P.; Piatkowska, J.; Pakulska, A.; Wisniewska, K.; Wierzbicka, A.; Rybak, K.; Pobiega, K. Plant-Based Fish Analogs- A Review. *Appl. Sci.* **2023,** *13*, 4509.
103. Ogunbusola, E. M.; Alabi, O. O.; Araoye, K. Y.; Sanni, T. A.; Jaiyeoba, C. N.; Adebayo-Alabi, I. B.; Akila, O. A. Impact of Extraction Methods on the Quality, Physicochemical, and Functional Properties of White Melon (*Cucumeropsis mannii*) Seed Protein Concentrates. *Food Chem. Adv.* **2022,** *1*, 100131.
104. Papoutsis, K.; Vuong, Q. V.; Golding, J. B.; Hasperué, J. H.; Pristijono, P.; Bowyer, M. C.; Scarlett, C. J., Stathopoulos, C. E. Pretreatment of Citrus by-Products Affects Polyphenol Recovery: A Review, *Food Rev. Int.* **2018,** *34*, 770–795.
105. Paul, A. A.; Kumar, S.; Kumar, V.; Sharma, R. Milk Analog: Plant Based Alternatives to Conventional Milk, Production, Potential and Health Concerns. *Crit. Rev. Food Sci. Nutr.* **2020,** *60*, 3005–3023.
106. Pathak, N.; Singh, S.; Singh, P.; Singh, P. K.; Singh, R.; Bala, S.; Thirumalesh, B. V., Gaur, R., Tripathi M. Valorization of Jackfruit Waste into Value Added Products and their Potential Applications. *Front. Nutr.* **2022,** *9*, 1061098.
107. Pattnaik, M.; Pandey, P.; Martin, G. J. O.; Mishra, H. N. Ashokkumar, M. Innovative Technologies for Extraction and Microencapsulation of Bioactives from Plant-Based Food Waste and their Applications in Functional Food Development. *Foods* **2021,** *10*, 279.
108. Pereira, J. A. M.; Berenguer, C. V.; Andrade, C. F. P.; Câmara, J. S. Unveiling the Bioactive Potential of Fresh Fruit and Vegetable Waste in Human Health from a Consumer Perspective. *Appl. Sci.* **2022,** *12*, 2747.
109. Pérez Saucedo, M. R.; Ulloa, J. A.; Rosas Ulloa, P.; Ramírez Ramírez, J. C.; Silva-Carrillo, Y.; Ulloa Rangel, B. E. Caracterización tecno-funcional de un concentrado proteínico obtenido de la semilla de mango (*Mangifera indica* L.). *Biotecnia* **2021,** *23*,120-126.

110. Pontonio, E.; Montemurro, M.; Dingeo, C.; Rotolo, M.; Centrone, C.; Carofiglio, V. E.; Giuseppe Rizzello, C. Design and Characterization of a Plant-Based Ice Cream Obtained from a Cereal/Legume Yogurt-Like. *LWT-Food Sci. Technol.* **2022,** *161*, 113327.
111. Rajpurohit, B.; Li, Y. Overview on Pulse Proteins for Future Foods: Ingredient Development and Novel Applications. *J. Future Foods* **2023,** *3*, 340–356.
112. Rahman, M. M.; Ho, K.; Netravali, A. N. Bio-Based Polymeric Resin from Agricultural Waste, Neem (*Azadirachta indica*) Seed Cake, for Green Composites. *J. Appl. Polym. Sci.* **2014,** *132*, 1–11.
113. Rahman M. M.; Lamsal, B. P. Ultrasound-Assisted Extraction and Modification of Plant-Based Proteins: Impact on Physicochemical, Functional, and Nutritional Properties. *Compr. Rev. Food Sci. Food Saf.* **2021,** *20*, 1457–1480.
114. Rao, M. V.; Sunil C. K.; Rawson, A.; Chidanand D. V.; Venkatachlapathy, N. Modifying the Plant Proteins Techno-Functionalities by Novel Physical Processing Technologies: A Review. *Crit. Rev. Food Sci. Nutr.* **2023,** *63*, 4070–4091.
115. Ray, R.; Dubey, K. K.; Marathe, S. J.; Singhal, R. Supercritical Fluid Extraction of Bioactives from Fruit Waste and its Therapeutic Potential. *Food Biosci.* **2023,** *52*, 102418.
116. Resendiz-Vazquez, J. A.; Ulloa, J. A.; UrÍas-Silvas, J. E.; Bautista-Rosales, P. U.; Ramírez-Ramírez, J. C.; Rosas-Ulloa, P.; González-Torres, L. Effect of High-Intensity Ultrasound on the Technofunctional Properties and Structure of Jackfruit (*Artocarpus heterophyllus*) seed Protein Isolate. *Ultrason. Sonochem.* **2017,** *37*, 436–444.
117. Rivera, J. E.; Chará, J. CH4 and N2O Emissions from Cattle Excreta: A Review of Main Drivers and Mitigation Strategies in Grazing Systems. *Front. Sustain. Food Syst.* **2021,** *5*, 657936.
118. Rivera, J.; Siliveru, K.; Li, Y. A Comprehensive Review on Pulse Protein Fractionation and Extraction: Processes, Functionality, and Food Applications. *Crit. Rev. Food Sci. Nutr.* [Online] Nov 3, 2022. https://www.tandfonline.com (accessed Aug 2, 2023).
119. Rousta, L. K.; Yazdi, A. P. G.; Amini. M. Optimization of Athletic Pasta Formulation by D-Optimal Mixture Design. *Food Sci. Nutr.* **2020,** *8*, 4546–4554.
120. Rosas-Ulloa, P.; Ulloa, J. A.; Ulloa-Rangel, B. E.; López-Mártir, K. U. Protein Isolate from Orange (*Citrus sinensis* L.) Seeds: Effect of High-Intensity Ultrasound on its Physicochemical and Functional Properties. *Food Bioproc. Technol.* **2023,** *16*, 589–602.
121. Scoones, I. Livestock, Methane, and Climate Change: The Politics of Global Assessments. *Wiley Interdiscip. Rev. Clim. Change* **2023,** *14*, e790.
122. Serrano-Sandoval, S. N.; Guardado-Felix, D.; Gutiérrez-Uribe, J. A. Legumes in Human Health and Nutrition (Fourth Edition). In *Encyclopedia of Human Nutrition*; Caballero, B., Eds.; Academic Press: New York, 2023; pp 430–437.
123. Shanthakumar, P.; Klepacka, J.; Bains, A.; Chawla, P.; Dhull, S. B.; Najda, A. The Current Situation of Pea Protein and its Application in the Food Industry. *Molecules* **2022,** *27*, 5354.
124. Shen, Y.; Babu, K. S.; Amamcharla, J.; Li, Y. Emulsifying Properties of Pea Protein/Guar Gum Conjugates and Mayonnaise Application. *Inter. J. Food Sci. Technol.* **2022,** *57*, 3955–3966.
125. Shewry, P. R. Chapter 6; Cereal Grain Proteins. In *ICC Handbook of 21st Century Cereal Science and Technology*; Shewry, P. R., Koksel, H., Taylor, J. R. N., Eds.; Academic Press, 2023; pp 63–70.

126. Shi, H.; Li, J.; Xu, E.; Yang, H.; Liu, D.; Yin, J. Microscale 3D Printing of Fish Analogues using Soy Protein Food Ink. *J. Food Eng.* **2023,** *347*, 111436.
127. Siddeeg, A.; Xu, Y.; Jiang, Q.; Xia, W. Physicochemical and Functional Properties of Flour and Protein Isolates Extracted from Seinat *(Cucumis melo* var. *tibish)* Seeds. *Food Sci. Biotechnol.* **2014,** *23*, 345–353.
128. Sinha, V.; Parmar, H. Development and Validation of Multidimensional Scale on Indian Consumer's Acceptance of Functional Food (FFS) - The Sustainable Option. *Clean. Responsible Consum.* **2023,** *10*, 100128.
129. Soares Mateus, A. R.; Pena, A.; Sendón, R.; Almeida, C.; Almeida Nieto, G.; Khwaldia, K.; Sanches Silva, A. By-Products of Dates, Cherries, Plums and Artichokes: A Source of Valuable Bioactive Compounds. *Trends Food Sci. Technol.* **2023,** *131*, 220–243.
130. Sridhar, A.; Ponnuchamy, M.; Kumar, P. S.; Kapoor, A. Food Preservation Techniques and Nanotechnology for Increased Shelf Life of Fruits, Vegetables, Beverages and Spices: A Review. *Environ. Chem. Lett.* **2021,** *19*, 1715–1735.
131. Su, J.; Cavaco-Paulo, A. Effect of Ultrasound on Protein Functionality. *Ultrason. Sonochem.* **2021,** *76*, 105653.
132. Suchintita Das, R.; Tiwari, B. K.; Chemat, F.; Garcia-Vaquero, M. Impact of Ultrasound Processing on Alternative Protein Systems: Protein Extraction, Nutritional Effects and Associated Challenges. *Ultrason. Sonochem.* **2022,** *91*, 106234.
133. Suri, A., Singh, A., Nema, P. K. Current Applications of Citrus Fruit Processing Waste: A Scientific Outlook. *Appl. Food Res.* **2022,** *1*, 100050.
134. Tanger, C.; Engel, J.; Kulozik, U. Influence of Extraction Conditions on the Conformational Alteration of Pea Protein Extracted from Pea Flour. *Food Hydrocoll.* **2020,** *107*, 105949.
135. Tian, R.; Zhu, G.; Feng, J.; Tian, B.; Zhang, Y.; Jiang, L.; Sui, X. Ultrasound Driven Conformational and Physicochemical Changes of Soy Protein Hydrolysates. *Ultrason. Sonochem.* **2020,** *68*, 105202.
136. Ulloa, J. A.; Villalobos Barbosa, M. C.; Resendiz Vazquez, J. A.; Rosas Ulloa, P.; Ramírez Ramírez, J. C.; Silva Carrillo, Y.; González Torres, L. Production, Physico-Chemical and Functional Characterization of a Protein Isolate from Jackfruit *(Artocarpus heterophyllus)* Seeds. *CyTA-J. Food* **2017,** *15*, 497–507.
137. United Nations. Sustainable Development Goals [Online] 2023a. https://www.un.org/sustainabledevelopment (accessed Aug 2, 2023a)
138. United Nations. Food and Climate Change: Healthy Diets for a Healthier Planet [Online] 2023b. https://www.un.org/en/climatechange/science/climate-issues/food (accessed Aug 5, 2023b).
139. UNPF. United Nations Population Fund [Online] 2023. https://www.unfpa.org (accessed Aug 6, 2023).
140. Van Zanten, H. H. H.; Herrero, M.; Van Hal, O.; Röös, E.; Muller, A.; Garnett, T.; Gerber, P. J.; Schader, C.; De Boer, I. J. M. Defining a Land Boundary for Sustainable Livestock Consumption. *Glob. Change Biol.* **2018,** *24*, 4185–4194.
141. Vardanega, R.; Salinas Fuentes, F.; Palma, J.; Bugueño-Muñoz, W.; Cerezal-Mezquita, P.; Ruiz-Domínguez, M. C. Valorization of Granadilla Waste (*Passiflora ligularis,* Juss.) by Sequential Green Extraction Processes Based on Pressurized Fluids to Obtain Bioactive Compounds. *J. Supercrit. Fluids* **2023,** *194*, 105833.
142. Villacís-Chiriboga, J.; Elst, K.; Van Camp, J.; Vera, E.; Ruales, J. Valorization of Byproducts from Tropical Fruits: Extraction Methodologies, Applications,

Environmental and Economic Assessment– A Review (Part 1: General Overview of the Byproducts, Traditional Biorefinery Practices and Possible Applications). *Compr. Rev. Food Sci. Food Saf.* **2020,** *19*, 405–447.

143. Villacís-Chiriboga, J.; Vera, E.; Van Camp, J.; Jenny Ruales; J., Elst, K. Valorization of Byproducts from Tropical Fruits: A Review, Part 2: Applications, Economic, and Environmental Aspects of Biorefinery via Supercritical Fluid Extraction. *Compr. Rev. Food Sci. Food Saf.* **2021,** *20*, 2305–2331.
144. Villacís-Chiriboga, J.; Prandi, B.; Ruales, J.; Van Camp, J.; Sforza, S.; Elst, K. Valorization of Soursop (*Annona muricata*) Seeds as Alternative Oil and Protein Source using Novel De-oiling and Protein Extraction Techniques. *LWT-Food Sci. Technol.* **2023,** *1*, 114777.
145. Vinayashree, S.; Vasu, P. Biochemical, Nutritional and Functional Properties of Protein Isolate and Fractions from Pumpkin (*Cucurbita moschata* var. Kashi Harit) Seeds. *Food Chem.* **2021**, *340*, 128177.
146. Vovk, H.; Karnpakdee, K.; Ludwig, R.; Nosenko, T. Enzymatic Pretreatment of Plant Cells for Oil Extraction. *Food Technol Biotechnol.* **2023,** *61*, 160–178.
147. Warmate, D.; Onarinde, B. A. Food Safety Incidents in the Red Meat Industry: A Review of Foodborne Disease Outbreaks Linked to the Consumption of Red Meat and its Products, 1991 to 2021. *Int. J. Food Microbiol.* **2023,** *398*, 110240.
148. Wen, C.; Zhang, J.; Yao, H.; Zhou, J.; Duan, Y.; Zhang, H.; Ma, H. Advances in Renewable Plant-Derive Protein Source: The Structure, Physicochemical Properties Affected by Ultrasonication. *Ultrason. Sonochem.* **2019,** *53*, 83–98.
149. Yousefi, N.; Abbasi, S. Food Proteins: Solubility & Thermal Stability Improvement Techniques. *Food Chem. Adv.* **2022,** *1*, 100090.
150. Yuliarti, O.; Kiat Kovis, T. J.; Yi, N. J. Structuring the Meat Analogue by using Plant-Based Derived Composites. *J. Food Eng.* **2021,** *288*, 110138.
151. Yücetepe, A.; Saroğlu, Ö.; Özçelik, B. Response Surface Optimization of Ultrasound-Assisted Protein Extraction from *Spirulina platensis*: Investigation of the Effect of Extraction Conditions on Techno-Functional Properties of Protein Concentrates. *J. Food Sci. Technol.* **2019,** *56*, 3282–3292.
152. Zhang, X.; Wang, Q.; Liu, Z.; Zhi, L.; Jiao, B.; Hu, H.; Ma, X.; Agyei, D.; Shi, A. Plant Protein-Based Emulsifiers: Mechanisms, Techniques for Emulsification Enhancement and Applications. *Food Hydrocoll.* **2023,** *144*, 109008.
153. Zheng, T.; Li, X.; Taha, A.; Wei, Y.; Hu, T.; Fatamorgana, P. B.; Zhang, Z.; Liu, F.; Xu, X.; Pan, S.; Hu, H. Effect of High Intensity Ultrasound on the Structure and Physicochemical Properties of Soy Protein Isolates Produced by Different Denaturation Methods. *Food Hydrocoll.* **2019,** *97*, 105216.
154. Zhou, H.; Vu, G.; McClements, D. J. Formulation and Characterization of Plant-Based Egg White Analogs using RuBisCO Protein. *Food Chem.* **2022,** *397*, 133808.
155. Zhu, Y.; Luan, Y.; Zhao, Y.; Liu, J.; Duan, Z.; Ruan, R. Current Technologies and Uses for Fruit and Vegetable Wastes in a Sustainable System: A Review. *Foods* **2023,** *12*, 1949.
156. Zou, Y.; Wang, L.; Li, P.; Cai, P.; Zhang, M.; Sun, Z.; Sun, C.; Gen, Z.; Xu, W.; Xu, X.; Wang, D. Effects of Ultrasound Assisted Extraction on the Physiochemical, Structural and Functional Characteristics of Duck Liver Protein Isolate. *Process Biochem.* **2019,** *52*, 174–182.

CHAPTER 6

Biotechnological Valorization of Cocoa Waste: Advances in the Production of Biomaterials and Beverages, and their Impact on Environmental Sustainability

MANUEL VARGAS-ORTÍZ[1], AARÓN FERNANDO GONZÁLEZ-CÓRDOVA[1], BELINDA VALLEJO-CORDOBA[1], ADRIÁN HERNÁNDEZ-MENDOZA[1], LILIA MARÍA BELTRÁN-BARRIENTOS[1], ALINE REYES-DÍAZ[2], JOSÉ ISIDRO MÉNDEZ-ROMERO[1*], and EMANUEL RIVAS-ROBLES[3]

[1]*CONAHCYT-Centro de Investigación en Alimentación y Desarrollo, Laboratorio de Calidad, Autenticidad y Trazabilidad de los Alimentos, Coordinación de Tecnología de Alimentos de Origen Animal, Hermosillo, Sonora, México*

[2]*Universidad Estatal de Sonora (UES), Unidad Académica Hermosillo, Av. Ley Federal del Trabajo, Sin Asignar en Nombre del Acuerdo, Sonora, México*

[3]*Universidad de Ciencias y Artes de Chiapas, Facultad de Ciencias de la Nutrición y Alimentos, Ingeniería en Agroalimentos, subsede Acapetahua, México*

*Corresponding author.

ABSTRACT

Cocoa, a crucial crop in many regions of the world, produces significant amounts of waste during its processing. Traditionally, these wastes have been discarded, but recent biotechnological advancements have unveiled enormous potential in their reuse and valorization. This chapter discusses how biotechnology has enabled the transformation of cocoa waste into useful biomaterials

Agroindustrial Waste Management and Natural Resources Conservation: Environmental Challenges and Transformation into Value-Added Products. José Juan Buenrostro-Figueroa, Juan Manuel Tirado-Gallegos, Mónica L. Chávez-González, A. K. Haghi, and Shaheemath Suhara K. K. (Eds.)

and new beverages. Biomaterials derived from cocoa have applications in areas such as medicine, where they can be used in the manufacture of prostheses or biocompatible materials, and in industry, where they can replace nonbiodegradable materials. On the contrary, beverages made from cocoa waste offer a sustainable and nutritious alternative to conventional drinks, and their production can generate economic opportunities in cocoa-producing regions. The impact on environmental sustainability is significant. The valorization of cocoa waste reduces greenhouse gas emissions, minimizes soil and water pollution, and promotes a circular economy (CE) where the use of all available resources is maximized. Additionally, reusing these wastes decreases the demand for landfills and other disposal facilities.

6.1 INTRODUCTION

Cocoa, a treasured gift that the Amazon rainforest has bestowed upon the world since prehistoric times, has been the cornerstone of many civilizations and cultures, from ancient American tribes to modern food industries.[1] However, over the years, the processing of this coveted bean has led to the generation of significant amounts of waste, posing both economic and environmental challenges.[2]

While the global demand for chocolate has experienced steady growth, boosting the economies of cocoa-producing countries such as Ivory Coast, Ghana, and Nigeria, waste generation has been a persistent shadow in the sector.[3] In 2021, with the production of more than 5.5 million tons of dried cocoa beans, there arise significant concerns about waste management.[3,4] These residues, mainly composed of pod husks, bean shells, and pulp, have historically been underutilized.[5]

Fortunately, recent research is shedding light on the intrinsic value of these residues. With their antioxidant properties, dietary fiber content, and polyphenols, these by-products have untapped potential in the food, pharmaceutical, and cosmetic sectors.[6–8] For instance, pod husks, once regarded as mere waste, are now seen as potential sources of dietary fiber.[6] Similarly, the cocoa mucilage, which surrounds the cocoa bean, has potential applications in food products such as juices and wines.[9,10]

From an economic perspective, the shift toward a circular economy (CE), which promotes resource reuse, not only has the power to mitigate costs associated with waste disposal but also opens up new economic opportunities. With the global chocolate market projected to surpass €169 billion by 2026, the biotechnological valorization of cocoa by-products could offer a significant boost to the industry.[11]

However, we cannot ignore the environmental challenges posed by the cocoa industry. For every ton of dried cocoa beans produced, approximately 10 tons of wet waste are generated.[12] If not properly managed, these residues can release greenhouse gases, such as methane, exacerbating the issue of climate change.[13] Therefore, it is essential to develop and adopt strategies that prevent pollution and reduce greenhouse gas emissions.[14]

Finally, in the context of a globalized and competitive world, intellectual property has become an essential tool for the protection of innovations in the field of biotechnological valorization of cocoa.[15] The increase in patents related to the optimization and commercialization of these residues evidences the growing relevance and interest in this field.[16–18]

This chapter seeks to provide a comprehensive view of biotechnology applied to cocoa waste, highlighting its advancements, opportunities, and challenges from a technical, economic, and environmental perspective.

6.2 ORIGIN AND DESCRIPTION OF COCOA FRUIT

Cocoa (*Theobroma cacao*) is indigenous to the American continent—specifically to the Amazonian rain forest where Brazil, Colombia, and Venezuela converge—where evidence of its existence since 10,000–15,000 years ago was found.[1] However, it was in Mexico and Guatemala that cocoa was first domesticated as a crop for feeding, ceremonial, and economic purposes. The word "cacao" (cocoa in Spanish) derives from the Mayan word Ka'kau.[2] Despite the fact that cocoa is indigenous to America, the main cocoa-producing countries in the world are Ivory Coast, Ghana, Indonesia, Nigeria, and Cameroon, followed by Brazil and Ecuador.[3]

During 2021, 5,580,432.37 tons of dry cocoa beans were produced in 58 countries, mainly distributed in a zone between 10° North and 10° south of the equator, where the weather conditions for its growth are predominant.[3,4] However, to achieve that amount of dry beans, about 55,804,323.7 tons of fresh fruit were required because only 10% of the total weight of the fruit corresponds to the bean, which represents a large generation of solid waste.[19] It is estimated that, for each ton of dry cocoa beans, ten tons of waste are generated.[5]

Chocolate is derived from the processing of cocoa beans, which must be extracted from the cocoa fruit (Fig. 6.1). Cocoa fruits have a thick husk known as the Cocoa pod husk (CPH), which represents between 67 and 80 % of the fruit weight.[5,19,20] Inside the fruit are the seeds that are covered by the mucilage, which consists of a white spongy covering; this mucilage represents between 3 and 5% of the weight of the fresh fruit.[21] The seeds are

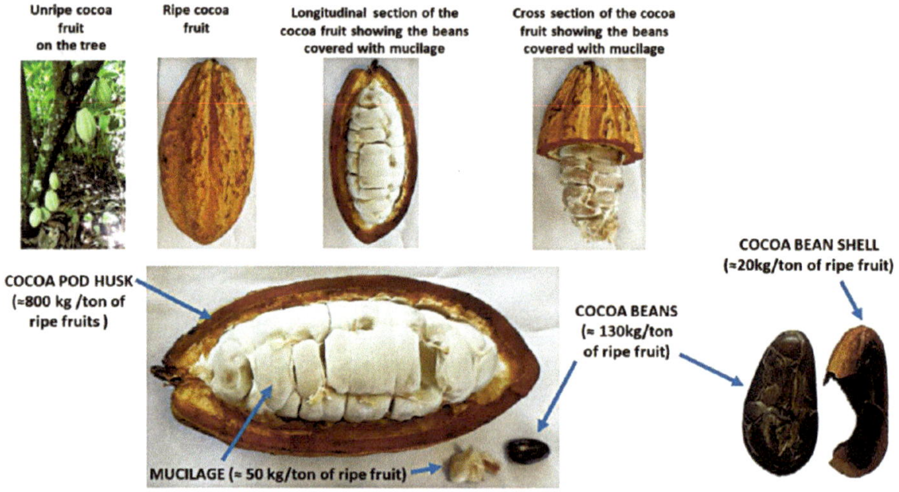

FIGURE 6.1 Description of the cocoa fruit.

covered by the cocoa bean shell (CBS), which represents between 1 and 2% of the total weight of the fresh fruit.[22] While the seeds represent between 10 and 15% of the total weight of fresh fruit.[19]

The transformation of cocoa into chocolate involves a complex process that can be divided into two major phases. The first phase (Fig. 6.2) comprises the entire process carried out very close to the cocoa plantations (cocoa mucilage is rich in fermentable sugars that reduce the fruit's shelf life, which requires processing as quickly as possible).[23] When the fruit reaches maturity, it is harvested and taken to nearby processing plants, where the fruits are opened, and the CPH is discarded (being the first waste of the process). The mucilage-covered cocoa beans are placed in containers for 4–7 days to let them ferment by the native microbiota, predominantly yeasts at the beginning and acidophilus bacteria at the end of the fermentative process,[24] reaching temperatures between 30 and 50°C. The final pH of the fermented beans is lower than 4.0[25]; during this fermentation, the mucilage is degraded and discarded in liquid form (about 100–150 liters per ton of fresh bean).[26] After fermentation, the beans are sun-dried until they reach a humidity level of 6–8%,[27] at which point the first phase of the process ends.

The dried beans are stored until they are transported to the industrial facilities where the second phase (Fig. 6.3) of the process takes place, which starts with roasting (10–35 min at temperatures between 120 and 140°C) of the dried beans.[28] After roasting, dehulling is performed, which consists of removing the CBS, which becomes a waste product. Subsequently, the roasted

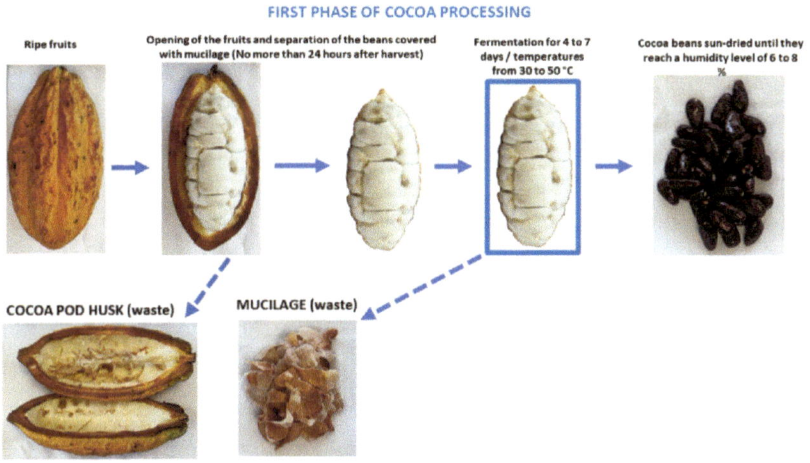

FIGURE 6.2 First phase of cocoa processing (normally carried out very close to the cocoa plantations).

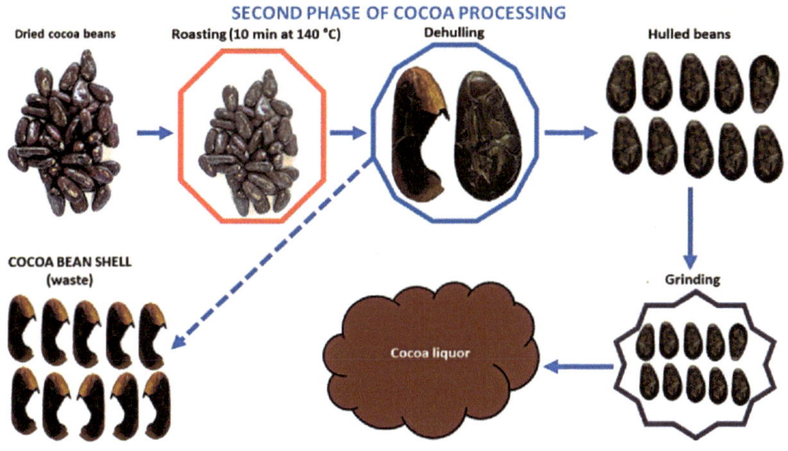

FIGURE 6.3 Second phase of cocoa processing (carried out in industrial facilities far from cocoa plantations).

and cleaned beans are ground to generate a semi-solid paste commonly known as cocoa liquor, from which different products can be generated.[25]

6.3 ECONOMIC IMPORTANCE OF COCOA WASTE

As previously mentioned, cocoa processing generates a large amount of waste, especially in the farmlands, at the first stage of processing. The accumulation of waste leads to the proliferation of fungi and insects that damage the cocoa crop. Thus, its reduction by composting has not proved to be a good solution.[20,29,30] However, all wastes generated during cocoa processing are considered important sources of compounds of interest for the pharmaceutical, cosmetic, food, and chemical industries, highlighting bioactive compounds, complex carbohydrates, fiber, aromatic compounds, fermentable sugars, lignin cellulosic complex, minerals, and phenolic compounds, in addition, these wastes have high potential.[19,20,31,32]

The main cocoa-producing countries are located in Africa (generating about 22.8% of world production), while the main cocoa-bean-processing countries (those that transform cocoa beans into cocoa-based products) are located in Europe (except for the Ivory Coast) and generate 36% of ground cocoa production.[34] As previously mentioned, cocoa industry wastes are of interest to many industries, and a niche market has developed for these wastes. Derived from the distribution of cocoa-producing and processing countries, the trade of cocoa wastes is concentrated between African and European countries. During 2021, cocoa waste (Cocoa shells, husks, skins, and waste) ranked 4404th of the most traded products in the world, representing an economic value of $39 million, with the main exporters being Nigeria ($8.56 million), Germany ($5.72 million), France ($5.61 million), Ghana ($4.76 million), and The Netherlands ($4.43 million),[22] while the main importers of cocoa processing wastes are The Netherlands ($10.7 million), Germany ($7.59 million), Malaysia ($3.01 million), Belgium ($2.99 million), and France ($2.6 million).[22]

6.4 SELECTION OF BIOTECHNOLOGICAL TECHNOLOGIES FOR COCOA WASTE VALORIZATION

Over the last decade, the valorization of by-products has become of greatest importance in the food industry. In this regard, there is a need to reduce waste generation by aiming for maximum reuse and recycling of materials and food components.[36] As previously mentioned, cocoa processing generates three main by-products, which include pod husks, bean shells, and pulp (mucilage). These by-products are usually underutilized and considered waste in the cocoa/chocolate industry[6]; in fact, 90% of these by-products are generally discarded.[7] Therefore, several studies have explored the valorization of cocoa by-products.[38] Although these by-products have been used mainly in agro-industrial, such as in animal feed, fertilizer, and biofuels,; recent studies have inquired to research and develop innovative food, nutraceutical, medical, and pharmaceutical products.[6,8] Moreover, several approaches have been made to exploit the bioactive compounds from these by-products and develop high value-added products with low costs and environmental risks.[6]

6.4.1 COCOA POD HUSKS

The CPH represents 67–76% of the weight of the whole fruit. They may vary in color and shape, depending on their variety and location. It consists of three layers: epicarp (outer layer), mesocarp (inner layer), and endocarp (innermost layer). In general, once the fresh cocoa pod is opened and seeds are extracted, the pod is known as a husk. In general, the CPH moisture content is approximately 90%. Hence, CPH may become a source of pathogens that may affect cocoa bean plantations. Therefore, several efforts in the food industry and research have been made to exploit this by-product.[6]

In this sense, a quick drying process (to reach 65% moisture) is recommended to prevent the deterioration of CPH. Also, grinding CPH to pellets and drying to reach 10% moisture before storage may also be suggested. The chemical composition of CPH has been widely reported. In this regard, CPH consists of moisture (6.4–14.1%), protein (2.1–9.1%), ash (5.9–13%), lipid (0.6–4.7%), carbohydrates (17.5–46%), and total dietary fiber (18.3–59%).[30] However, the chemical composition of CPH may vary depending on the variety, location, and pretreatment used to process it.[30] In general, the incorporation of CPH into food products has been challenging; since the undesirable texture and the perception of consumers, several approaches have been made.

In this sense, several studies have established that CPH may be an excellent source of dietary fiber, which includes soluble fiber (11%) and insoluble fiber (48%).[40] Since the content of insoluble fiber is higher than soluble fiber, it has been proposed that a chemical pretreatment to hydrolyze this fiber may be suitable before adding CPH to food products.[20] In this regard, it has been suggested the incorporation of CPH into food products with poor fiber content. Nevertheless, special attention must be paid to the technofunctional properties that CPH may provide to food products since they may possess water-holding and swelling properties.[41]

On the contrary, other studies have shown that CPH is a valuable source of antioxidants. In this regard, phenolic compounds, such as catechin, quercetin, gallic acid, coumaric acid, and theobromine, have been previously reported.[20] Moreover, drying methods are recommended to guarantee the stability of CPH; however, it has been suggested that these processes may affect phenolic compounds.[42] Contrary to this, Delgado-Ospina et al.[41] (reported that the inactivation of enzymes with rapid heat treatment on CPH powder protected polyphenolic compounds. Also, the authors observed a significant decrease in these compounds in frozen samples without enzyme inactivation.

Additionally, the antioxidant activity of CPH has been previously determined, and this effect was attributed to phenolic acids.[8,41] In fact, other beneficial health effects from CPH have been studied. In this regard, the anti-obesity and cardiometabolic risk factors lowering effects (blood pressure, cholesterol, and triglycerides) were reported in a murine model administered with pellets and extracts made with CPH.[43] Similarly, a hypolipidemic effect was observed in cholesterol-fed rats after the administration of a soluble fiber CPH product obtained after enzymatic treatment of cocoa husks. Interestingly, the main effect was attributed to fiber present in the CPH product; however, polyphenols were not overlooked.[44]

6.4.2 COCOA MUCILAGE

Cocoa mucilage is the substance that covers the cocoa bean inside the cocoa husk, also known as cocoa pulp. This mucilage is mainly composed of sugars (10–13%), salts (8–10%), and citric acid (1–2%).[45] In general, this cocoa mucilage is naturally removed during fermentation and may be hydrolyzed by fermentative microorganisms. The chemical composition of cocoa mucilage is 93.5% of moisture, 4.10% of carbohydrates, 0.53% of protein, 0.87% of ash, 1% of fat, and 0.47% of crude fiber.[10] Interestingly, 95% of the fiber in cocoa mucilage is soluble.[46]

Also, these mucilages range in pH of 3.6–3.75, where citric, malic, acetic, oxalic, lactic, fumaric, citric, and ascorbic acids are present. Interestingly, it has been reported that 18.3 mg/L of ascorbic acid was observed.[47] On the contrary, the total flavonoid content and antioxidant activity has been previously reported in cocoa mucilage.[6,48] Although CBS showed higher antioxidant activity, cocoa mucilage also presented this effect.[48] Few studies have reported the use of cocoa mucilage in food products. In this regard, this by-product mixed in fruit juice drinks,[10] kefir beverages,[49] wine,[9] and marmalade[47] has been explored.

6.4.3 COCOA BEAN SHELL

CBS is the outer cover of the cocoa bean, and its chemical composition may be variable since it depends on the origin and processing process.[50] In this regard, it has been reported that 13.1% of protein, 13% of crude fiber,

8.7% of fat, and 9.1% of dried CBS.[45] Also, phenolic compounds have been reported in CBS. In this regard, catechins, epicatechins, quercetin, and citric acid have been reported. Furthermore, the antioxidant activity of CBS has also been determined.[41]

Overall, the addition of CBS to food products has been aimed at enhancing the flavor, aroma, and color characteristics of chocolate.[41] Also, infusions of CBS have been widely used in some countries.[22] Moreover, corn-extruded snacks fortified with CBS have been shown to possess a high content of phenolic compounds.[51] Interestingly, the addition of CBS to other food products increased antioxidant activity and reduced lipid oxidation in soy oil[52] and cooked beef.[53]

Recent studies have evidenced the beneficial health effects that CBS may provide. In this sense, the neuroprotective,[54] antidiabetic,[55] anticarcinogenic,[56–58] antiobesity,[59] cardioprotective effects,[60] and hypolipidemic[61] have been reported. Similarly, all these potential effects were attributed to polyphenols. Hence, CBS has been demonstrated to be a promising by-product for the development of functional foods.

In general, the introduction of CPH and CBS during the development of food products with beneficial effects must comply with different strategies to achieve a tasteful food product. Also, these strategies must guarantee to be safe for human consumption. Moreover, in-depth knowledge of the physicochemical, technofunctional, and beneficial health effects of the bioactive compounds from CPH and CBS is crucial for the development of food products. Therefore, the food industry may use technology and "green" practices to overcome the difficulties that may present during the development of these food products, as well as to improve the sustainability of the cocoa production industry.

6.5 ECONOMIC AND FINANCIAL EVALUATION OF BIOTECHNOLOGICAL VALORIZATION OF COCOA WASTE

6.5.1 COST-BENEFIT ANALYSIS AND FINANCIAL EVALUATION OF VALORIZATION SYSTEMS

Cocoa is a food product with a high economic potential for the chocolate processing industry. The global size of the chocolate confectionery market was estimated at €116.79 billion in 2017 to exceed €169 billion in 2026.[11]

There are, however, a number of problems associated with the cocoa production sector, as well as its further processing and consumption. Considering the market volume of cocoa (*Theobroma cacao* L.), it is clear that the cocoa/chocolate industry generates significant amounts of waste.[62] Approximately 10 tons of wet waste and by-products are disposed of for each ton of dry beans, making waste management extremely challenging.[12] Moreover, the cocoa industry has been facing sustainability issues for decades as resource scarcity, and the demand for cocoa products has been increasing year by year.[62]

Mankind is now on the brink of a new challenge: the constant demand for natural, nonrenewable resources leads to them being over-exploited, making the traditional linear economic model of "take, make, consume, dispose" unsustainable.[14,64] It is well known that resource deficiency can lead to economic problems, with manufacturing industries facing rising raw material prices and uncertainty of supply.[14]

With the aforesaid issues, it is hardly surprising, with a growing world population and dwindling resources, that there is a growing awareness of the need to conserve and reuse materials that are treated as waste.[65]

An alternative model is the CE, which promises to be an efficient option in the medium and long term to reuse or recover natural resources and derived byproducts.[66] CE is a system that currently accounts for 8.6% of the global economy.[67] It involves optimizing the production and consumption model, as materials can be circulated in a closed loop, thus avoiding the disposal of end-of-life goods in landfills.[67] This model aims to focus on better use of resources and reduce our dependence on the exploitation of natural resources. This makes CE economically viable as their processing requires fewer resources and less energy than conventional material recycling.[14,68]

In addition, CE promotes the use of waste and by-products to recover valuable compounds.[68] By following the CE principles, the valorization of cocoa by-products can reduce environmental impact and lead to the development of profitable commodities, creating new business opportunities and consequent economic benefits.[14,68,69]

6.5.2 ESTIMATION OF INVESTMENT COSTS, OPERATING COSTS, INCOME GENERATED FROM THE SALE OF BIOTECHNOLOGICAL COMPOUNDS, AND ADDITIONAL ECONOMIC BENEFITS

CE focuses on making better use of resources by optimizing the production and consumption model.[67] The valorization of waste and by-products is one of the key steps in closing material flows according to CE.[70] As 67% of cocoa fruit is essentially a by-product, its valorization adds value to the entire cocoa production process. It reduces the costs associated with the disposal of lost or wasted food.[14,71] Alternative models, such as CE, are driving the need to adopt disruptive technologies that can lead to more sustainable production processes.[14]

CE has not yet been fully adopted in manufacturing and supply chains, partly due to insufficient information on product life cycles and material processing by innovative cleaner technologies. Economic uncertainty is caused by investment costs, return on profit, and the time required for implementation, which is another barrier to the implementation of CE principles.[14,72]

It is estimated that the implementation of the CE principles could be responsible for a 17–24% reduction in resource demand by 2030, with a total savings potential of €630 billion per year for European industry. In addition, greenhouse gas emissions could be reduced by up to 70%, and employment could increase by 4%.[14]

6.5.3 CONSIDERATION OF THE LONG-TERM FINANCIAL SUSTAINABILITY OF VALORIZATION PROCESSES

Food waste management is a major issue worldwide, and as a result, several new processes and methodologies are being developed and applied to reduce its impact while increasing the valorization of food by-products.[71] Residual biomass, especially that of cocoa, has the particular characteristics of being cheap, renewable, and abundant, and these characteristics can, therefore, become an important economic advantage if appropriate technologies are developed to recover value-added compounds from these wastes.[32,73] To this end, consideration should be given to the adoption of integrated technologies, such as those referred to as the "fourth industrial revolution" or "Industry 4.0."[74]

The core technologies of Industry 4.0 are the Internet of Things, cloud manufacturing, big data, cyber–physical systems, and additive manufacturing.[72] CE business models can be integrated and enhanced by key elements of Industry 4.0.[14] Emerging digital Industry 4.0 technologies aim to increase productivity, create new products and services, and facilitate communication between producers and customers, while CE focuses on the sustainability of manufacturing processes and supporting the creation of new products that could provide added value.[14]

Industry 4.0 technology tools can help track raw materials as well as postoperational by-products and waste, generating useful information about the production process, individual machines, and product postconsumption, paving the way for the adoption of CE concepts.[69] Data analysis can be used to optimize resource and energy efficiency, control manufacturing emissions, and increase productivity.[14]

The principles of CE and Industry 4.0 should be key elements in ensuring sustainable production in the cocoa industry, which is facing several sustainability challenges as demand for cocoa products outstrips supply.[14]

6.6 ENVIRONMENTAL CONSIDERATIONS IN THE BIOTECHNOLOGICAL VALORIZATION OF COCOA WASTE

6.6.1 EVALUATION OF ENVIRONMENTAL IMPACTS ASSOCIATED WITH BIOTECHNOLOGICAL VALORIZATION PROCESSES

The main raw material for the production of all types of cocoa products is dried and fermented cocoa beans.[65] The beans are usually processed into four semi-finished products, namely cocoa liquor, cocoa butter, cocoa powder, and cake, and one finished product, chocolate.[75] In this sense, it is worth noting that the cocoa field processing chain includes harvesting, fermentation, roasting, shelling, grinding, and grinding.[14]

Cocoa beans are an important raw material for the global food industry; however, cocoa production generates huge amounts of residual biomass at the "on-farm" and "postprocessing" levels, such as cocoa pod shells, CBSs, cocoa pulp, and sweat.[13,32] In fact, for every ton of dry cocoa beans, approximately 10 tons of wet waste and by-products are generated, which makes waste management extremely challenging, especially in terms of its correct use in the CE.[12,32,76]

Unfortunately, this waste has increased with the growing demand and use of cocoa products. Typically, solid agricultural wastes are generally buried in landfills, and liquids are discharged into sewers, resulting in air, water, and soil pollution. Others are recycled, mainly as animal feed and compost.[71,78]

However, without proper treatment, such waste releases methane, a greenhouse gas far more potent than carbon dioxide, into the atmosphere and contributes to climate change. In addition, the high organic content of these wastes makes them suitable for microbial growth.[13]

In addition, there are several elements in the different stages of cocoa processing that have also been linked to environmental impacts and should be taken into account. These stages include the extraction of raw materials (e.g., fossil fuels and minerals), the production of agricultural inputs (e.g., fertilizers and pesticides), and all agricultural operations in the field (e.g., tillage, application of fertilizers and pesticides, harvesting, etc.). The transport of the beans to the factory and the industrial processing of the beans into cocoa butter, liquor, cocoa cake, and cocoa powder should also be considered.[13,75]

The inappropriate disposal of these wastes represents crops and health risks to the population due to environmental contamination or the proliferation of pathogens. A better destination for these wastes is undoubtedly important because it promotes their valorization, in addition to ensuring a sanitary quality of the environment as a whole.[79,80]

6.6.2 IMPLEMENTATION OF WASTE MANAGEMENT SYSTEMS, INCLUDING WASTEWATER TREATMENT AND ORGANIC MATTER MANAGEMENT

Due to the economic and environmental problems caused by the accumulation of organic waste, there is a growing interest in the implementation of waste management systems.

Nowadays, waste management is exceptionally challenging, and it is therefore necessary raising awareness for the recovery of waste. The European Union (EU) is promoting the reduction of food wastes, and the search for new end-uses for food by-products to decrease these problems.[81,71,76]

Minimizing waste is one of the principles of the CE, which aims to transform products that are at the end of their production chain into resources for other purposes and thus close loops in industrial ecosystems.[82] This dissemination of sustainability awareness means that ever more attention is being paid to the reduction of waste and pollution in the agro-food industry. The EU Waste Framework Directive (EU (2008) Directive 2008/98/EC)[83] on food waste gives priority to waste reduction at the source, followed by recycling and reuse, leaving elimination as the last option. The majority of industries worldwide are, therefore, now involved in the development of environmentally friendly protocols that reduce waste production.[73]

Therefore, from an environmental perspective, several positive impacts are expected in reducing food loss and waste, such as in water management, climate change, marine resources, terrestrial ecosystems, forestry, and biodiversity.[71]

6.6.3 STRATEGIES FOR PREVENTING SOIL AND WATER POLLUTION AND REDUCING GREENHOUSE GAS EMISSIONS

In 2017, the value of exported cocoa waste was approximately $244 million, with Côte d'Ivoire, Sierra Leone, and The Netherlands as the main exporting countries, and Spain, The Netherlands, and Malaysia were the main importers.[84] Recently, greater interest has been given from companies around the world to the use of cocoa by-products to develop various products on an industrial scale (e.g., feed and complements for animal nutrition and fertilizers).[73]

Beyond, the by-products of the agri-food industry (peels, seeds, shells, pomace, and leaves) are useful due to their content of bioactive compounds (phenols, peptides, carotenoids, anthocyanins, and fatty acids), fibers, and enzymes of great interest, because of antioxidant, antihypertensive, anti-atherogenic, anti-inflammatory, hypoglycemic, and hypocholesterolemic properties.[76] Henceforth, these by-products can be used for the production of functional foods and drugs against acute and chronic diseases that can be applied to the food, cosmetic, or pharmaceutical sectors.[64,65,76] It is worth mentioning that certain cocoa by-products may contain some undesirable pollutants, such as mycotoxins, which could complicate the incorporation of this by-product into food formulations.[76]

Additionally, cocoa by-products have been employed in several applications, such as adsorbents to remove emerging contaminants from wastewater, a substrate for biopolymer production, corrosion inhibitors, and the obtention of biofuels.[13,76]

Until now, extraction techniques (such as extraction with organic solvents) have been used, although they are often laborious, time-consuming, and require large amounts of solvent.[73] However, greener and more environmentally friendly techniques are more desirable than conventional ones for intensifying the extraction process.[85]

By implementing these latter strategies, environmental and economic benefits can be ensured, in accordance with the principles of CE, by implementing emerging technologies, such as pressurized liquid extraction (PLE), subcritical water extractions (SWE), supercritical fluid extraction (SFE), Ultrasound-assisted extractions (UAE), hydrodynamic cavitation (HC), microwave-assisted extractions (MAE), electrical-disruption techniques, could all be key innovations for greener and cost-effective production.[14]

6.7 PATENTING AND INTELLECTUAL PROPERTY IN THE BIOTECHNOLOGICAL VALORIZATION OF COCOA WASTE

6.7.1 CONSIDERATIONS REGARDING PATENTING PROCESSES AND PRODUCTS DERIVED FROM COCOA WASTE VALORIZATION

Biotechnological valorization of cacao by-products has garnered attention in the patent and intellectual property industry. A growing collection of patents demonstrates efforts to capitalize and optimize these residues, ranging from the extraction of carcinogenesis-inhibiting compounds[15] to the production of soluble fibers.[16] Noteworthy innovations include the creation of teas made from cacao husks[86] and foods enriched with alkalized husks,[87] as well as drinks enhanced with alkalized CBS.[18]

Cacao beans have traditionally taken the spotlight, but its by-products, especially the honey and pulp, offer a promising future. Major corporations such as Société des Produits Nestlé and Mars, Incorporated, have ventured into and patented unique applications targeting both gastronomy and dermatology. This trend is evidenced by a French patent regarding cacao pulp.[88] With its distinctive organoleptic profile, cacao honey has been utilized in a variety of products, from jams to spirits, as proposed by Scampini (2020).[89] Renowned Brazilian universities, such as the University of São Paulo and the Federal University of Bahia, have pioneered patents involving cacao honey.[17,90] A review of existing patents suggests that only a fraction focus on biotechnological processes, indicating a vast field yet to be explored. This points toward a rising potential for research that merges biotechnology with waste valorization, turning these by-products into high-value goods.

Given the worldwide interest in capitalizing on cacao by-products on an industrial scale, opportunities are arising in the patentability of extraction and transformation techniques. Conventional techniques, such as extraction with organic solvents, are being overshadowed by more sustainable and eco-friendly methods such as PLE, SWE, SFE, and others. These methods represent the future of the industry and pave the way for the development of cacao-derived products.[14] Ensuring rigorous analysis of transformation and recovery processes is vital, leading to the creation of new production and commercialization avenues. The ever-evolving cacao industry is distinguished by patent creation and the establishment of specialized R&D centers. To enhance the entire cacao production chain, from cultivation to the factory, it is imperative to pursue advanced research, strengthen research networks, generate patents, and craft industrial applications. Collaborating with academic and research entities further boosts the path to innovation and sustainability.[91]

6.7.2 STRATEGIES TO PROTECT AND MAXIMIZE INTELLECTUAL PROPERTY IN THE SECTOR

The rising importance and diversity in the biotechnological management of cacao waste necessitate the implementation of robust strategies for the protection and maximization of intellectual property in this field.[92,93] Effective intellectual property management not only safeguards innovations but also acts as a catalyst for further research and development.[94] The landscape of innovations in cacao waste valorization is both diverse and continuously evolving. Patents ensure the protection of these advancements and recognize leading entities in this area, as observed with various institutions worldwide.[17,90]

Trademarks become paramount when innovations find traction in the market, ensuring the identity and perception of creations derived from cacao waste. Protecting names, logos, and other distinctive symbols is essential to ensure brand integrity and value remain intact. Additionally, with a wide range of potential applications, copyrights and industrial designs offer protection for aspects outside the purview of patents, covering areas from software developments to specialized machinery. The importance of trade Secrets in the global competitive landscape cannot be understated. Companies must have the option to keep certain processes or techniques confidential, preventing competitors from gaining an edge by accessing this crucial information.[95]

Training and raising awareness about intellectual property are paramount in this emerging field. All involved parties, from researchers to producers, need education to foster a consistent understanding and respect for intellectual property rights. Licensing and technology transfers stand out as strategic opportunities in this domain. Collaborating across different entities enhances the impact of innovations, fosters adaptability, and champions sustainability in the industry. With the swift pace of advancements in cacao waste valorization, actively monitoring the market and patent databases is essential to ensure the integrity of protected innovations remains uncompromised. Collaborations with Research Centers are pivotal, bolstering the innovation ecosystem and facilitating joint protection of discoveries.

6.8 INDUSTRY INTERESTS IN THE BIOTECHNOLOGICAL VALORIZATION OF COCOA WASTE

6.8.1 IDENTIFICATION OF THE INTERESTS AND NEEDS OF THE COCOA INDUSTRY REGARDING WASTE VALORIZATION

The cocoa industry, such as other agro-industries, faces challenges in waste management, which accounts for about 60% of plant biomass.[96] The treatment of these materials is intricate due to their volume and structure.[97] In

2017, cocoa waste had an export value of $244 million, with key players such as Côte d'Ivoire and Sierra Leone[98] indicating a rise in its utilization for industrial products. Other agro-industrial residues have found varied applications, from biofuels to biomaterials.[99,100] Cocoa has been valued for over 3000 years by civilizations such as the Mayans and Aztecs for its nutritional properties.[101] Despite its recognized potential, the valorization of residual cocoa biomass is still in its early stages. The industry is seeking strategies to capitalize on these by-products, reducing environmental impacts and generating new economic opportunities. The focus on the revalorization of cocoa waste not only provides a sustainable solution to the waste issue but also creates employment opportunities in emerging areas of the industry. With the establishment and expansion of these new businesses, there's potential for job creation in research, production, distribution, and other related fields, thus benefiting the labor community connected to cocoa.

6.8.2 SUSTAINABILITY AND CIRCULAR ECONOMY IN THE INDUSTRY

Sustainability is characterized by three key elements: social equity, economic viability, and environmental preservation. This means ensuring social justice, maintaining a healthy economy, and safeguarding the environment for present and future generations.[102] Within the context of the cocoa industry, the social component takes on unique significance. Since this sector largely relies on family-based operations, with between 40 and 50 million people economically dependent on cocoa farming, it is paramount to ensure farmers enjoy equitable and dignified living conditions.[103] Currently, according to the International Cocoa Organization, Latin America alone supplies 20% of the world's cocoa and over 80% of organic and aroma cocoa. This results in production across over 400,000 family farms, representing a significant source of rural employment for producing countries.[104] Globally, in the 2015–2016 harvest, family-managed cocoa cultivation played a crucial role, yielding a worldwide production of 3.9 million tons and exports valued at approximately $47 billion.[98] (International Cocoa Organization, 2018) Therefore, it is evident that measures promoting education and ongoing development in cocoa-growing communities are needed to ensure enduring social equity in this industry.[91]

However, it is clear that cocoa production is not optimized due to various factors. Labor saturation; land scarcity and related disputes; limitations on fertilizer use; and climatic and political challenges negatively impact cocoa production in countries such as Côte d'Ivoire. This highlights the importance of fostering measures that promote education and continuous development in cocoa-farming communities, not just to ensure lasting social equity in this sector but also to address the direct and indirect socioeconomic challenges faced by farmers.[63] Cocoa production also confronts social sustainability and governance challenges, often overshadowed by environmental and economic concerns. While eco-friendly agricultural practices are prioritized, the direct and indirect socioeconomic benefits must be considered. Cocoa-chocolate businesses should diversify their sustainability mechanisms, considering both the farm level and systemic challenges. Inclusivity of all stakeholders and a comprehensive understanding of sustainability dimensions are key to addressing challenges such as child labor and deforestation in the supply chain.[105]

On the economic front, cocoa prices are notoriously volatile, influenced by factors ranging from extreme weather conditions to pest outbreaks and diseases, speculative practices, and political instability in producing nations. This is evident in Ghana's cocoa industry between 1975 and 2019, which faced various challenges, including political instability and the negative impact of the El Niño climate phenomenon, causing droughts and periods of production uncertainty. Key recommendations have been put forth in light of this, such as strengthening Ghana's Meteorological Agency, setting up stabilization funds in the Cocoa Board, and promoting reforestation as measures to bolster the sustainability of the cocoa industry.[63,103,105,106]

In economic terms, a fundamental challenge in the cocoa industry is achieving a balance between productivity increase and minimizing negative impacts on the environment and health. This equilibrium is vital to ensure the industry's long-term sustainable growth. Furthermore, protecting local ecosystems should be a priority, as they play an essential role in preserving biodiversity and preventing environmental degradation. Adapting to changing climate conditions, such as droughts and floods, is a pressing necessity to maintain consistent cocoa production. This calls for the implementation of climate-resilient agricultural practices and efficient management strategies.[105]

On the contrary, appropriate waste management generated during cocoa production is essential to minimize the entire industry's environmental footprint. This is where the valorization of cocoa waste comes into play. Not only does it have a positive economic impact by adding value to these by-products, but it also brings significant ecological benefits by preventing them from polluting the environment and being wasted. Leveraging these residues becomes an environmentally responsible and economically beneficial practice for the cocoa industry.[102]

6.9 CONCLUSIONS AND FUTURE PERSPECTIVES

Cocoa, a crop native to America, is at the heart of a complex production chain that transcends its borders, reaching major producers in Africa and processing centers in Europe. However, the transformation of cocoa into chocolate produces significant amounts of waste, with only 10% of the total fruit weight being used. This figure highlights the magnitude of the challenge but also the potential for utilization.

Biotechnology has emerged as a crucial pillar in valorizing cocoa waste. Husks, mucilage, and shells, previously considered as waste, are now revealed as sources of valuable compounds with applications in food, nutraceuticals, and medicine. On the contrary, CE and sustainability have taken center stage in the global debate, with cocoa at the focal point. The industry has the chance to transition from a linear model to a CE, where by-products are valued, reducing costs and environmental impact. This shift can lead to significant savings while promoting environmental and economic sustainability.

Moreover, from an environmental perspective, proper management of cocoa waste is essential to prevent the release of greenhouse gases, such as methane. Strategies to reduce these emissions include reusing by-products for food and medicine production, as well as for wastewater treatment and biopolymer production.

The realm of patents and intellectual property is showing growth in relation to the biotechnological valorization of cocoa, with large companies and universities leading the way. However, there is much ground to cover in terms of biotechnological processes, highlighting an untapped potential in the field.

Finally, the cocoa industry faces a crossroads: continue with historical practices or adopt an innovative approach focused on sustainability and a CE. Adapting and valorizing waste is not just an option but a necessity. Besides benefiting the environment, it can unveil new economic opportunities and improve the living conditions of millions who depend on cocoa.

In perspective, it is essential to continue research and promote collaborations between the industry, research centers, and governments to maximize the potential of cocoa waste. By focusing on sustainability, the CE, and innovation, we can turn challenges into opportunities, steering the cocoa industry toward a greener and more prosperous future.

KEYWORDS

- **cocoa**
- **biotechnological valorization**
- **solid waste**
- **biomaterials**
- **fermentation**

REFERENCES

1. Young, A. M. In *The Chocolate Tree: A Natural History of Cacao*; Smithsonian Institution Press: Washington, DC, 1994.
2. De La Cruz-Medina, J.; Vargas-Ortiz, M.; Del Angel-Coronel, O. CACAO: Operaciones poscosecha. Editado por Danilo Mejía; Food and Agriculture Organization (FAO) [Online] 2012. https://www.fao.org/in-action/inpho/crop-compendium/es/
3. FAOSTAT. Crops and Livestock Products. Cocoa, Production by Region [Online] 2023. http://www.fao.org/faostat/en/#data/QC (accessed Aug 3, 2023).
4. ICCO (International Cocoa Organization). Origins of Cocoa and its Spread Around the World [Online] 2023. Retrieved from. https://www.icco.org/growing-cocoa/ (accessed Aug 8, 2023).
5. Acosta, N.; De Vrieze, J.; Sandoval, V.; Sinche, D.; Wierinck, I.; Rabaey, K. Cocoa Residues as Viable Biomass for Renewable Energy Production through Anaerobic Digestion. *Bioresour. Technol.* **2018,** *265*, 568–572.
6. Nieto-Figueroa, K. H.; Mendoza García, N. V.; Campos Vega, R. Cocoa By-Products. In *Food Wastes and By-Products: Nutraceutical and Health Potential*, First Edition; Campos-Vega, R., Dave Oomah, B., Vergara-Castañeda, H. A., Eds.; John Wiley & Sons Ltd.: USA, 2020; pp 373–403.
7. Indiarto, R.; Raihani, Z. R.; Dewi, M. P.; Aqila, Z. A.; Efendi, M. Y. A Review of Innovation in Cocoa Bean Processing By-Products. *I.J.E.T.E.R.* **2021,** *9* (8), 1162–1169.
8. Cádiz-Gurrea, M. L.; Fernández-Ochoa, A.; Leyva-Jiménez, F. J.; Guerrero-Muñoz, N.; Villegas-Aguilar, M. C.; Pimentel-Moral, S.; Ramos-Escudero, F.; Segura-Carretero A. LC-MS and Spectrophotometric Approaches for Evaluation of Bioactive Compounds from Peru Cocoa By-Products for Commercial Applications. *Molecules* **2020,** *25*, 3177.
9. Duarte, W. F.; Dias, D. R.; Oliveira, J. M.; Teixeira, J. A.; de Almeida e Silva, J. B.; Schwan, R. F. Characterization of Different Fruit Wines made from Cacao, Cupuassu, Gabiroba, Jaboticaba and Umbu. *LWT - Food Sci. Technol.* **2010,** *43* (10), 1564–1572.
10. Afolabi, O.; Ibitoye, O.; Agbaje, F. Evaluation of Nutritional and Sensory Properties of Cocoa Pulp Beverage Supplemented with Pineapple Juice. *J. Food Res.* **2015,** *4* (6), 58.
11. Misevic, P.; Volarevic, H.; Peric, M. In *The Trend Analysis of Less Sugar Chocolate Confectionery Markets*, Proceedings of the Economic and Social Development (Book

of Proceedings), 52nd International Scientific Conference on Economic and Social Development, Porto, Portugal, April 16–17, 2020; p 655.
12. Ramos, L. H.; Cisneros-Yupanqui, M.; Santisteban Soto, D. V.; Lante, A.; Favaro, L.; Casella, S.; Basaglia, M. Exploitation of Cocoa Pod Residues for the Production of Antioxidants, Polyhydroxyalkanoates, and Ethanol. *Fermentation* **2023,** *9*, 843.
13. Belwal, T.; Cravotto, C.; Ramola, S.; Thakur, M.; Chemat, F.; Cravotto, G. Bioactive Compounds from Cocoa Husk: Extraction, Analysis and Applications in Food Production Chain. *Foods* **2022,** *11*, 798.
14. Mariatti, F.; Gunjević, V.; Boffa, L.; Cravotto, G. Process Intensification Technologies for the Recovery of Valuable Compounds from Cocoa by-Products. *Innov. Food Sci. Emerg. Technol.* **2021,** *68*, 102601.
15. Jingxuan, K.; Yihuan, P.; Fuzhong, L.; Jizhong, L.; Jiyuan, L.; Jindong, L. J. Z., Y., J. Extracts of Cacao Bean and Cacao Bean Husk with Inhibitory Effects on Carcinogenesis, 100394860, 2008.
16. Nakamura, A.; Yoshida, R.; Maeda, H.; Nagaoka, S. Cacao Husk-Derived Watersoluble, Dietary Fiber, Process for its Production, Foods and Beverages Containing It and Methods for their Preparation, 2010.
17. Santos, F. C. G. D.; Kalid, R. D. A. Technological Prospecting: A Study of Technologies Applied to Cocoa Processing and Derivatives, 2020.
18. Bernaert, H.; Ruysscher, I. Process, Use, and Product. US Patent 2016135478A1, May 2016.
19. Soares, T. F.; Oliveira, M. B. P. Cocoa by-Products: Characterization of Bioactive Compounds and Beneficial Health Effects. *Molecules* **2022,** *27* (5), 1625.
20. Lu, F.; Rodriguez-Garcia, J.; Van Damme, I.; Westwood, N. J.; Shaw, L.; Robinson, J. S.; Warren, G.; Chatzifragkou, A.; McQueen Mason, S.; Gomez, L.; Faas, L.; Balcombe, K.; Srinivasan, C.; Picchioni, F.; Hadley, P.; Charalampopoulos, D. Valorisation Strategies for Cocoa Pod Husk and its Fractions. *Curr. Opin. Green Sustain. Chem.* **2018,** *14*, 80–88.
21. Scheuer, P. M.; Southgate, A. N. N.; Martelli, M. F.; Dias, C.; da Silva, M. E.; Coelho, A. A.; de Francisco, A. Quality Properties of a Bread made with Levain and Cocoa Waste. *J. Culin. Sci. Technol.* **2022,** *20* (5), 409–420.
22. Rojo-Poveda, O.; Barbosa-Pereira, L.; Zeppa, G.; Stévigny, C. Cocoa Bean Shell—A by-Product with Nutritional Properties and Biofunctional Potential. *Nutrients* **2020,** *12* (4), 1123.
23. Mendoza-Meneses, C. J.; Feregrino-Pérez, A. A.; Gutiérrez-Antonio, C. Potential Use of Industrial Cocoa Waste in Biofuel Production. *J. Chem.* **2021,** 1–11.
24. Oddoye, E. O. K.; Agyente-Badu, C. K.; Gyedu-Akoto, E. Cocoa and its Byproducts: Identification and Utilization. In *Chocolate in Health and Nutrition*; Watson, R., Preedy, V., Zibadi, S., Eds.; Springer: New York, 2013; vol 7; pp 23–37.
25. Perez, M.; Lopez-Yerena, A.; Vallverdú-Queralt, A. Traceability, Authenticity and Sustainability of Cocoa and Chocolate Products: A Challenge for the Chocolate Industry. *Crit. Rev. Food Sci. Nutr.* **2021,** *62* (2), 475–489.
26. Dwapanyin, A. O.; Adomako, D.; Tetteh, J. P. The Sugar Content of Cocoa Sweatings and the Effect of Pressing the Sweatings Prior to Fermentation on Bean Quality. *J. Biochem. Mol. Biol.* **1991,** *1*, 109–120.
27. Afoakwa, E. O. In *Chocolate Science and Technology*; Wiley-Blackwell Publishers: UK, 2010.

28. Barišić, V.; Kopjar, M.; Jozinović, A.; Flanjak, I.; Ačkar, Đ.; Miličević, B.; Šubarić, D.; Jokić, S.; Babić, J. The Chemistry behind Chocolate Production. *Molecules* **2019,** *24* (17), 3163.
29. Watson, R. R.; Preedy, V. R.; Zibadi, S. In *Chocolate in Health and Nutrition In Nutrition and Health*; Bendich, A., Ed.; Humana Press: Heidelberg, Germany, 2013; p 23.
30. Campos-Vega, R.; Nieto-Figueroa, K. H.; Oomah, B. D. Cocoa (*Theobroma cacao* L.) Pod Husk: Renewable Source of Bioactive Compounds. *Trends Food Sci. Technol.* **2018,** *81*, 172–184.
31. Galanakis, C. M. Separation of Functional Macromolecules and Micromolecules: From Ultrafiltration to the Border of Nanofiltration. *Trends Food Sci. Technol.* **2015,** *42* (1), 44–63.
32. Vásquez, Z. S.; de Carvalho Neto, D. P.; Pereira, G. V.; Vandenberghe, L. P.; de Oliveira, P. Z.; Tiburcio, P. B.; Rogez, H. L. G.; Neto, A. G.; Soccol, C. R. Biotechnological Approaches for Cocoa Waste Management: A Review. *Waste Manage.* **2019,** *90*, 72–83.
33. Mariatti, F.; Gunjević, V.; Boffa, L.; Cravotto, G. Process Intensification Technologies for the Recovery of Valuable Compounds from Cocoa by-Products. *Innov. Food Sci. Emerg. Technol.* **2021,** *68*, 102601.
34. STATISTICA. Share of Cocoa Bean Grindings Worldwide in 2022/23, by Region [Online] 2023. https://www.statista.com/statistics/655435/global-share-cocoa-bean-grindings-by-region/ (accessed Aug 23, 2023).
35. OEC (Economic Complexity Observatory). Cocoa Shells, Husks, Skins and Waste [Online] 2023. https://oec.world/en/profile/hs/cocoa-shells-husks-skins-and-waste (accessed Aug 3, 2023).
36. Camponio, F.; Piga, A.; Poiana, M. Valorization of Food Processing By-Products. *Foods* **2020,** *11*, 3246.
37. Indiarto, R.; Raihani, Z. R.; Dewi, M. P.; Aqila, Z. A.; Efendi, M. Y. A Review of Innovation in Cocoa Bean Processing By-Products. *I.J.E.T.E.R.* **2021,** *9* (8), 1162–1169.
38. Chico, M. F. Valorization of Cocoa by Products: Applications and Perspectives in the Food Industry. *Rev. Alime. Ciencia Ing.* **2022,** *29* (2), 57–101.
39. Nguyen, V. T. Potential of Agricultural Wastes: Uses and Future Perspectives. In *Recovering Bioactive Compounds from Agricultural Wastes*; Nguyen, V. T., Ed.; John Wiley & Sons: Hoboken, NJ, USA, 2017; pp 1–32.
40. Yapo, B. M.; Besson, V.; Koubala, B. B.; Koffi, K. L. Adding Value to Cacao Pod Husks as a Potential Antioxidant-Dietary Fiber Source. *Am. J. Food Nutr.* **2013,** *1* (3), 38–46.
41. Delgado-Ospina, J.; Lucas-González, R.; Viuda-Martos, M.; Fernández-López, J.; Pérez-Álvarez, J. A.; Martuscelli, M.; Chaves-López, C. Bioactive Compounds and Techno-Functional Properties of High-Fiber Co-products of the Cacao Agro-Industrial Chain. *Heliyon* **2021,** *7* (4), e06799.
42. Valadez-Carmona, L.; Plazola-Jacinto, C. P.; Hernández-Ortega, M.; Hernández-Navarro, M. D.; Villarreal, F.; Necoechea-Mondragón, H.; Ortiz-Moreno, A.; Ceballos, G. Effects of Microwaves, Hot Air and Freeze-Drying on the Phenolic Compounds, Antioxidant Capacity, Enzyme Activity and Microstructure of Cacao Pod Husks (*Theobroma cacao* L.). *Innov. Food Sci. Emerg. Technol.* **2017,** *41*, 378–386.
43. Hidalgo, I.; Ortiz, A.; Sanchez-Pardo, M.; Garduño-Siciliano, L.; Hernandez-Ortega, M.; Villarreal, F.; Meaney, E.; Najera, N.; Ceballos, G. M. Obesity and Cardiovascular Risk Improvement using Cacao by-Products in a Diet-Induced Obesity Murine Model. *J. Med. Food* **2019,** *22*, 567–577.

44. Ramos, S.; Moulay, L.; Granado-Serrano, A. B.; Vilanova, O.; Muguerza, B.; Goya, L.; Bravo, L. Hypolipidemic Effect in Cholesterol-Fed Rats of a Soluble Fiber-Rich Product Obtained from Cocoa Husks. *J. Agric. Food Chem.* **2008,** *56*, 6985–6993.
45. Jayeola, C. O.; Adebowale, B. A.; Yahaya, L. E.; Ogunwolu, S.; Olubamiwa, O. Production of Bioactive Compounds from Waste. In *Therapeutic, Probiotic, and Unconventional Foods*; Grumezescu, A. M., Holban, A. M., Eds.; Academic Press, 2018; pp 317–340.
46. Martínez, R.; Torres, P.; Meneses, M. A.; Figueroa, J. G.; Pérez-Álvarez, J. A.; Viuda-Martos, M. Chemical, Technological and In Vitro Antioxidant Properties of Cocoa (*Theobroma cacao* L.) Co-Products. *Food Res. Int.* **2012,** *49* (1), 39–45.
47. Anvoh, K. Y. B.; Bi, A. Z.; Gnakri, D. Production and Characterization of Juice from Mucilage of Cocoa Beans and its Transformation into Marmalade. *Pakistan J. Nutr.* **2009,** *8* (2), 129–133.
48. Llerena, W.; Samaniego, I.; Vallejo, C.; Arreaga, A.; Zhunio, B.; Coronel, Z.; Quiroz, J.; Angós, I.; Carrillo, W. Profile of Bioactive Components of Cocoa (*Theobroma cacao* L.) By-Products from Ecuador and Evaluation of Their Antioxidant Activity. *Foods* **2023,** *12*, 2583.
49. Puerari, C.; Magalhães, K. T.; Schwan, R. F. New Cocoa Pulp-Based Kefir Beverages: Microbiological, Chemical Composition and Sensory Analysis. *Food Res. Int.* **2012,** *48* (2), 634–640.
50. Martín-Cabrejas, M. A.; Valiente, C.; Esteban, R. M.; Mollá, E.; Waldron, K. Cocoa Hull: A Potential Source of Dietary Fiber. *J. Sci. Food Agric.* **1994,** *66* (3), 307–311.
51. Netania, G.; Permana, T.; Effendy, J.; Santoso, F.; Oslo, E. A. A Review on the Potential Applications of Cocoa Shell in Food Industry. *Adv. Biol. Sci. Res.* **2022,** *16*, 126–131.
52. Manzano, P.; Hernández, J.; Quijano-Avilés, M.; Barragán, A.; Chóez-Guaranda, I.; Viteri, R.; Valle, O. Polyphenols Extracted from *Theobromoa cacao* Waste and its Utility as Antioxidant. *Emir. J. Food Agric.* **2017,** *29* (1), 45–50.
53. Ismail, A.; Chew, L. Y. Antioxidative Effects of Extracts of Cocoa Shell, Roselle Seeds and a Combination of Both Extracts on the Susceptibility of Cooked Beef to Lipid Oxidation. *J. Food Technol.* **2006,** *4*, 10–15.
54. Ruesgas-Ramón, M.; Figueroa-Espinoza, M. C.; Durand, E.; Suárez-Quiroz, M. L.; González-Ríos, O.; Rocher, A.; Reversat, G.; Vercauteren, J.; Oger, C.; Galano, J. M. Identification and Quantification of Phytoprostanes and Phytofurans of Coffee and Cocoa by-and Co-Products. *Food Funct.* **2019,** *10*, 6882–6891.
55. Lecumberri, E.; Mateos, R.; Izquierdo-Pulido, M.; Rupérez, P.; Goya, L.; Bravo, L. Dietary Fiber Composition, Antioxidant Capacity and Physico-Chemical Properties of a Fiber-Rich Product from Cocoa (*Theobroma cacao* L.). *Food Chem.* **2007,** *104*, 948–954.
56. Zainal, B.; Abdah, M.; Taufiq-Yap, Y.; Roslida, A.; Rosmin, K. Anticancer Agents from Non-Edible Parts of *Theobroma cacao*. *Nat. Prod. Chem. Res.* **2014,** *2*, 134–142.
57. Zainal, B.; Abdah, M.; Taufiq Yap, Y.; Roslida, A.; Mohd Redzuan, S.; Kasran, R. Bioactivity-Guided Fractionation of Potent Anti-Cancer Properties from Non-Edible Tissues of *Theobroma cacao*. *Malasyan Cocoa J.* **2016,** *9*, 170–181.
58. Baharum, Z.; Akim, A.; Taufiq-Yap, Y.; Hamid, R.; Kasran, R. In Vitro Antioxidant and Antiproliferative Activities of Methanolic Plant Part Extracts of *Theobroma cacao*. *Molecules* **2014,** *19*, 18317–18331.

59. Rebollo-Hernanz, M.; Zhang, Q.; Aguilera, Y.; Martín-Cabrejas, M. A.; de Mejia, E. G. Cocoa Shell Aqueous Phenolic Extract Preserves Mitochondrial Function and Insulin Sensitivity by Attenuating Inflammation between Macrophages and Adipocytes In Vitro. *Mol. Nutr. Food Res.* **2019,** *63*, 1801413.
60. Nsor-Atindana, J.; Zhong, F.; Mothibe, K. J. In Vitro Hypoglycemic and Cholesterol Lowering Effects of Dietary Fiber Prepared from Cocoa (*Theobroma cacao* L.) Shells. *Food Funct.* **2012,** *3*, 1044–1050.
61. Braojos, C.; Benitez, V.; Rebollo-Hernanz, M.; Cañas, S.; Aguilera, Y.; Arribas, S. M.; Martin-Cabrejas, M. A. Evaluation of the Hypolipidemic Properties of Cocoa Shell after Simulated Digestion Using In Vitro Techniques and a Cell Culture Model of Non-Alcoholic Fatty Liver Disease. *Proceedings* **2020,** *61*, 19.
62. Souza, F. N. S.; Rocha Vieira, S.; Leopoldina Lamounier Campidelli, M.; Abadia Reis Rocha, R.; Milani Avelar Rodrigues, L.; Henrique Santos, P.; de Deus Souza Carneiro, J.; Maria de Carvalho Tavares, I.; Patrícia de Oliveira, C. Impact of Using Cocoa Bean Shell Powder as a Substitute for Wheat Flour on Some of Chocolate Cake Properties. *Food Chem.* **2022,** *381*, 132215.
63. Coulibaly, S. K.; Erbao, C. An Empirical Analysis of the Determinants of Cocoa Production in Cote d'Ivoire. *J. Econ. Struct.* **2019,** *8*, 5.
64. Del Rio Osorio, L.; Flórez-López, E.; David Grande-Tovar, C. The Potential of Selected Agri-Food Loss and Waste to Contribute to a Circular Economy, Applications in the Food, Cosmetic and Pharmaceutical Industries. *Molecules* **2021,** *26*, 515.
65. Panak Balentić, J.; Ačkar, Đ.; Jokić, S.; Jozinović, A.; Babić, J.; Miličević, B.; Šubarić, D.; Pavlović, N. Cocoa Shell: A By-Product with Great Potential for Wide Application. *Molecules* **2018,** *23*, 1404.
66. Esposito, B.; Sessa, M. R.; Sica, D.; Malandrino, O. Towards Circular Economy in the Agri-Food Sector. A Systematic Literatura Review. *Sustainability* **2020,** *12*, 7401.
67. Villegas-Yarlequé, M.; Tirado Kulieva, V. A.; Seminario-Sanz, R.; Camacho-Orbegoso, W.; Calderón-Castillo, B.; Bruno-Coveñas, P. Bibliometric Analysis and Text Mining to Reveal Research Trends on Fruit by-Products under Circular Economy Strategies. *Sustain. Chem. Pharm.* **2023,** *35*, 101232.
68. Korhonen, J.; Honkasalo, A.; Seppälä, J. Circular Economy: The Concept and its Limitations. *Ecol. Econ.* **2018,** *143*, 37–46.
69. Rosa, P.; Sassanelli, C.; Urbinati, A.; Chiaroni, D.; Terzi, S. Assessing Relations between Circular Economy and Industry 4.0: A Systematic Literature Review. *Int. J. Prod. Res.* **2020,** *58*, 1662–1687.
70. Mishra, B.; Mohanta, Y. K.; Reddy, C. N.; Reddy, S. D. M.; Mandal, S. K.; Yadavalli, R.; Sarma, H. Valorization of Agro-Industrial Biowaste to Biomaterials: An Innovative Circular Bioeconomy Approach. *Circ. Econ.* **2023,** *2*, 2773–1677.
71. Socas-Rodríguez, B.; Alvarez-Rivera, G.; Valdés, A.; Ibañez, E.; Cifuentes A. Food by-Products and Food Wastes: Are they Safe Enough for their Valorization? *Trends Food Sci. Technol.* **2021,** *114*, 133–147.
72. de Lopes Sousa Jabbour, A. B.; Jabbour, C. J. C.; Godinho Filho, M.; Roubaud, D. Industry 4.0 and the Circular Economy: A Proposed Research Agenda and Original Roadmap for Sustainable Operations. *Ann. Oper. Res.* **2018,** *270*, 273–286.
73. Grillo, G.; Boffa, L.; Binello, A.; Mantegna, S.; Cravotto, G.; Chemat, F.; Dizhbite, T.; Lauberte, L.; Telysheva, G. Cocoa Bean Shell Waste Valorisation; Extraction from Lab to Pilot-Scale Cavitational Reactors. *Food Res. Int.* **2019,** *115*, 200–208.

74. Machado, C. G.; Winroth, M. P.; Ribeiro da Silva, E. H. D. Sustainable Manufacturing in Industry 4.0: An Emerging Research Agenda. *Int. J. Prod. Res.* **2020,** *58*, 1462–1484.
75. Ntiamoah, A.; Afrane, G. Environmental Impacts of Cocoa Production and Processing in Ghana: Life Cycle Assessment Approach. *J. Clean. Prod.* **2008,** *16*, 1735–1740.
76. Sánchez, M.; Laca, A.; Laca, A.; Díaz, M. Cocoa Bean Shell: A By-Product with High Potential for Nutritional and Biotechnological Applications. *Antioxidants* **2023,** *12*, 1028.
78. Nair, L. G.; Agrawal, K.; Verma, P. An Overview of Sustainable Approaches for Bioenergy Production from Agro-Industrial Wastes. *Energy Nexus* **2022,** *6*, 100086.
79. Sato, M. K.; de Lima, H. V.; Costa, A. N.; Rodrigues, S.; Pedroso, A. J. S.; de Freitas Maia, C. M. B. Biochar from Acai Agroindustry Waste: Study of Pyrolysis Conditions. *Waste Manag.* **2019,** *96*, 158–167.
80. Veloso, M. C. R. A.; Pires, M. R.; Villela, L. S.; Scatolino, M. V.; Protásio, T. P.; Mendes, L. M.; Guimarães Júnior, J. B. Potential Destination of Brazilian Cocoa Agro-Industrial Wastes for Production of Materials with High Added Value. *Waste Manag.* **2020,***118*, 36–44.
81. Wunder, S.; Keighley, M. F.; Martin, H. G.; et al. *Food Waste Prevention and Valorisation: Relevant EU Policy Areas*; Report of the REFRESH Project, D3.3 Review of EU Policy Areas with Relevant Impact on Food Waste Prevention and Valorization, 2018.
82. Stahel, W. R. The Circular Economy. *Nature* **2016,** *531*, 435–438.
83. EU. Directive 2008/98/EC of the European Parliament and of the Council of 19 November 2008 on Waste and Repealing Certain Directives. Official Journal of the European Union, 2008.
84. International Trade Center. List of Exporters for the Selected Product: 1802 Cocoa Shells, Husks, Skins and Other Cocoa Waste; [WWW Document] 2019.
85. More, P. R.; Jambrak, A. R.; Arya, S. S. Green, Environment-Friendly and Sustainable Techniques for Extraction of Food Bioactive Compounds and Waste Valorization. *Trends Food Sci. Technol.* **2022,** *128*, 296–315.
86. Marcolini, P. Tea Made of Cacao-Bean Peels, and Method for Producing Same. Patent 130278, 2011.
87. Brandstetter, B.; Chronopoulos, D.; Jung, C.; Zuurbier, R. Food Comprising Alkalized Cocoa Shells and Method Therefor. Patent EP2174557A1, 2008.
88. Derhy, Y. Cosmetic and/or dermatological composition based on cocoa pulp. Patent FR2849374B1. 2002. https://patents.google.com/patent/FR2849374B1/en?oq=FR+2849374B1.
89. Scampini, A. L. C. Processo de produção de aguardente pela preparação, fermentação e destilação do mel de cacau. BR 102018016061-3 A2. 2020.
90. Lannes, S. C. S.; Da Silva, M. V.; Da Silva, E. N.; Ramos, D. D. C.; Su, F. Composiciones alimenticias de chocolate y helado comestible que contienen "miel de cacao". Universidad de São Paulo (USP, Sao Paulo, Brasil) y Universidad Estadual del Suroeste de Bahía (UESB, Bahía, Brasil), 2013. https://patents.google.com/patent/BR102013005053B1
91. Vásquez, Z. S.; de Carvalho Neto, D. P.; Pereira, G. V.; Vandenberghe, L. P.; de Oliveira, P. Z.; Tiburcio, P. B.; Soccol, C. R. Biotechnological Approaches for Cocoa Waste Management: A Review. *Waste Manage.* **2019,** *90*, 72–83.
92. Okiyama, D. C.; Navarro, S. L.; Rodrigues, C. E. Cocoa Shell and its Compounds: Applications in the Food Industry. *Trends Food Sci. Technol.* **2017,** *63*, 103–112.

93. Sufficiency, E.; Qamar, S. A.; Ferreira, L. F. R.; Franco, M.; Iqbal, H. M.; Bilal, M. Emerging Biotechnological Strategies for Food Waste Management: A Green Leap Towards Achieving High-Value Products and Environmental Abatement. *Energy Nexus* **2022,** *6*, 100077.
94. Guirlanda, C. P.; da Silva, G. G.; Takahashi, J. A. Cocoa Honey: Agro-Industrial Waste or Underutilized Cocoa by-Product?. *Fut. Foods* **2021,** *4*, 100061.
95. O'Donnell, R. W.; O'Malley, J. J.; Huis, R. J.; Halt, G. B. In *Intellectual Property in the Food Technology Industry: Protecting your Innovation*; Springer Science & Business Media, 2008.
96. Rodrigues, C.; Woiciechowski, A. L.; Letti, L. A. J.; Karp, S. G.; Goelzer, F. D.; Sobral, K. C. A.; Soocol, C. R. Materiais lignocelulósicos como matéria-prima para a obtenção de biomoléculas de valor comercial. *Biotecnologia aplicada a agroindústria* **2017,** *4*, 283–314.
97. Martínez, R.; Torres, P.; Meneses, M. A.; Figueroa, J. G.; Pérez-Álvarez, J. A.; Viuda-Martos, M. Chemical, Technological and In Vitro Antioxidant Properties of Cocoa (*Theobroma cacao* L.) Co-Products. *Food Res. Int.* **2012,** *49* (1), 39–45.
98. International Trade Center. List of Exporters for the Selected Product: 1802 Cocoa Shells, Husks, Skins and Other Cocoa Waste, 2019.
99. Islam, M. A.; Yousuf, A.; Karim, A.; Pirozzi, D.; Khan, M. R.; Wahid, Z. A. Bioremediation of Palm Oil Mill Effluent and Lipid Production by *Lipomyces starkeyi*: A Combined Approach. *J. Clean. Prod.* **2018,** *172*, 1779–1787.
100. Ngaosuwan, K. Solid Acid Catalyst Derived from Coffee Residue for Biodiesel Production. In *Renewable Energy in the Service of Mankind*; Sayigh, A., Ed.; Springer: Cham, 2015; pp 47–55.
101. Scapagnini, G.; Davinelli, S.; Di Renzo, L.; De Lorenzo, A.; Olarte, H. H.; Micali, G.; Cicero, A. F.; Gonzalez, S. Cocoa Bioactive Compounds: Significance and Potential for the Maintenance of Skin Health. *Nutrients* **2014,** *6*, 3202–3213.
102. Franzen, M.; Borgerhoff Mulder, M. Ecological, Economic and Social Perspectives on Cocoa Production Worldwide. *Biodivers. Conserv.* **2007,** *16*, 3835–3849.
103. Beg, M. S.; Ahmad, S.; Jan, K.; Bashir, K. Status, Supply Chain and Processing of Cocoa-A Review. *Trends Food Sci. Technol.* **2017,** *66*, 108–116.
104. Instituto Interamericano de Cooperación para la Agricultura. [Online] 2023. https://iica.int/en/press/news/international-cocoa-organisation-and-iica-launched-initiative-address-cadmium-levels
105. Tennhardt, L.; Lazzarini, G.; Weisshaidinger, R.; Schader, C. Do Environmentally-Friendly Cocoa Farms Yield Social and Economic Co-Benefits?. *Ecol. Econ.* **2022,** *197*, 107428.
106. Benneh, E. Y.; Anaman, K. A. Economic Shocks and the Growth of the Ghanaian Cocoa Industry from 1975 to 2019. *Ghana J. Dev. Stud.* **2022,** *19* (2), 57–82.

CHAPTER 7

Analysis of Spatiotemporal Changes in Vegetation of Lower Bhavani Basin Using Google Earth Engine: A Case Study for Natural Resource Management

JANANI, N.[1], BALAJI KANNAN[2], NIRMALA DEVI M[2], and SHAHEEMATH SUHARA K. K.[1*]

[1]*Department of Soil and Water Conservation Engineering, Agricultural Engineering College and Research Institute, Tamil Nadu Agricultural University, Coimbatore, Tamil Nadu, India*

[2]*Department of Physical Science and Information Technology, Agricultural Engineering College and Research Institute, Tamil Nadu Agricultural University, Coimbatore, Tamil Nadu, India*

*Corresponding author.

ABSTRACT

Vegetation has undergone phenomenal changes globally due to the effects of climate change and anthropogenic interruptions. Vegetation plays a vital role in crop production and food security at regional and national scales, and it influences the global climate, surface temperature, and crop phenology. To generate large-scale vegetation maps, Landsat 8 images at 30-m spatial resolution were used for the Lower Bhavani basin, and a time series graph was generated for 2015–2021 using the Google Earth Engine (GEE) platform. In GEE, vegetation values were calculated using the Normalized Difference Vegetation Index (NDVI), and the algorithm generated a vegetation map over the region. The vegetation values were compared with climatic factors such as precipitation, temperature, and soil surface wetness. The notable correlation was

Agroindustrial Waste Management and Natural Resources Conservation: Environmental Challenges and Transformation into Value-Added Products. José Juan Buenrostro-Figueroa, Juan Manuel Tirado-Gallegos, Mónica L. Chávez-González, A. K. Haghi, and Shaheemath Suhara K. K. (Eds.)

found between vegetation and precipitation over the basin. The annual and seasonal intervariability analyses were studied. In the annual NDVI pattern, the lowest NDVI value was observed in 2016, and the highest was in 2020. Similarly, the month-wise NDVI pattern represented that the month of April may show the lowest NDVI and the month of October showed higher NDVI values.

7.1 INTRODUCTION

Monitoring and assessment of earth surface is a crucial requirement for global surface change detection.[19] The growing population influences land use changes by reducing the natural vegetation in many countries. Urbanization brings socioeconomic changes and expresses a distinct set of land use and human behaviors. It can affect the production of crops indirectly by changing the vegetation environment.[4] Vegetation is a component in soil–water–atmosphere system that serves as a good indicator corresponding to the changes in the environmental, climatic, and hydrological factors such as precipitation, temperature, evapotranspiration, and soil moisture through various monitoring indices.[11]

Mapping vegetation is a technical task for natural resource management and plays an essential role in affecting global climate change.[23,32,33] Satellite-derived vegetation indices monitor the Earth's vegetative surface and plant ecology. There are many remote sensing indices available for calculating vegetation over the Earth surface. One such popular index is the normalized difference vegetation index (NDVI), which is widely implemented for earth surface vegetation monitoring. The density of green in the area can be estimated with NDVI, and its value ranges from − 1 and +1. NDVI time series are increasingly used to monitor vegetation dynamics.[1] When the index is related to climatic variables, important details of climate-ecosystem interactions can be inferred.

The effect of variations in precipitation influences the vegetation, especially in semi-arid regions. Water availability is a principle and ultimate determinant of the distribution and composition of vegetation. The spatial and temporal distribution of rainfall is highly unpredictable in semi-arid regions, and it is not easy to generalize the relationship between precipitation and vegetation.[2,16] Unlike precipitation, temperature range varies seasonally and it is complex to build a connection between temperature and vegetation.[1,34] The function of surface soil wetness determines the hydrological properties and soil moisture holding capacity.[25] Since the climatic factor affects the vegetation over the region, it's hard to determine which variable has highly influenced the plant phenology.

7.2 MATERIALS AND METHODS

7.2.1 STUDY AREA DESCRIPTION

The study area was the Lower Bhavani basin (Fig. 7.1), covering the area of 2424 km^2 in Erode district of Tamil Nadu state, India. The basin lies between 11°15′ and 11°45′ North latitudes and 77°00′ and 77°40′ East longitudes. The average annual temperature of the Basin varies from 22 to 40°C. The area has an average annual rainfall of about 740 mm, of which most of the rainfall occurs from Southwest monsoon. The elevation ranges from 66 to 1473 m. The basin has the major land uses of forest/hills, agricultural land, built-up areas, and riverbed regions. The study area was dominated by sandy clay loam, sandy clay, and loamy sand texture.

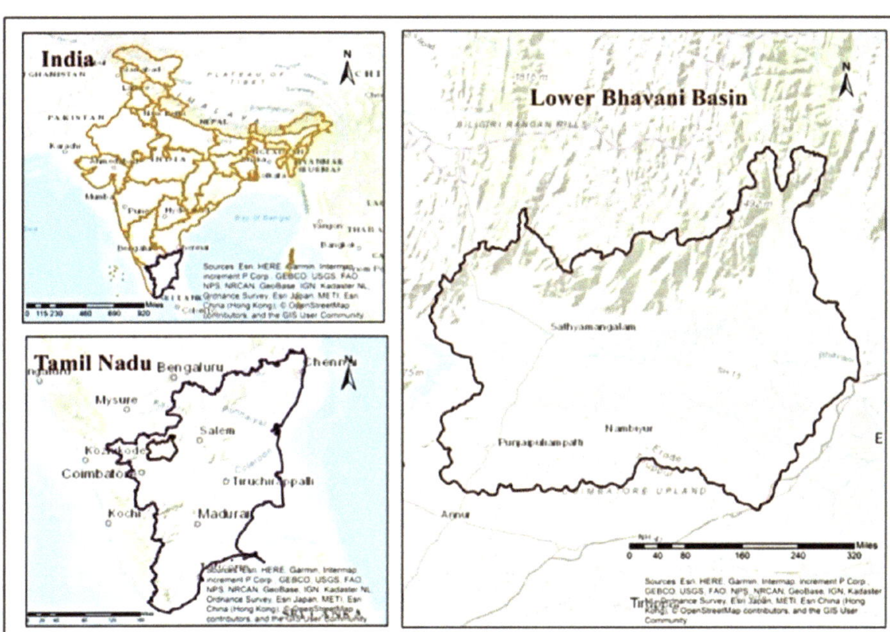

FIGURE 7.1 Study area map of Lower Bhavani Basin.

7.2.2 GOOGLE EARTH ENGINE (GEE)

Google Earth Engine (GEE) is a cloud-based platform that makes it easy to access high-performance computing resources for processing very large geospatial datasets. It helps to store and process enormous remote sensing datasets for open-source use and other ready-to-use products.[17,24] GEE has wide areas of application in monitoring and mapping vegetation, water resource, land use, agriculture, environment and earth science, forest cover change, and crop-yield estimation.[5,8,12,22,26,28,29] The GEE platform makes use of Google's computational infrastructure to enable parallel geospatial data processing to reduce computational time.[24] The code editor is a feature of GEE, which is available in a web-based Integrated Development Environment (IDE) designed for writing, developing, and running complex scripts applying JavaScript application programing interface (API).[10,31] In this study, the GEE platform was used for the estimation of vegetation indices and to chart the variations of vegetation value over the time period.

7.2.3 VEGETATIVE INDEX (VI)

Vegetative index is the property with two or more spectral reflectance wavelength combinations to accentuate vegetation over the region. NDVI is one such index that helps to understand the vegetation phenology response to short-term meteorological conditions.[18] The NDVI value varies between –1 to +1 depending upon the different land use classes captured in the sensor. Generally, higher NDVI stands for healthier vegetation with high chlorophyll content; NDVI more than 0.2 shows vegetation; positive vegetation up to 0.2 shows urban and barren land; negative NDVI values indicate the presence of water bodies.[3,7] NDVI for any satellite images is computed using near-infrared (NIR) band and RED band, and the formula for calculating NDVI is expressed as follows:

$$\text{NDVI} = \frac{NIR - RED}{NIR + RED} \tag{7.1}$$

7.2.4 WORKFLOW

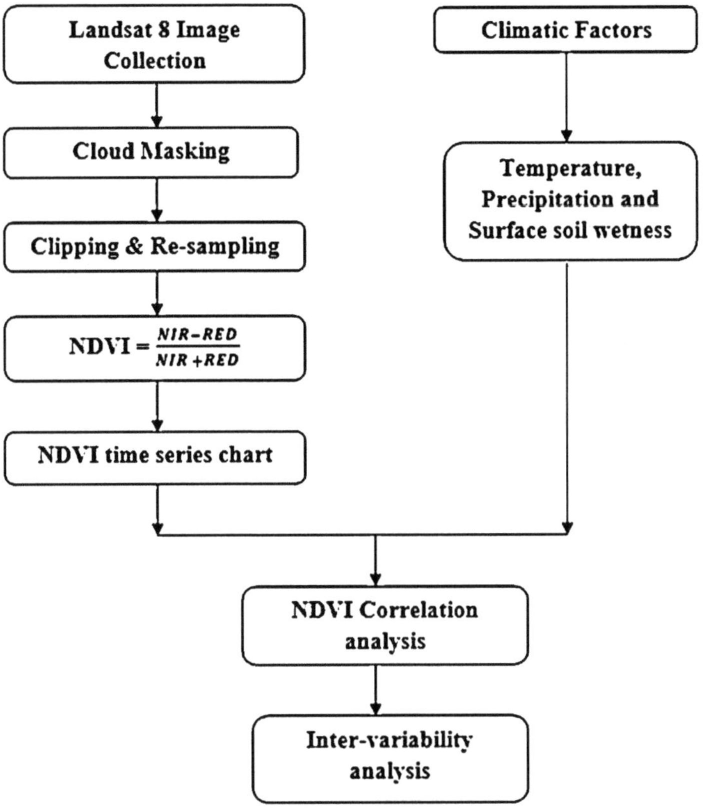

FIGURE 7.2 Methodology workflow.

7.2.5 NDVI TIME SERIES ANALYSIS IN GOOGLE EARTH ENGINE

NDVI analysis for the Lower Bhavani basin was carried out for the period 2015–2021. Monthly satellite images were taken, and NDVI values were calculated for the time period. The values of NDVI vary based on the precipitation, crop sowing, and harvest stages.[13,30] In GEE, the Landsat 8 image collection was imported, and with the help of a specific band collection (eq 7.1), NDVI has been calculated. After calculating NDVI, the commands have been provided to create a time series graph for each image collection. Finally, commands have been added for NDVI data and plot as map layers to the Google Maps base map.

7.2.6 HYDROCLIMATIC FACTORS

The hydroclimatic factor tends to influence the vegetation of the area. The variability of NDVI is directly related to climatic change, especially rainfall and temperature.[6] In this study, hydroclimatic factors such as precipitation, temperature, and surface soil wetness have been considered and correlated with NDVI values. The changes in NDVI values based on the variations in climatic factors on technological advances in remote sensing data were studied.

Figure 7.3 shows the interannual variability of mean monthly precipitation, temperature, and surface soil wetness for 7 years (2015–2021). The values of climatic factors were based on the mean monthly values of the Lower Bhavani Basin. Observing at a seasonal scale, the total maximum temperature was exhibited in the months of March and April, and maximum rainfall in the months of October, November, and December. However, surface soil moisture was higher in the maximum rainfall period.

7.3 RESULTS AND DISCUSSION

7.3.1 NDVI

In GEE, the NDVI was mapped for the period of 2015–2021. The Landsat 8 image collection was imported to the GEE. The Lower Bhavani basin was given as the geometry of extent for calculating NDVI. The images of every year were calculated by extracting the median values of each image from 2015 to 2021 and the images were classified. The range of classification of NDVI varied from 0 to 1. Figure 7.4 shows the spatial plots of long-term mean yearly NDVI images.

7.3.2 NDVI TIME SERIES ANALYSIS

Monthly NDVI values were plotted as a chart (Fig. 7.5) for every year in GEE. GEE provides the continuous line chart for the monthly NDVI values. The NDVI values were higher every year during the monsoon period. This is because the study area experiences both North East and South West monsoon. Comparing the year-wise composite NDVI data, the duration between 2016 and 2017 registered lower values, and 2015, 2019, 2020, and 2021 registered higher range of NDVI values especially during monsoon.

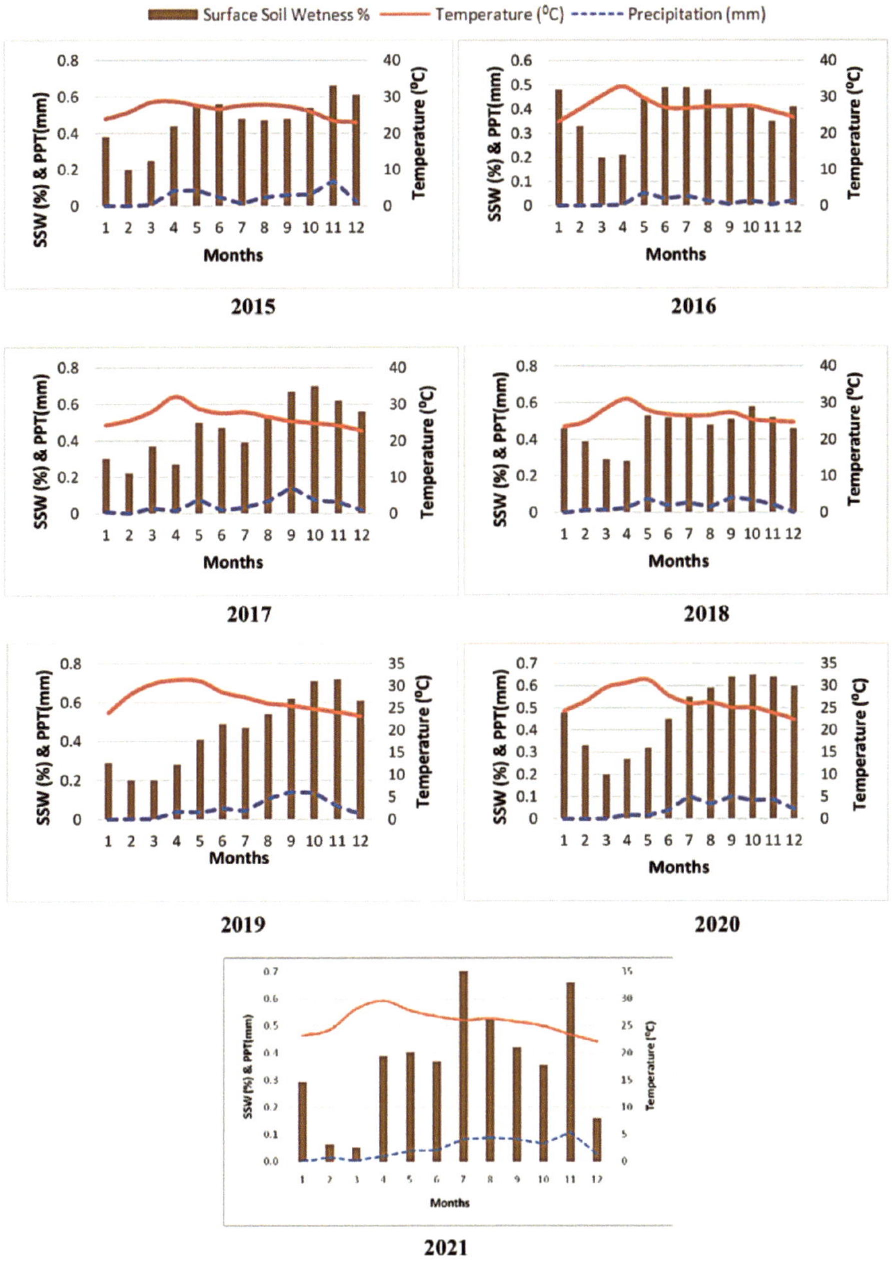

FIGURE 7.3 Mean monthly precipitation, temperature, and surface soil wetness for the years 2015–2021.

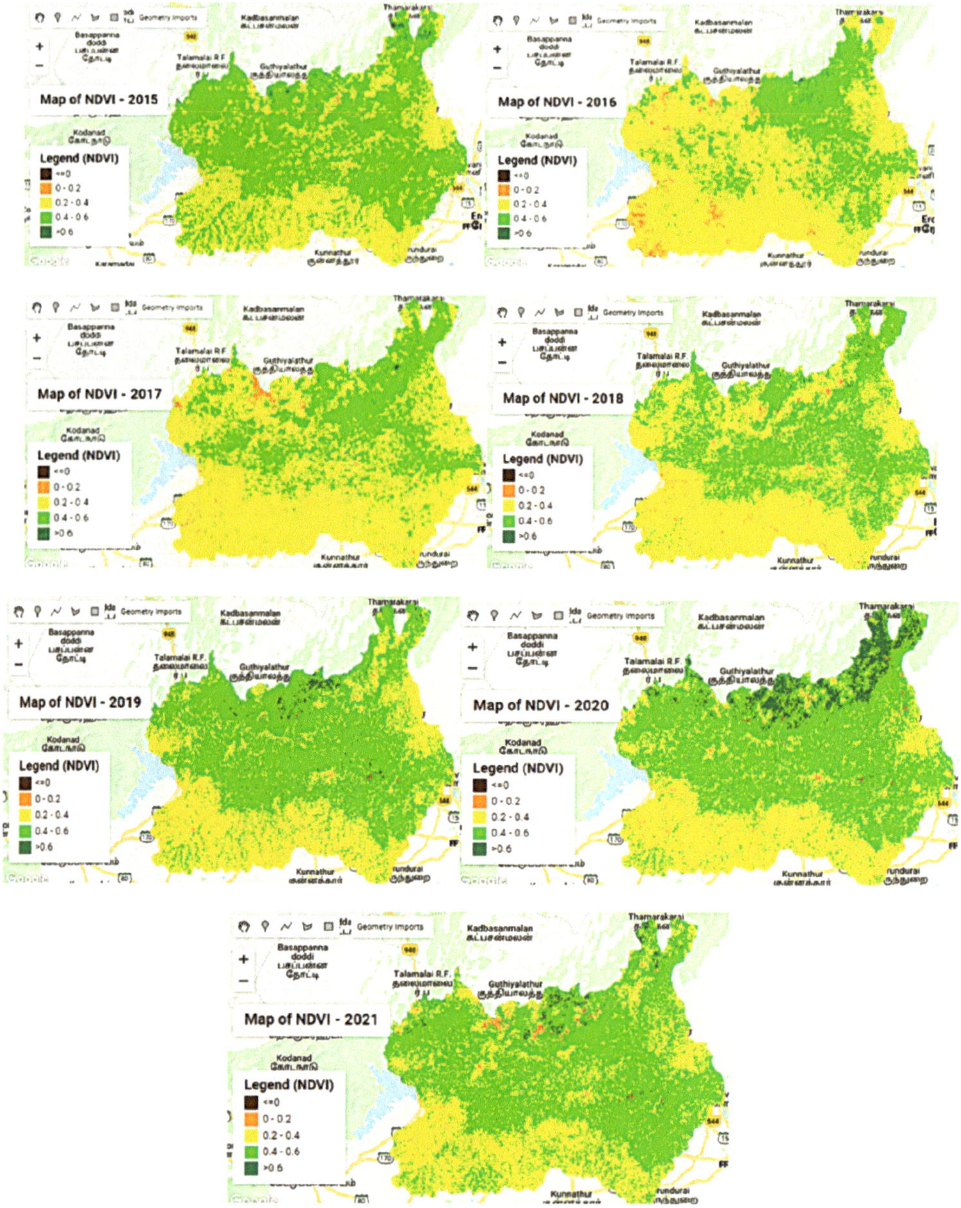

FIGURE 7.4 Yearly mean NDVI images.

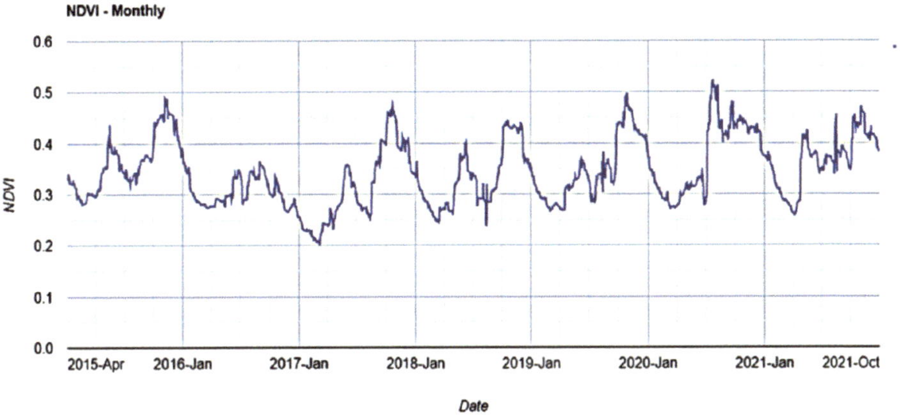

FIGURE 7.5 NDVI time series chart.

7.3.3 NDVI AND CLIMATIC FACTOR

The relationship of NDVI with precipitation, temperature, and surface soil wetness was plotted. The precipitation and NDVI yielded very little correlation of $r^2 = 0.174$ (Fig. 7.6), as there was fluctuation and unpredictable rainfall in the last few decades. Studies observed that there was variability of rainfall and extreme rainfall events in Tamil Nadu and India over the years.[15,20,21] The temperature and NDVI correlation showed a satisfactory value of $r^2 = 0.506$ (Fig. 7.7). The strength of the correlation between temperature and NDVI varies based on the season, time of day, and land cover.[14] On the contrary, surface soil wetness and NDVI showed a moderate correlation value of $r^2 = 0.444$ (Fig. 7.8). Surface soil wetness is otherwise soil moisture in the soil has influenced the vegetation over the study area.

7.3.4 INTERVARIABILITY ANALYSIS

7.3.4.1 INTERANNUAL VARIABILITY OF NDVI

The interannual variability of vegetation index was plotted for the duration 2015–2021. The temporal variation of NDVI was represented in the form of a box plot for every year (Fig. 7.9). The boxplot corresponding to the annual NDVI temporal patterns shows the NDVI fluctuation between the years 2015–2021. The NDVI for 2016 and 2017 was lower compared to the other years. The reason was that 2016 and 2017 faced severe drought in the state

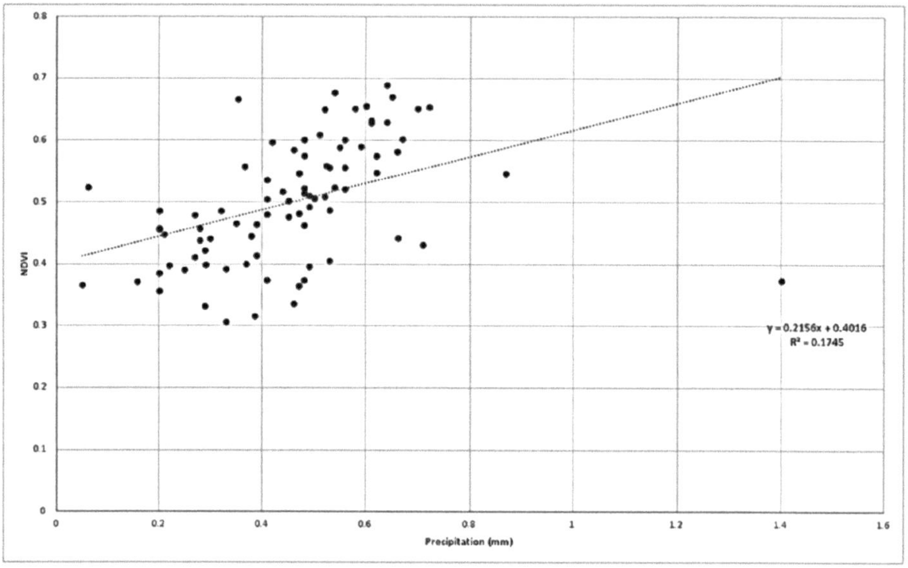

FIGURE 7.6 Relationship of mean temporal NDVI plotted and precipitation.

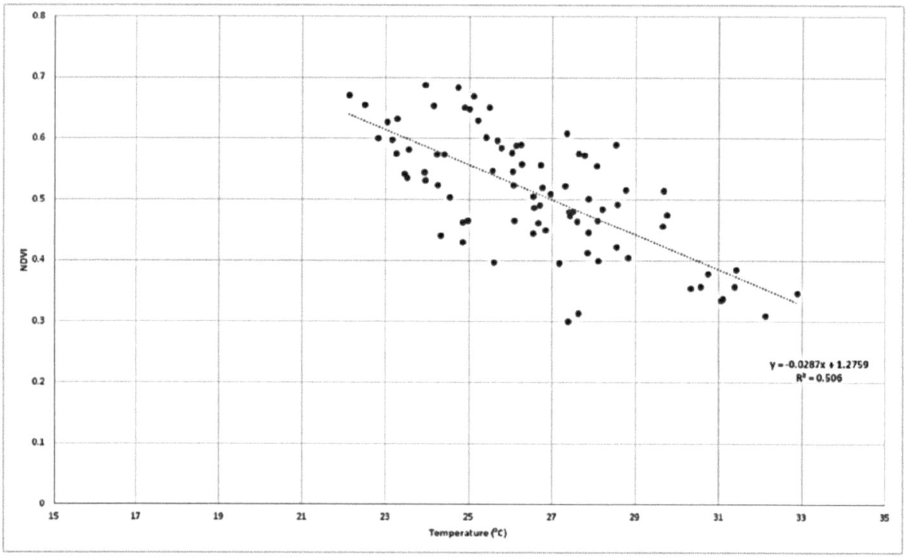

FIGURE 7.7 Relationship of mean temporal NDVI plotted and temperature.

of Tamil Nadu.[9,27] It was observed that the maximum (~0.7) and minimum NDVI values (~0.33) occurred in the years 2019 and 2016, respectively. The 25th percentile of NDVI values and the 75th percentile of NDVI values

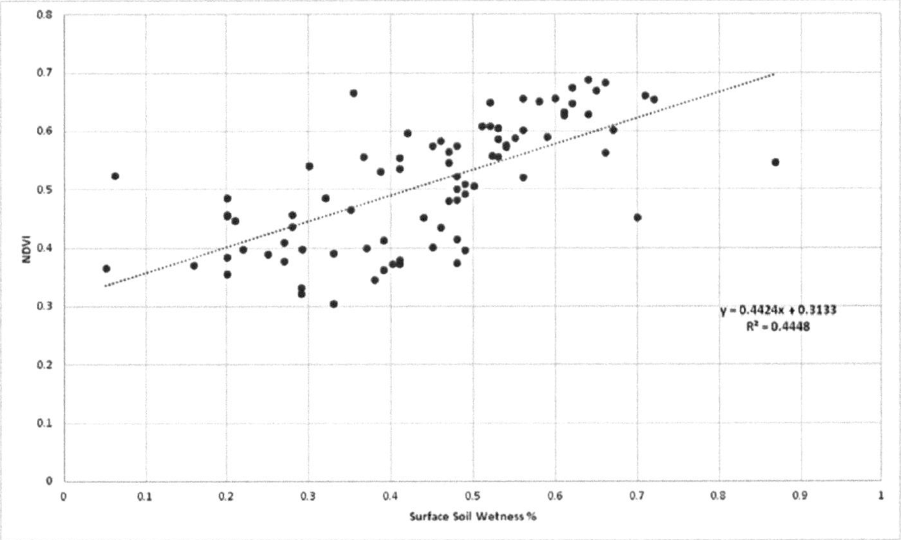

FIGURE 7.8 Relationship of mean temporal NDVI plotted and Surface soil wetness.

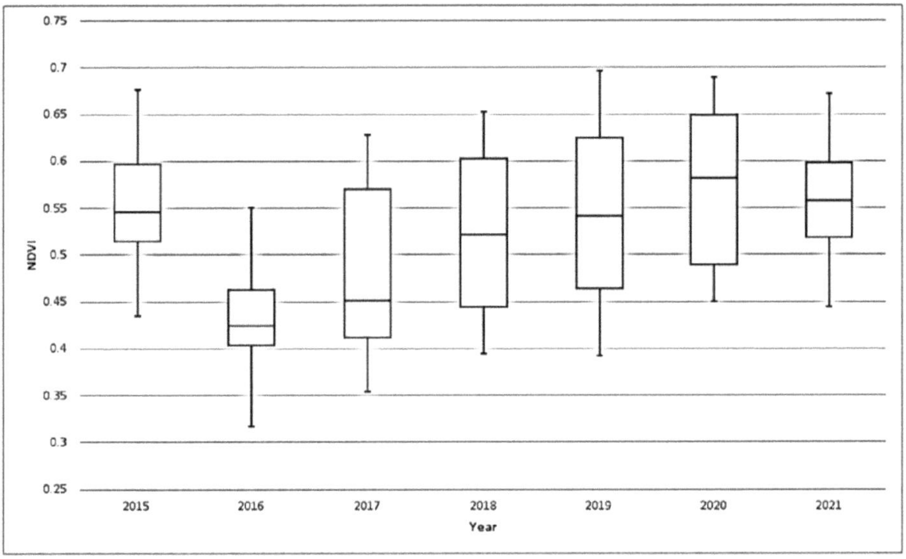

FIGURE 7.9 Boxplot for the interannual variability of NDVI over the duration 2005–2021.

varied from ~0.41 to ~0.65, respectively, during 2015–2021. In addition, the 25th percentile of the NDVI ranges from ~0.46 to 0.65, while the 75th percentile of the enhanced vegetation index (EVI) ranges from ~0.35 to

~0.43. On the contrary, the time series shown in Figure 7.5 indicates that NDVI values were always higher than ~0.4, mostly suggesting that the basin had sufficiently higher vegetation greenness throughout those years.[11]

7.3.4.2 SEASONAL VARIABILITY OF NDVI

The seasonal variability of NDVI for the Lower Bhavani basin was represented in the form of a boxplot from 2015–2021. The mean monthly NDVI values are shown in Figure 7.10. It was observed that the NDVI values have their seasonal peak in October, while March, April, and May were observed to have the lowest NDVI mean values in the basin. It was also observed that the vegetation greening usually starts from June and reaches its maximum in October, and there after starts decreasing till the next March. Monthly NDVI values have fewer fluctuations, especially in the summer and winter seasons.

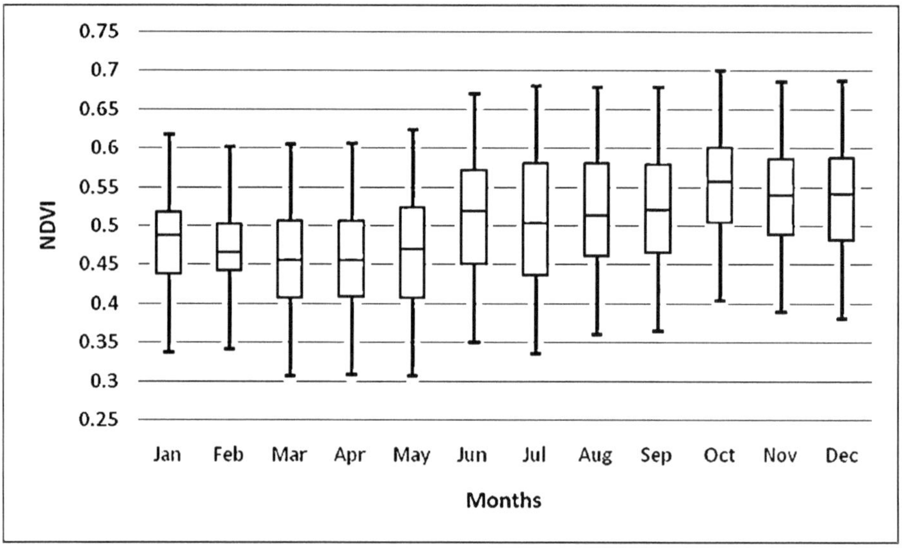

FIGURE 7.10 Boxplot for the seasonal variability of NDVI over the period 2005–2021.

7.4 CONCLUSION

In this study, variations in vegetation values were studied for the Lower Bhavani basin using GEE for the period 2015–2021. The study also focused on the influences of climatic factors on vegetation values over the time period. The monitoring of vegetation over the larger basin is made possible through a remote sensing approach. For the Lower Bhavani basin, the Landsat data were used for calculating the NDVI in GEE and thereafter used the datasets such as precipitation, temperature, and surface soil wetness to understand the influence of hydroclimatic conditions over the vegetation in the basin. In the annual NDVI pattern, the lowest NDVI value was observed in 2016, and the highest was observed in 2020. Similarly, the month-wise NDVI pattern represented that the months of April may show the lowest NDVI and the month of October showed higher NDVI values. The result of the study helps in ecological environment protection and water resource management in the Lower Bhavani basin.

KEYWORDS

- **vegetation**
- **Landsat 8**
- **Lower Bhavani basin**
- **Google Earth Engine**
- **hydroclimatic factors**

REFERENCES

1. Bradley, B. A.; Blumenthal, D. M.; Wilcove, D. S.; Ziska, L. H. Predicting Plant Invasions in an Era of Global Change. *Trends Ecol. Evol.* **2010,** *25*, no. 5, 310–318.
2. Buyantuyev, A.; Wu, J. Urbanization Diversifies Land Surface Phenology in Arid Environments: Interactions Among Vegetation, Climatic Variation, and Land Use Pattern in the Phoenix Metropolitan Region, USA. *Landsc. Urban Plan.* **2012,** *105*, nos. 1/2, 149–159.
3. Chen, X. L.; Zhao, H. M.; Li, P. X.; Yin, Z. Y. Remote Sensing Image-Based Analysis of the Relationship Between Urban Heat Island and Land Use/Cover Changes. *Remote Sens. Environ.* **2006,** *104*, no. 2, 133–146.
4. Chen, Y.; Huang, B.; Zeng, H. How Does Urbanization Affect Vegetation Productivity in the Coastal Cities of Eastern China? *Sci. Total Environ.* **2022,** *811*, 152356.
5. Feizizadeh, B.; Omarzadeh, D.; Kazemi Garajeh, M.; Lakes, T.; Blaschke, T. Machine Learning Data-Driven Approaches for Land Use/Cover Mapping and Trend Analysis Using Google Earth Engine. *J. Environ. Plan. Manag.* **2021,** *66*, 665–697.
6. Ghebrezgabher, M. G.; Yang, T.; Yang, X.; Sereke, T. E. Assessment of NDVI Variations in Responses to Climate Change in the Horn of Africa. *Egypt. J. Remote Sens. Space Sci.* **2020,** *23*, no. 3, 249–261.

7. Guha, S.; Govil, H. An Assessment on the Relationship Between Land Surface Temperature and Normalized Difference Vegetation Index. *Environ. Dev. Sustain.* **2021,** *23*, no. 2, 1944–1963.
8. Hird, J. N.; DeLancey, E. R.; McDermid, G. J.; Kariyeva, J. Google Earth Engine, Open-Access Satellite Data, and Machine Learning in Support of Large-Area Probabilistic Wetland Mapping. *Remote Sens.* **2017,** *9*, no. 12, 1315.
9. Kumar, K. A.; Reddy, G. O.; Masilamani, P.; Turkar, S. Y.; Sandeep, P. Integrated Drought Monitoring Index: A Tool to Monitor Agricultural Drought by Using Time-Series Datasets of Space-Based Earth Observation Satellites. *Adv. Space Res.* **2021,** *67*, no. 1, 298–315.
10. Kumar, L.; Mutanga, O. Google Earth Engine Applications Since Inception: Usage, Trends, and Potential. *Remote Sens.* **2018,** *10*, no. 10, 1509.
11. Kumari, B.; Tayyab, M.; Ahmed, I. A.; Baig, M. R. I.; Khan, M. F.; Rahman, A. Longitudinal Study of Land Surface Temperature (LST) Using Mono-and Split-Window Algorithms and its Relationship with NDVI and NDBI Over Selected Metro Cities of India. *Arab. J. Geosci.* **2020,** *13*, no. 19, 1–19.
12. Langner, A.; Miettinen, J.; Kukkonen, M.; Vancutsem, C.; Simonetti, D.; Vieilledent, G.; Verhegghen, A.; Gallego, J.; and Stibig, H. J. Towards Operational Monitoring of Forest Canopy Disturbance in Evergreen Rain Forests: A Test Case in Continental Southeast Asia. *Remote Sens.* **2018,** *10*, no. 4, 544.
13. Liu, W. T.; Kogan, F. Monitoring Brazilian Soybean Production Using NOAA/AVHRR Based Vegetation Condition Indices. *Int. J. Remote Sens.* **2002,** *23*, no. 6, 1161–1179.
14. Marzban, F.; Sodoudi, S.; Preusker, R. The Influence of Land-Cover Type on the Relationship Between NDVI–LST and LST-T Air. *Int. J. Remote Sens.* **2018,** *39*, no. 5, 1377–1398.
15. Mertens, A.; Balakrishnan, K.; Ramaswamy, P.; Rajkumar, P.; Ramaprabha, P.; Durairaj, N.; Hubbard, A. E.; Khush, R.; Colford Jr, J. M.; and Arnold, B. F. Associations Between High Temperature, Heavy Rainfall, and Diarrhea Among Young Children in Rural Tamil Nadu, India: A Prospective Cohort Study. *Environ. Health Perspect.* **2019,** *127*, no. 4, 047004.
16. Mo, K.; Chen, Q.; Chen, C.; Zhang, J.; Wang, L.; Bao, Z. Spatiotemporal Variation of Correlation Between Vegetation Cover and Precipitation in an Arid Mountain-Oasis River Basin in Northwest China. *J. Hydrol.* **2019,** *574*, 138–147.
17. Mutanga, O.; Kumar, L. Google Earth Engine Applications. *Remote Sens.* **2019,** *11*, no. 5, 591.
18. Nagai, S.; Nasahara, K. N.; Muraoka, H.; Akiyama, T.; Tsuchida, S. Field Experiments to Test the Use of the Normalized-Difference Vegetation Index for Phenology Detection. *Agric. Forest Meteorol.* **2010,** *150*, no. 2, 152–160.
19. Pail, R.; Bingham, R.; Braitenberg, C.; Dobslaw, H.; Eicker, A.; Güntner, A.; Horwath, M.; Ivins, E.; Longuevergne, L.; Panet, I.; Wouters, B. Science and User Needs for Observing Global Mass Transport to Understand Global Change and to Benefit Society. *Surv. Geophys.* **2015,** *36*, no. 6, 743–772.
20. Pattanaik, D. R.; Rajeevan, M. Variability of Extreme Rainfall Events Over India During Southwest Monsoon Season. *Meteorol. Appl.* **2010,** *17*, no. 1, 88–104.
21. Sanap, S. D.; Priya, P.; Sawaisarje, G. K.; Hosalikar, K. S. Heavy Rainfall Events Over Southeast Peninsular India During Northeast Monsoon: Role of El Niño and Easterly Wave Activity. *Int. J. Climatol.* **2019,** *39*, no. 4, 1954–1968.

22. Shelestov, A.; Lavreniuk, M.; Kussul, N.; Novikov, A.; Skakun, S. *Large Scale Crop Classification Using Google Earth Engine Platform*. 2017 IEEE International Geoscience and Remote Sensing Symposium (IGARSS), IEEE, 2017; pp. 3696–3699.
23. Tallis, H.; Polasky, S. Mapping and Valuing Ecosystem Services as an Approach for Conservation and Natural-Resource Management. *Ann. N. Y. Acad. Sci.* **2009,** *1162*, no. 1, 265–283.
24. Tamiminia, H.; Salehi, B.; Mahdianpari, M.; Quackenbush, L.; Adeli, S.; Brisco, B. Google Earth Engine for Geo-Big Data Applications: A Meta-Analysis and Systematic Review. *ISPRS J. Photogramm. Remote Sens.* **2020,** *164*, 152–170.
25. Terleev, V. V.; Topaj, A. G.; Mirschel, W. The Improved Estimation for the Effective Supply of Productive Moisture Considering the Hysteresis of Soil Water-Retention Capacity. *Russ. Meteorol. Hydrol.* **2015,** *40*, no. 4, 278–285.
26. Tsai, Y. H.; Stow, D.; Chen, H. L.; Lewison, R.; An, L.; Shi, L. Mapping Vegetation and Land Use Types in Fanjingshan National Nature Reserve Using Google Earth Engine. *Remote Sens.* **2018,** *10*, no. 6, 927.
27. Venkadesh, S.; Pazhanivelan, S.; Ragunath, K. P.; Kumaraperumal, R.; Panneerselvam, S.; Sathy, R. Assessment of Agricultural Drought Using MODIS NDVI Based Vegetation Status for Different Agro Climatic Zones of Tamil Nadu. *Int. J. Curr. Microbiol. Appl. Sci.* **2017,** *8*, no. 5, 2019.
28. Venkatappa, M.; Sasaki, N.; Shrestha, R. P.; Tripathi, N. K.; Ma, H. O. Determination of Vegetation Thresholds for Assessing Land Use and Land Use Changes in Cambodia Using the Google Earth Engine Cloud-Computing Platform. *Remote Sens.* **2019,** *11*, no. 13, 1514.
29. Wang, C.; Jia, M.; Chen, N.; Wang, W. Long-Term Surface Water Dynamics Analysis Based on Landsat Imagery and the Google Earth Engine platform: A Case Study in the Middle Yangtze River Basin. *Remote Sens.* **2018,** *10*, no. 10, 1635.
30. Wang, R.; Cherkauer, K.; Bowling, L. Corn Response to Climate Stress Detected with Satellite-Based NDVI Time Series. *Remote Sens.* **2016,** *8*, no. 4, 269.
31. Wimberly, M. C.; Nekorchuk, D. M.; Kankanala, R. R. Cloud-Based Applications for Accessing Satellite Earth Observations to Support Malaria Early Warning. *Sci. Data* **2022,** *9*, no. 1, 1–11.
32. Wu, D.; Zhao, X.; Liang, S.; Zhou, T.; Huang, K.; Tang, B.; Zhao, W. Time-lag Effects of Global Vegetation Responses to Climate Change. *Global Change Biol.* **2015,** *21*, no. 9, 3520–3531.
33. Xiao, X.; Zhang, Q.; Braswell, B.; Urbanski, S.; Boles, S.; Wofsy, S.; Moore, B.; and Ojima, D. Modeling Gross Primary Production of Temperate Deciduous Broadleaf Forest Using Satellite Images and Climate Data. *Remote Sens. Environ.* **2004,** *91*, no. 2, 256–270.
34. Yu, F.; Price, K. P.; Ellis, J.; Shi, P. Response of Seasonal Vegetation Development to Climatic Variations in Eastern Central Asia. *Remote Sens. Environ.* **2003,** *87*, no. 1, 42–54.

CHAPTER 8

Plastic Waste Generation and Management in Agroindustial Packaging Industries: Perspectives on Emerging Approaches

YURIDIA ORTIZ-RIVERA[1], CLAUDIA CAROLINA HERNÁNDEZ-PEÑA[1], COYOLXAUHQUI FIGUEROA[2*], and ANLUI RODRÍGUEZ-CARREÓN[3]

[1]*Programa de Biotecnología, Departamento de Ciencias Químico Biológicas, Instituto de Ciencias Biomédicas, Universidad Autónoma de Ciudad Juárez, Ciudad Juárez, Chihuahua, México*

[2]*Programa de Biología. Herbario UACJ, Unidad de Producción de plantas, México*

[3]*Independent Researcher, Av. Benjamín Franklin #4650, Zona PRONAF, Ciudad Juárez, 32310, Chihuahua, México*

*Corresponding author.

ABSTRACT

Plastic supplies are widely used in modern life due to their versatility and low cost; however, a deficient control in their disposal and the lack of effective treatments for most of them have led to severe environmental problems. In the world, only 9% of plastic is recycled; therefore, the search for new and more sustainable plastic waste management solutions is now a priority. In this regard, biological treatment has become an attractive option. The use of living organisms for plastic degradation shows notable advantages such as utilizing waste as a substrate for biomass production, avoiding the use of potentially toxic chemicals, obtaining value-added by-products and reducing treatment costs. This chapter aims to analyze and discuss the most

Agroindustrial Waste Management and Natural Resources Conservation: Environmental Challenges and Transformation into Value-Added Products. José Juan Buenrostro-Figueroa, Juan Manuel Tirado-Gallegos, Mónica L. Chávez-González, A. K. Haghi, and Shaheemath Suhara K. K. (Eds.)

recent advances in bacterial, fungal, and arthropodal treatment technologies, emphasizing on their economic and technical feasibility for large-scale applications.

8.1 INTRODUCTION

It is hard to imagine our modern life without plastic since this material is the main component of millions of daily-use products. This exaggerate dependence on plastic polymers has caused an excessive increase in their production, generating negative impacts on human and environmental health.[40] The presence of macro and micro plastics has affected marine fauna through the ingestion of these compounds, which will consequently enter to the trophic chains. Moreover, the amount of greenhouse gas (GG) released to the atmosphere during plastics' life cycle represents an important contribution to global warming.[93]

Currently, plastic waste management is implemented through different strategies, being landfill disposal the most common one. Due to their low degradation rate, plastics tend to accumulate in landfills, reaching the open environment through lixiviation or mechanical traction. In order to mitigate these accumulation problems and its significant use of land, plastic management can also be addressed by different techniques. Incineration offers the advantage of not using space for plastic waste storage and the possibility of energy production, but the high grade of technology required and the emission of toxic compounds and GG of the process are issues that must be taken into consideration. On the other side, plastics can be recycled, although the presence of additives and impurities in plastic items can complicate this procedure, making it more expensive; therefore, only a small percentage of plastic waste is suitable for recycling. Finally, biological treatment, also known as biodegradation or biorecycling, is another promising emergent technique for plastic waste management, and it is perhaps the most environmentally friendly option nowadays.[126]

The development of new plastic materials with better characteristics, and the enhancement of biorecycling techniques of conventional synthetic polymers seems to be the optimal path for environmental conservation. However, despite the numerous reports that corroborate the technical feasibility of biological treatments, large-scale applications have been scarce and most studies have been applied in controlled laboratory conditions. Thus, in spite of its high potential, the need of specialized manpower, plus the still elevated operation cost have made of biodegradation a technology that requires much more research efforts before it becomes present in the plastic waste markets.[78]

The main goal of this chapter is to discuss the status of plastic pollution issues and their impacts in environmental and human health, analyzing and comparing the available options for its treatment. Special emphasis will be made in plastic biorecycling since it is considered the most ecologically favorable alternative. Finally, the importance of scientific research and technology development will be highlighted through the analysis of several studies that propose biodegradation as the potential solution for the contemporary plastic waste crisis.

8.2 PLASTIC USAGE AND ACCUMULATION

Plastic-based products are present in virtually every area of our lives, and they have become indispensable in many cases. However, our high grade of dependency for plastics has exacerbated their production, leading to severe environmental problems and to a major increase of human health risks.

Plastics can help us store food safely, isolate buildings and constructions, protect and use electronics, make clothing, among many other applications. There are literally millions of daily-use articles that can be fabricated with plastic materials. Despite the undeniable benefits that plastic polymers can offer, our society's carbon footprint has risen significantly as a consequence of the enormous amount of plastics that are consumed on a daily basis. This carbon footprint is mainly related to the highly contaminating plastic productions processes and the generally deficient plastic waste management systems.[87]

Plastic pollution persists for decades or even centuries, producing negative impacts on wildlife and ecosystems. Contamination by plastic waste has been documented in all major rivers, lakes, beaches, ocean basins, and even in remote locations like the Artic Polar Circle. According to 2019 reports, the estimated global amount of plastic waste in oceans and soil is around 22 million metric tons (MMt); this value is expected to increase up to 44 MMt in 2060. This dataset is particularly disturbing since the application of intelligent and efficient plastic waste treatments has not grown at the same rate.[87]

Plastic production has exponentially increased over the last decades, from 2 MMt in 1950 to 380 MMt in 2015.[40] Recent data indicates that 139 MMt of single-use plastic waste were produced in 2021; this represents 6 MMt more than the quantity reported in 2019. The amount of virgin polymers produced from petroleum derivative substances was 15 times higher than the number of plastics fabricated from recycled raw materials; this disparity suggests that recycling is not advancing fast enough in our societies.[93]

Most of the plastic waste is represented by macro-plastics, which are common and recognizable items such as disposable bottles, food containers, and many other single-use products. Most of the plastic waste reached the environment due to a deficient recollection or disposal. Another important concern related to an incorrect management of plastic waste is the formation of micro-plastics (MP). MP are synthetic solid particles or polymeric matrices, with regular or irregular shape and with size ranging from 1 μm to 5 mm, of either primary or secondary manufacturing origin, which are insoluble in water.[39] Primary MP are designed that way for commercial use, for example, as raw materials in textile and cosmetic industries. Secondary MP result from the breakdown of larger plastic items. This breakdown is caused by exposure to environmental factors like the sunlight, ocean waves, wind, rain, or general mechanic stress. As previously mentioned, MP have been detected worldwide in a large range of shapes, sizes, and concentrations in the environments of marine water,[21] freshwater,[21] agroecosystems,[100] atmosphere,[95] food,[131] drinking-water,[92] biota,[98] and other remote locations.[30]

The existing plastic environmental crisis urges the necessity to improve the current techniques of plastic waste management, focusing the efforts not only on the treatment of domestic waste but also on the plastic pollution of the oceans. Plastic waste not only is affecting terrestrial and marine ecosystems, but also is also endangering important.

Plastic waste is undoubtedly altering marine and terrestrial ecosystems, and, at the same time, it is creating substantial risks for economic activities that rely on the integrity of those ecosystems, such as tourism or fishing. 109 MMt of plastic waste went into rivers around the world during 2019 but only 1.7 MMt ended in the oceans; this is because the process in which plastics are driven through the rivers and reach the oceans can take years or even decades. It is expected that the increasing rate of plastic consumption generates a considerable increase of plastic waste accumulation in marine environments, from 140 MMt in 2019 to 493 MMt in 2060. To clean up this pollution becomes harder and more expensive with the pass of the years as these materials start to fragment.

Plastic polymers significantly contribute to global warming through greenhouse gas emissions (GGE). Plastics generated 1800 MMt of GGE in 2019; this represents around 3.4% of global GGE. 90% of these emissions came because of fossil fuel consumption during their production. According to 2021 estimations, GGE from single-use plastics were ~450 MMt CO_2 equivalents; this exceeds the annual total GGE of the UK. At the current rate, 2060 GGE from plastics' life cycle will duplicate, reaching 4300 MMt of GGE.[93]

Moreover, MPs have been found in the air of remote regions, including the Artic Pole, where they can enhance global warming by absorbing sunlight and reducing the superficial albedo of snow.

A deficient plastic waste management of both, land and marine activities are the major causes of global plastic accumulation. Single-use plastics have been a matter of public concern for years. "Single-use" is a generic term that encompasses plastic items designed to be used once and be discarded immediately after, for example, grocery plastic bags, food containers, disposable bottles, drinking straws, etc.

Although plastic pollution is usually referred to as a management problem, the whole concept of plastic pollution covers the full spectrum of harmful effects, toxic emissions, and resource consumption that follows along the complete life cycle of plastics.

A complete life cycle assessment of synthetic polymers can help us identify and evaluate all the possible effects of plastic-related goods and services, thus developing alternatives of use that allow us to obtain higher performances, economic gains, and recycling potentials. Only this way, can we mitigate and ultimately decrease the ever-growing accumulation rate of plastics in the environment.

More than 13,000 chemical compounds can be associated with plastics as they act as additives and plasticizers. Many of these molecules possess high toxicity and represent a human health and environmental hazard. Some of these substances can lixiviate during the plastic's life cycle, reaching the air, water, or soil. Pregnant women, elderly, and children are especially vulnerable to the effects of toxic additives and chronic exposition can result in long-term health issues. For instance, phthalates and bisphenol-A are directly linked to estrogenic activity.[137]

As mentioned, most plastic pollution in the oceans was derived from plastic waste that was originally discarded in soils.[34] Therefore, it is believed that the concentration of plastics in continental land is 4–23 times higher than the concentration in the oceans.

Plastic ingestion is another great concern related to plastic pollution. Accidental plastic and MP ingestion can lead to a loss of reproductive capacity in wildlife or even to the death of organisms by mechanical obstruction of the respiratory and/or digestive tract.[29]

Plastic reuse and recycling is now more relevant than ever; hence, research efforts in bio-recycling of plastics offer an auspicious alternative to mitigate and diminish the worrying accumulation of plastic waste in our surroundings.

8.3 OPTIONS OF PLASTIC WASTE TREATMENTS

The broad applicability of synthetic polymers in many human activities, and the comfort of using them as disposable products have led to numerous problems.[126] Plastic waste is highly insoluble and possesses minimal chemical reactivity; as a result, it has a very low bioavailability, and it is considered nondegradable under most environmental conditions.[78]

The growing global population and the changes in life habits of many modern societies have caused a significant increase in the use of disposable plastics.[112] COVID-19 pandemic induced our society into new social dynamics, many of which involved the use of plastic products for personal protection or food storage, generating an intense accumulation of plastics waste in the environment. This raise in the disposal of synthetic polymers combined with the still limited treatment options are the main reasons for the current plastic waste crisis, which everyday becomes more visible (e.g., the detection of floating plastic islands).

Despite the advances that many countries have made in regards of stabilizing regulatory frameworks for plastic use and disposal,[126] many political, social, and economic obstacles have yet to be addressed. To implement an integral and efficient plastic waste management system, a combination of adequate treatment and recycling techniques must be applied.[49] Currently, several treatment techniques can be applied on plastic waste. In a global scale, landfill disposal is the most used treatment, as 45% of plastic waste receives this treatment. Incineration (25%) and mechanical recycling (16%) are also globally implemented. Despite the existence of previously mentioned options, it is believed that 19% of plastic waste ultimately ends in terrestrial and aquatic ecosystems, contaminating them. It has been estimated that nearly 80% of plastic materials are discarded as waste with no second-use opportunity.[11] Table 8.1 summarizes the main treatment strategies implemented for plastic waste management.

As mentioned earlier, not all kinds of plastic can be recycled; hence, a bad management or disposal will inevitably result in their accumulation in the environment. Therefore, an efficient management system of plastic waste

TABLE 8.1 Advantages and Disadvantages of Strategies of Plastic Management.

Strategies	Advantages	Disadvantages	Reference
Landfill disposal	If applied correctly, it can significantly reduce the environmental impact of plastic waste.	There are risks of leaks or migration of the plastics into the ecosystems.	[126,110,88]
	Some studies suggest that a controlled accumulation of plastics in a landfill has less negative side effects than other treatments.	Its capacity is limited by space.	
		Because of their stability, plastics will be accumulated permanently within the landfill.	
		Additives and plasticizers can be hazardous and migrate into the soil or water.	
Incineration	The produced calorific energy can be converted into electricity.	Toxic substances, such as dibenzo-dioxins, dibenzo-furans, and polychlorinated biphenyls, can be produced.	[6,88,110]
	Good option if landfill space is saturated or limited.	Some plastics contain heavy metals which cannot be destroyed and will persist as pollutants.	
	Environmental impacts can be reduced through the treatment of combustion gasses.		
Mechanical recycling	It is environmentally friendly	It becomes complex because plastic waste usually contains a mixture of plastic types and many of them contain additives.	[36,110,112,126]
		Different types of plastic will have different fusion and processing temperatures.	
		Just a few types of plastics can be recycled. Impurities such as paint, dyes, and other compounds should be removed for an efficient recycling of the polymers.	
		Separation and selection of the target plastics can be laborious and expensive.	

requires individual cooperation as well as governmental and corporative investments and actions.[110] At-home plastic separation from other types of solid waste is crucial because it reduces time and efforts from the personnel in charge of its classification and cleaning.[99]

Despite being technically feasible for some type of pollutants, conventional techniques for plastic waste management are not sustainable in the long run nor at global scale. Technological capacity, capital investment potential, and regulations vary between countries, making plastic waste treatment a nonachievable option. In order to stop plastic from being continuously accumulated, local governments should apply tax incentives and other economic enticements for companies of the recycling sector and stricter fiscal regulations or tax payments for high plastic-producing corporations.[48] Bioplastic and biodegradable plastic-like materials are promising options as these materials are degraded significantly more rapidly in the environment. Although the use of bioplastics emerges as an encouraging idea, more research is needed in order to produce them in a less expensive manner.[88]Actions that induce a reduction in the consumption of plastic products, promoting an increase in plastic reuse and recycling are more necessary nowadays than ever because they generate impacts (economic, social, environmental) by themselves. More research is crucial to determine which waste management option is the more optimal for every specific scenario, in example, recycling is usually the eco-friendlier treatment option; but, if possible, it will always be better to not produce plastic waste in the first place.[44]

One of the key alternatives for dealing with plastic pollution in a global scale is biodegradation. Biodegradation of plastic waste occurs when hydrolysis and oxidation–reduction reactions take place within polymer's backbone, shortening the long hydrocarbon chains and producing the subsequent fragmentation–assimilation and mineralization of the compound. In most cases, these reactions are catalyzed by microbial enzymes, although environmental factors like UV light and high temperatures can preoxidize the pollutant, making the job easier for the biomolecules involved.[36]

Biodegradation of plastic can happen in 3 different scenarios: aerobic degradation that needs oxygen and would occur in an open environment, anaerobic degradation will likely take place in landfills, and semi-aerobic degradation would occur during compost processes. Biodegradation is mainly done by microorganisms such as bacteria and fungi, although recent reports have discovered that insects are also able to biodegrade complex plastic polymers. Some factors can significantly affect biodegradation efficiency, for instance polymer type, inoculum type, time of exposure, and the use of pretreatments. Pretreatments are necessary since they can form short oligomers or monomers that are more easily assimilated by microorganisms. Most common pretreatments are UV radiation, acidic treatment, and heat treatment.[78]

8.4 BIODEGRADATION OF PLASTIC

Plastics can undergo degradation processes, such as hydrolysis, electrochemical, thermal oxidative, photo and biodegradation when released into the environment.[3] The degradation of plastic is frequently categorized based on any detrimental change in its physical, mechanical, chemical, or structural properties. Plastic degradation is a complex process, as it cannot be measured and controlled easily, especially under open environmental conditions.[66] Despite environmental conditions, the polymer degradation rate depends on the composition of polymeric materials.[65] Extreme environmental conditions such as weathering, sunlight, aging, water, and soil burial can accelerate plastic degradation. Temperature additionally plays an important role in the plastic degradation process.[7] The thermal properties of different plastic materials such as glass transition temperature, crystalline/amorphous ratio, and melting temperature affect the rate of plastic degradation. Temperature in some landfills and composters has been reported to reach 80–100°C, which accelerates the degradation rates if sufficient moisture and oxygen is present for thermal oxidation and hydrolysis.[19] Plastic biodegradation refers to the degradation of plastic using enzymes or microbes during a specific time. It involves catabolic processes that result in the chemical modification of plastics.[123] Biodegradation has been considered one of the key alternatives to effectively deal with plastic pollution.[68] During biodegradation, organisms use organic materials as carbon sources or their enzymes are applied directly to the pollutants; this may not only produce nontoxic by-products but also provide energy or transform organic compounds into other useful products.[46] Biodegradation can effectively reduce the harm of additives to plastics and achieve a better ecological environment.[23] While knowledge of the environmental degradation of plastics is still limited, it has been proved that the fate of plastics is strongly related to the type and strength of the degradation they suffer.[146] Therefore, exploring the ability of different organisms and the interaction between their enzymes and plastics is critical to obtain feasible biological treatment options.[20]Biodegradation seems to be a promising solution to tackle the plastic waste issue, but it requires a thorough understanding of the efficient bacteria, fungi, or insects and the gene clusters, pathways, and enzymes involved in the process[2,124,134]The following sections aim to examine the viability of the biological treatment of plastic waste, with a particular emphasis on the most recent studies of various insects, microorganisms, and their enzymes in the biodegradation of synthetic plastics. It also points out the existing biodegradation challenges and corresponding methods. Emphasis has also been placed on the biodegradation mechanisms and the degradation ability and effectiveness of bio-catalytic methods and their application potential.

8.4.1 CHALLENGES IN BIODEGRADATION OF PLASTICS

As a result of its manufacturing performance, conventional plastics consist of a long chain molecule with a repetitive functional group, elevated molecular weight (MW), high hydrophobicity, and crystallinity. Because of this, the ability of microorganisms to deteriorate them is usually very limited. Plastics are not rich in polar functional groups or hydrolysable bonds; thus, they are not a suitable substrate for microbial attachment and enzymatic reaction.[5] On the opposite side, biodegradable plastics have susceptible functional groups and enzyme-sensitive ester linkages that might be attacked by enzymatic and abiotic oxidation.[83]Conventional polymers are more difficult to disintegrate in comparison to other plastics, such as emerging biodegradable plastics, due to their short chains, low molecular weights, and flexible, abundant functional groups.[83] The problem and difficulty of plastic degradation relies on traditional plastics with a unique chemical composition, for example, polymers like PE have a fairly simple chemical structure, with the monomer (-CH-CH-) repeated tens of thousands of times and the extent and spatial arrangement of ramifications within the chain define the different types of PE (low density, linear low density, or high density). Chemical substitution of a lateral group to CH_3 or a benzene ring changes the polymer from polyethylene (PE) to polypropylene (PP) or polystyrene (PS), respectively.[41,133]

The sturdy structure of polymers like PE, PS, and PP and their highly hydrophobic nature impede the colonization of these plastics by many bacteria or fungi, decreasing the ability of these microorganisms to grow and use such molecules to satisfy their energetic needs. To date, intensive research efforts in the field of biodegradation have resulted in the identification of very few microbes with the enzymatic machinery to break down these durable polymers, with a common major difficulty being the oxidation of the long hydrocarbon chains.[17,77,79,52] In general, microorganisms can only metabolize and ultimately degrade synthetic polymers after these have undertaken preliminary abiotic oxidation. One apparent exception is represented by polyethylene terephthalate (PET), the material of which disposable bottles are made. A few microorganisms are in fact capable of breaking it down[115,121] ; however, this process is not as straightforward as it seems, and crystalline PET, one of the most used plastics, is not as degradable as it seems to be. In addition to the fact that effective plastic biodegradation using microorganisms has not been accomplished so far, it is important to highlight that the metabolic use of plastics as a source of energy releases CO_2. Therefore, it is in fact a form of slow biologically mediated incineration. For these reasons, microbial degradation techniques must be improved, or more suitable and efficient strains should be found.[67] Currently, there are still undeveloped technologies and cost issues associated with biological degradation technologies, as well as capital consumption not in line with the sustainable development our society urgently needs.

8.4.2 MECHANISMS OF PLASTIC BIODEGRADATION

Biodiversity is the key condition for biodegradation. Organisms capable of degrading plastics include microorganisms such as bacteria, fungi, a few algae and insects (whose degradation ability is mostly attributed to their gut microbiome). Usually, a complete degradation of plastics cannot be achieved by biological means alone and a combination of biotic and abiotic factors is required. The basic process of plastic degradation in the environment includes steps like the rupture of polymers to smaller particles, followed by the degradation of the smaller polymers to oligomers, dimers, and monomers.[109] Researchers have made a thorough research on the process of surface colonization of plastics by degrading consortium forming a biofilm on the particles. During this colonization, microorganisms bind to the surface of plastic material and secrete degrading enzymes that convert complex polymers into their simpler forms.[12] The attachment of microorganisms follows several steps including biofouling and degradation of plasticizers, and enzymatic attack on the backbone of the polymer followed by hydration and penetration of microbes into the polymeric structure. After this, mineralization takes place through various metabolic mechanisms. On complete mineralization, carbon dioxide is produced along with the formation of several intermediate compounds that are used as a source of energy to promote the growth of microbes. Microbial enzymes that play important roles during the degradation of plastics include hydrolases (e.g., lipase, esterase, and cutinase) which degrade plastics by catalyzing the chain cleavage reactions on the surface of the polymer, generating cracks. Other important groups of enzymes consist of oxidases such as lignin peroxidase, manganese peroxidase, and laccase; these proteins increase the hydrophilicity of plastics and convert them to carbonyl or alcohol residues.[119] The monomers generated are absorbed into the cytoplasm of microorganisms and enter into different metabolic pathways. Extensive research has already been done using extracellular enzymes in plastic biodegradation, but very little information is available on the role of intracellular enzymes. Although the pathways involved in the uptake of monomers are still not clear, it has been proposed that after the fragmentation–oxidation steps of some plastics like PET, the metabolic intermediates with carbonyl and hydroxyl groups are processed inside the cell using the tricarboxylic acid cycle and β-oxidation pathway.[119]

8.4.3 INSTRUMENTAL TECHNIQUES TO DETECT PLASTIC BIODEGRADATION

As mentioned before, it is generally accepted that biological degradation of plastic is divided into three stages[79]: (1) Bio-deterioration, which involves initial modifications on the surface and cracking of plastics; (2) Bio-fragmentation, which takes place when long hydrocarbon chains of the polymer suffer oxidation and hydrolysis, realizing intermediate products of degradation; (3) Bio-assimilation of monomers and short chain metabolite molecules by microbes and their incorporation to the cell with concomitant degradation of plastic to completely oxidized and reduced molecules (mineralization).

It is possible to use a variety of techniques to evidence the changes that occur to the plastic for each of these stages. The main methods utilized to identify plastic biodegradation are shown in Table 8.2. Information on the alterations of the plastic surface occurring in the first stage can be acquired using several microscopic techniques, for example, scanning electron microscopy (SEM) and atomic force microscopy (AFM) are used to compare specimens from which the attached cells have been washed off to control samples incubated in the absence of microorganisms.[109] The presence of microscopic holes, cracks, and fractures on the plastic would indicate degradative activity in the samples. Other methods can be applied to analyze structural changes of plastic: for example, tensile strength, extension at break, elastic constant, and yield point.[76]

Bio-fragmentation can be recognized based on the decrease of mean molecular mass (Mn) of polymeric carbon chains and a wider molecular mass distribution.[128] Both parameters can be measured using gel permeation chromatography (GPC); intermediate products of plastic bio-fragmentation can be detected through gas chromatography with mass spectrometry (GC-MS), high-performance liquid chromatography (HPLC), Fourier-transform infrared spectroscopy (FTIR), and nuclear magnetic resonance (NMR) imaging. In particular, FTIR can provide valuable information in the abundance of many types of atomic bonds in the samples, for instance C-C, C-Cl, C-O, or C=O.[60] The formation or rupture of these atomic bonds may indicate oxidation, dechlorination (for polyvinyl chloride PVC), or degradation of polymer chains.

At this stage, thermo-gravimetric analysis is another powerful tool that allows researchers to detect polymer chain shortening or breaking due to a decrease in the thermal stability of the material.[89]

Bio-assimilation can be measured by the increase in microbial biomass of those microorganisms cultivated in the presence of a plastic substrate, for example, by amount of yielded protein produced in comparison to control.

Finally, mineralization of plastics is assessed based on the amount of released CO_2 or, in the case of anaerobic processes, products of the terminal reductive reactions, such as CH_4, H_2S, or reduced nitrogen compounds. As a supplemental method, assessing of microbial isolates can be implemented on a solid medium containing fine particles of the polymer, where colonies of plastic-degrading microorganisms would be surrounded by a clear zone. Finally, high accuracy can be achieved by using plastics labeled with ^{14}C isotope and determining accumulation of the label in the end products of degradation: $^{14}CO_2$ or $^{14}CH_4$.[60] Although this method presents the advantage of being not sensitive to the presence of biodegradable impurities or additives in the substrate, its application is limited due to the complexity of the assay and the high cost of reactants.

Robust evidence of plastic biodegradation is more likely to be obtained by employing several techniques (See Table 8.2) to register the changes within polymer structure, variation of biomass and its characteristics, and formation of metabolites or end products from plastic degradation.[71,82,102] Still, a thorough examination of 408 studies revealed that of all those microorganisms reported to degrade plastics, 48% of reports used analytical methods relating to only one of these categories, 39% implemented assays that covered two categories, and just 10% used techniques that related to all three. Moreover, most reports on plastic biodegradation using microbial isolates do not deepen on the genes and enzymes responsible for the reported activity. Only around 14% of the strains capable of degrading plastic have their gene sequence conferring the degradation activity elucidated. This becomes an important limitation since deep understanding of the relevant biochemical and molecular data is vital to advance in the plastic biodegradation field because it allows the search for new plastic-degrading gene candidates in novel microbiomes by comparison to enzyme data banks.

TABLE 8.2 Analytical Techniques for the Appraisal of Plastic Degradation.

Technique	Changes in plastic's properties detected	Function of technique	Reference
SEM, AFM	Morphological and surface changes	SEM reveals the presence of cracks, holes, and erosion. AFM estimates the roughness of material.	[33,43]
High temperature GPC	Molecular weight	Detection of changes in average molecular weight and molecular weight distribution.	[52,117]
X-ray diffraction (XRD) and differential scanning calorimetry (DSC)	Crystallinity	Detect crystallinity changes in the polymer.	[22]
FTIR	Chemical properties	Detection of formation or depletion of certain functional groups.	[47]
Titration methods, trapping, sturm test, radiolabeling	CO_2 evolution	Used as an indication to prove biological degradation and mineralization.	[8,25]
Gas chromatography–Mass spectrometry	Metabolites formation	Detection of bio-fragments and the presence of saturated linear alkanes.	[64]
Gravimetric weight loss, thermo-gravimetry	Weight	Determination of mass loss of the polymer.	[113]

8.5 PLASTIC BIODEGRADATION BY INSECTS

The observation that microorganisms can degrade organic polymers has motivated the search for bacteria, fungi, and algae that are capable of breaking down sturdy fossil fuel–derived plastics[77,133]Unfortunately, despite being the center of attention in the quest to biologically treat plastics for a few decades, microbial degradation of plastics, in most cases, has presented considerable obstacles. As mentioned before, biodegradation process follows four steps: biodeterioration, biofragmentation, assimilation, and mineralization; microbes capable of assimilating and mineralizing short chain molecules derived from oxidation and depolymerization have been characterized, but the initial two steps are rarely achieved in the environment. In most cases, fungi and bacteria can only metabolize and fully degrade plastics after the polymers have undergone a preliminary phase of abiotic oxidation (one notable exception being PET). More recently, interest has switched to insects (see Table 8.3), whose ability to digest and consume plastics has been reported in Coleopterans species, being *Tenebrio molitor (T. molitor)*, *Tenebrio obscurus (T. obscurus)*, the super-worm, *Zophobas atratus (Z. atratus)*, *Tribolium castaneum*, and *Plesiophthalmus davidis* a few examples of this. Besides coleopterans larvae, lepidopteran fauna like *Plodia interpunctella*, *Galleria mellonella (G. mellonella)*, *Achroia grisella*, and *Spodoptera frugiperda* are also reported to digest synthetic polymers like PS, PE, PVC, and PP. These advances opened new paths in the pursuit of a sustainable way to dispose plastics, and given the current metabolic paradigm of plastic biodegradation, scientists rushed to analyze the gut microbiota of these insects in order to clarify whether or not these microorganisms are directly involved in such degradation processes. Although most studies suggest that gut microbes are mainly responsible for plastics degradation in most analyzed species, a new discovery in the larva of *G. mellonella* is now offering an alternative to this metabolic archetype.[106] Certain enzymes in the saliva of the caterpillars are able to fragment and break PE in a matter of only a few hours. Molecular and biochemical analyses have identified these enzymes as phenol-oxidases which add elemental oxygen to the substrate, producing a series of chain reactions that will result in depolymerization of the plastic. In this case, plastic is not used to produce energy, and no CO_2 is released. Biological oxidation of PE in the extracellular matrix at room temperature represents a new model within plastic degradation.[13] The following sections aim to analyze the state of the art in the use of insects for the biological treatment of polymers, detailing the proposed degradation mechanisms and discussing the possibilities and potential of these procedures for being applied in a larger scale.

8.5.1 MECHANISM OF PLASTIC DEGRADATION IN LARVAL GUT

Larvae of different insects use their mandibles to mechanically break down and ingest plastics or diets with a high amount of plastic. It has been observed that symbiont and commensal microbiota in these larvae is altered when

TABLE 8.3 List of Reported Insects with Plastic Consumption Capacity and their Associated Gut Microbiota and Enzymes.

Insect	Consumed plastic	Gut microbiota	Consumption associated enzymes	Reference
Lasioderma serricorne	PP, PE, and Polyester	*Symbiotaphrina kochi*	Cutinase-like enzyme	[35,101,127]
Rhyzopertha dominica	PE and PP	*Aeromonas liquifaciens*	Lipase Chitinase Protease	[101]
Tenebrio molitor	PS, PP, and PE	*Exiguobacterium* sp. YT2 *Kluyvera* sp. *Citrobacter* sp. *Kosakonia* sp.	Alkaline proteases Alkali-tolerant esterase Hydrolase	[139]
Tenebrio obscurus	PS	Spiroplasmataceae Enterococcaceae	Further research needed	[91]
Tribolium castaneum	PS	*Acinetobacter* sp. AnTc-1	Further research needed	[129,130]
Plesiophthalmus davidis	PS	*Serratia* sp.	Lipase Protease Chitinase	[135]
Alphitobius diaperinus	PS	*Pseudomonas* sp *Kocuria* sp *Cronobacter* sp	Hydroquinone Peroxidase	[32]
Zophobas atratus	PS, PE and PU	*Pseudomonas* sp. strain DSM 50071 *Klebsiella pneumoniae* *Alcaligenes* sp. *Acinetobacter* sp. *Citrobacter* sp. *Mangrovibacter* sp.	Monooxygenase Lipase Cutinase Esterase Polyurethanase	[139,58,72, 118]
Sitophilus oryzae	Nylon, PE and PP	*Bacillus subtilis* strain TLO3 *Staphylococcus* sp.	Hydrolase Lipase	[94,101]
Oryzaephilus surinamensis	Nylon, PE	Further research needed	Further research needed	[38,45]
Galleria mellonella	PS and PE	*Massilia* sp. FS1903 *Acinetobacter* sp. *Bacillus* sp. *Serratia* sp. *Enterobacter* sp. D1 *Aspergillus flavus*	Manganese peroxidase, Lacase, Lignin peroxidase, Lipase Protease Polyurethanase	[16,24,53,69, 97,146]

TABLE 8.3 *(Continued)*

Insect	Consumed plastic	Gut microbiota	Consumption associated enzymes	Reference
Plodia interpunctella	PE	*Enterobacter asburiae* YT1 *Bacillus* sp. YP1	Esterase	[138]
Achroia grisella	HDPE	Further research needed	Further research needed	[61]
Corcyra cephalonica	PE	*Staphylococcus saprophyticus*	Further research needed	[55]
Gryllus bimaculatus	Polyester, PU	*Aspergillus flavus* G10	Hydrolytic enzymes	[56]
Uloma sp.	PS	Further research needed	Further research needed	[62]

they are force-fed or co-fed with plastics. Independently of insect species or polymer type, consuming plastics will result in notable changes of the abundance or diversity of certain operational taxonomic units (OTUs) like Enterobacteriaceae, Enterococcaceae, and Streptococcaceae in comparison to larvae-fed natural, plastic-free diets.[91] These OTUs will initially generate biofilms either individually or in consortium; this colonization is assisted by various polysaccharides and/or proteins.[107] Consequently, polysaccharides and cysteine-rich proteins penetrate trough the surface and change the size of the polymeric pore.[42] The sturdiness of the polymer will decrease eventually, but its surface area will increase, giving microbes a larger superficial area to stick to. The microbial consortia in the insect's gut will excrete protein that will initiate the enzymatic degradation process with concomitant depolymerization of the plastic. After short chain oligomers and monomers are produced, bacteria and other microbes can uptake these products for energy production. Sometimes, the larvae can use the depolymerization products to synthetize different biomolecules.[13]

8.5.1.1 BIODEGRADATION OF PLASTIC BY COLEOPTERANS

Tenebrionidae family is the main representative of Coleopterans with the capacity for degrading synthetic polymers. Recent discoveries concerning these species and their interaction with different types of plastic are discussed and detailed below.

8.5.1.1.1 *Tenebrio molitor*

The yellow mealworm, *T. molitor*, is a globally distributed beetle species which is commonly found in grain and bran warehouses; thus, it is usually considered a pest. The larvae of yellow mealworms typically measure about 2.5 cm or more in length and have lighter body colors than adults, with long and slender structures. *T. molitor* larvae were able to efficiently degrade PS samples within a 24 h period, mineralizing approximately 50% of the ingested material, although only about 1% of the carbon was converted into biomass.[145] *Exiguobacterium* sp. (YT2), a bacterial strain with the ability to degrade PS, was isolated from yellow mealworm's gut. YT2 was noticed to cause surface topography changes on PS materials, and as a result, the hydrophobicity of PS decreased, and carbonyl groups formed.[144] It has also been opined that besides *Exiguobacterium* sp., various microorganisms play an essential role in the digestion process of mealworms.[125] Another strain of mealworm (strain CA) was reported to be capable of biodegrading seven PS wastes.[140] Further investigation has confirmed the hypothesis that the capability of mealworms to biodegrade PS is highly conserved within the specie regardless of their geographic origin and seems to be ubiquitous.[140] Scientists have proposed that chewing and consuming PS by *T. molitor* is their adaptive intrinsic behavior, as they feed upon decaying forest vegetation in the wild.

It has been observed that several factors can significantly alter PS consumption in *T. molitor.*[141] Degradation efficiency of PS increased up to 72% when yellow mealworm larvae were co-fed with bran and plastic compared to a sole plastic diet. Plastic composition and temperature also impact PS degradation rates as yellow mealworms seem to prefer low-density polymers and mild temperatures (25°C).

Yellow mealworms have also shown capacity for low-density polyethylene (LDPE) degradation. According to recent reports, *T. molitor* larvae were able to mineralize LDPE at similar rates than those of PS mineralization. Interestingly, GPC analysis demonstrated a significant reduction of LDPE's average Mw, and NMR/FTIR confirmed modifications of the internal chemical structure of the polymer; this suggests that long hydrocarbon chains underwent partial depolymerization and oxidation. In addition, intestinal microbiome of larvae was analyzed through second-generation genetic sequencing revealing that *Citrobacter* sp. and *Kosakonia* sp. OTUs were associated with LDPE and PS consumption, respectively.[18] Several other bacteria namely *Listeria* sp., *Nitrospira defluvii*, *Pedomicrobium* sp., *Aquihabitans* sp., unclassified Xanthomonadaceae, Saprospiraceae, and Burkholeriale*s* were found to be associated with PS degradation in the yellow mealworm's gut.[70] The information of various microbes and their activity within larval gut suggests that mixtures of PE and PS could be depolymerized simultaneously, thus making a potential application of this treatment even more robust.

Other investigations have proposed that *T. molitor* larvae can degrade other plastic polymers such as PVC[136] and vulcanized rubber,[1] although degradation rates prove to be low and further examinations of the excreted products are required in order to demonstrate efficient depolymerization.

8.5.1.1.2 *Tenebrio obscurus*

The dark mealworm *T. obscurus*, similar to yellow mealworms in appearance, is also known for being able to consume plastic. The size of dark mealworm larvae is usually between 1.5 and 2.5 cm in length and, in the wild, they tend to consume seeds, vegetables, flour, and oats. *T. obscurus* larvae have a higher light sensitivity than *T. molitor*.[91]

The larvae of dark mealworms were found to depolymerize PS at a higher rate than equally sized yellow mealworms. When *T. obscurus* were supplied with PS as sole diet, PS exhibited a mass loss of 55% in a month, but the amount of PS degradation increased by 67% when larvae received a conventional food supplemented co-diet. This indicates that even though larvae had access to a nutritional source capable of satisfying their energetic demands, their voracity toward PS seems to increase instead of decreasing. Additionally, according to GPC and FTIR analyses, excreted PS had a considerable Mw diminution and newly formed hydrophilic functional groups; this would probably raise its solubility reducing its toxicity and recalcitrance.

Before feeding PS, the gut microbiome of *T. obscurus* showed higher diversity than the one of *T. molitor*. Microbial community analysis indicated that bacteria from the Enterococcaceae, Spiroplasmataceae, and Enterobacteriaceae families were particularly linked to dark mealworm's intestines for PS depolymerization and degradation.[91]

8.5.1.1.3 *Zophobas atratus*

Z. atratus larvae are commonly known as super worms because of their larger size in comparison with other grain-eating Coleopterans. Super worms are similar in appearance to mealworms; however, super worms are larger (2.5–5 cm) and after attaining maturity, the beetles become darker and are then called "black beetles." Super worms have mandibulate mouthparts, which offer these species the capability to chew and eat plastic.

During a study conducted using PS as sole diet, super worms were able to digest and degrade the polymer with a specific rate (0.58 mg/day/worm) almost 5 times higher than yellow mealworms (0.12 mg/day/worm). The survival rate of *Z. atratus* eating PS exclusively was almost equal to that of a regular diet, which indicates that super worms can complete their life cycle by consuming PS diets.[143] Subsequently, a different study revealed that the bacterial strain *Pseudomonas* sp. plays an active role during PS degradation in the gut of super worms. Furthermore, scientists observed through RT-qPCR, that when cultivated with PS as sole carbon source, *Pseudomonas* sp. increased its expression of serine-hydrolase and formyl-glutathione hydrolase while the expression of aryl esterase and thioesterase decreased.[58]

Biodegradation of LDPE has also been observed in super worms where two strains from China showed a LDPE consumption rate (0.58 mg/worm/day) almost equal to PS consumption rates (0.61 mg/worm/day). FTIR, TGA, and GPC analyses indicated a limited gut-microbe-dependent depolymerization of the LDPE within *Z. atratus* .[90] Super worms and yellow mealworms have also evidenced capacity to degrade expanded polypropylene (EPP) via gut-microbe-dependent depolymerization with diversified microbiome. High-throughput 16S rRNA analysis discovered that *Citrobacter* sp. and *Enterobacter* sp. were the two main OTUs linked to EPP diets in the gut microbiome of superworms while *Kluyvera* was the major OTU in the yellow mealworm.

8.5.1.2 BIODEGRADATION OF PLASTIC BY LEPIDOPTERANS

The caterpillars of different moths belonging to Pyralidae family are commonly known as "wax worms." Wax worms have been actively studied because of their ability for plastic degradation. Among these Lepidopterans, three species stand out: The Indian meal moth (*Plodia interpunctella*), the lesser wax moth (*Achroia grisella*), and the greater wax moth (*Galleria mellonella*). In the following lines, the capability of moth larvae for degrading synthetic polymers will be discussed and detailed.

8.5.1.2.1 The Indian Meal-Moth

One of the first reports on insect's degradation of plastic came out when scientists discovered that caterpillars of *Plodia interpunctella (P. interpunctella)*, the Indian meal moth, were able to chew and ingest PE foam.[139]

During this investigation, two bacterial strains, *Enterobacter asburiae* and *Bacillus* sp., were isolated. Both organisms produced a biofilm and were able to chemically modify PE in laboratory conditions, creating holes and cracks in the surface and reducing the hydrophobicity of the material.

8.5.1.2.2 The Lesser Wax Moth

An adult lesser wax worm, *Achroia grisella (A. grisella)*, is light brown with golden highlights and black scales with long filiform antennae. Its length is usually 8–13 mm long. *A. grisella* is largely distributed in tropical and subtropical regions and is considered a serious pest of beehives as its larvae consume bee wax.

Like other members of this family, *A. grisella* has also been studied as a PE-degrading organism. The lesser wax moth has exhibited capacity to degrade high-density polyethylene (HDPE), a very recalcitrant plastic polymer. When left in direct contact with HDPE films, *A. grisella* makes holes in them within a few days and one organism can degrade nearly 2 mg of HDPE film every day.[61] In addition, it was observed that a co-diet of bee wax and plastic enhanced HDPE consumption and impacted favorably on *A. grisella*'s life cycle, making possible for the organisms to breed a second generation of plastic-degrading larvae. FTIR and NMR confirmed the presence of new carbonyl and alcoholic groups with the increase of unsaturated hydrocarbons, thus providing evidence for the biodegradation process of HDPE in the lesser wax worm. However, further research must be done to understand whether this PE biodegradation is gut-dependent or independent.

8.5.1.2.3 The Greater Wax Moth

The greater wax moth, *Galleria mellonella (G. mellonella)*, is a lepidopteran insect whose plastic-consuming capability has also been reported. *G. mellonella* larvae are creamy white, 3–30 mm length, sexually dimorphous and distributed worldwide.[63] The greater wax moth feeds on pollen, honey, wax, and broods. Its larval stage is highly dangerous for bees and beehives due to its voracious feeding habits.[80] It has been observed that, when confined in a plastic bag, 100 greater wax moth larvae consumed almost 100 mg of LDPE in 12 h.[16] The authors have hypothesized that because bee wax, an ethylene-rich compound, is one of the main food sources of *G. mellonella*; this organism possesses the enzymatic machinery to break down the synthetic polymer. Although more research into the molecular complexities of wax biodegradation is needed, it appears that one of the targets in its digestion is the C-C single bond of aliphatic molecules. The presence of holes in PE films exposed to *G. mellonella* caterpillars and the FTIR analysis of damaged PE revealed chemical disintegration of PE, including the breakage of C-C bonds.

Recent evidence has shown that bacterial genera *Enterobacter*,[97] *Acinetobacter*[24] and the fungus *Aspergillus flavus*[146] are directly involved in the biodegradation process of PE within the greater wax moth. Hence, a close relationship between *G. mellonella*'s gut microbiome and LDPE biodegradation is likely to exist.

The effects of a supplemented diet on PE and PS degradation by *G. mellonella* have also been studied. Significant mass loss of both plastics was observed when insects were incubated with PS and PE as a sole diet for 21 days with subsequent formation of fatty acids and long chain esters in the intestinal fraction of larvae. Nevertheless, an enhancement in PE and PS depolymerization rates as well as increases in larvae survival rates and microbial diversity were achieved when *G. mellonella* caterpillars received a supplement of bee wax and wheat bran in their diet.[69] The gram-positive lactic bacteria, *Enterococcus* sp., and facultatively anaerobic gram-positive bacteria, *Bacillus cereus,* were isolated from the *G. mellonella* larval gut and associated with PE degradation, while the gram-negative bacteria, *Serratia marcescens*, is suspected of participating in PS degradation.

G. mellonella has proven to degrade PE via the action of its saliva. After being exposed in aqueous solution at room temperature, saliva from the grater wax moth oxidized the polymer. Later on, it fragmented it into shorter molecules as products of degradation.[106] This previous study also showed that the saliva of larvae contained enzymes with phenol oxidase activity which were named PEases, becoming the first report of enzymes that can break down PE and make the polymer more susceptible to enzymatic attacks without any abiotic pretreatment. The recently identified PEases are secreted by the wax worms and oxidize PE outside of the insect's gut. Then, these enzymes are probably preparing food for the gut microbiota, meaning that now the polymer could be metabolized by no specific bacteria or fungi since the capacity to assimilate short-chain oligomers and hydrocarbons is vastly conserved within both kingdoms.

As previously noted, wax worms possess microbial communities in their guts, but the structure and composition of these communities is highly heterogeneous among individuals of the same species; in fact, it seems to be determined by the particular conditions in which each individual lives.[42]

Even when microbiota changes due to a plastic-based diet, promoting the enrichment of certain microbial genera, this does not necessarily mean that these augmented populations are capable to bio-transform the plastics; but, once plastics are preoxidized by salivary enzymes, the small molecules could be degraded by virtually any species present within the gut. This scenario hypothesizes a predigestive function of these PEases. However, we cannot eliminate any possibility, and only with in-depth biochemical, structural, and evolutionary studies we will be able to understand the real function inside the animal, and their full range of potential applications.

8.5.2 FUTURE PROSPECT ON BIODEGRADATION OF PLASTICS: CAN INSECTS DELIVER A SOLUTION FOR PLASTIC BIOTREATMENT?

The number of microorganisms with the ability to affect the structure of sturdy polymers such as PE, PS, and PP without abiotic pretreatment is fairly low, and, even in this case, it may take weeks for the effects to be noticeable. Without considering bacteria and fungi, insects have brought a renewed hope in the field of plastic biorecycling.

As we have seen, ingestion and consumption of plastic waste by insects is a reality, and, as we mentioned earlier, depolymerization and mineralization of these compounds is achieved in many cases. However, we must mention that the economic feasibility and environmental impact derived from implementing these techniques in a larger scale have raised questions that urgently need to be addressed. The discussion regarding this topic has focused on three points: (1) contrasting evidence about biotransformation of PS and PE within insects' gut, (2) short longevity shown by several larvae when fed with plastics as sole diet, and (3) high associated costs and little valuable products derived from biodegradation. In the following lines we will deepen in these points in order to bring a thorough vision on the future perspectives of biological remediation of plastic.

Some studies on *G. mellonella* have focused on the analysis of its excreted metabolites during LDPE degradation.[15] Their results suggest that a minimal amount of ethylene glycol (a degradation by-product of PE) is formed when the gut microbiota of the greater wax moth is in contact with LDPE films for 36 h. Thus, no significant depolimerization of PE is present or, if any, it is very slow. These findings support earlier scientific statements.[132] The information presented in the previous lines contraries aforementioned reports in this chapter,[24,69] but it could indirectly support the hypothesis that the first steps on PE depolymerization require active PEases contained in the saliva of wax worms.[106] Another possible explanation for these divergent results is that the ability to degrade PE might not be ubiquitous among *G. mellonella* strains.

Other investigation applied a Dynamic Energy Budget (DEB) model analysis to assess the technical feasibility of a plastic treatment system using mealworms fed with PS.[75] Results indicated that organisms reared with PS as sole diet showed considerable mass loss, minimal growth and average survival rate of only 4% after 98 days. Interestingly, *T. molitor* showed a significantly higher and very similar survival rate (≈50%) when larvae were fed with conventional or mixed (PS + oat diets) diets. This evidence proposes that energy input obtained from PS assimilation is not sufficient for mealworms to develop and complete its life cycle. Therefore, a mealworm treatment facility that exclusively uses PS as sole diet seems unsustainable.

Further projections using streamlined life cycle assessment and techno-economic analysis have indicated that it would be necessary to raise 4–10 metric ton of *T. molitor* larvae in order to effectively treat 1 metric ton of LDPE,[15] representing an estimated cost of 300–1000€ per LDPE metric ton treated. This considerably exceeds incineration costs for energy production in Europe.[50]

In summary, utilization of insects as either a remediation or management technology for plastic waste is very unlikely, given the conversion to microplastics, the absence of valuable products and the high energy input required. Even though this direct utilization of insects is not considered a sustainable option, we must not diminish our research efforts in this area. Accelerating the field of plastic biodegradation in the coming years will require considerable endeavors in enzyme discovery, microbial culturing, enzyme characterization, and protein engineering. Insect larvae and their gut microbiomes have been a treasure trove for the recovery of potential plastic-degrading enzymes and, once plastic-degrading enzymes are discovered, the computationally predicted functions of these enzymes have to be experimentally tested in the laboratory. We should expect more experimental validations and detailed enzyme characterizations in the coming years, generating a large arsenal of enzymes that will enhance bio-recycling approaches.

8.6 BACTERIAL TREATMENT

The use of microorganisms for plastic waste treatment has recently become relevant. Bacterial strains with capacity to degrade plastic resins such as PVC, PET, PE, LDPE, and HDPE have been reported over the last years.[51,74,122]

Metabolic versatility of microorganisms enables them to degrade a broad range of chemical compounds and use them as carbon sources and energy.[60] Two different approaches have been traditionally implemented to study synthetic polymer degradation: (1) using pure axenic cultures for their ability to consume plastic; (2) incubating plastic materials in the presence of natural samples or directly into natural environments. The second is used as a screening methodology that allows researchers to find strains that naturally possess the required ability.

Bacterial degradation of plastics has presented as an attractive alternative in the biorecycling field because it offers more economic and simpler operation conditions than other treatments like incineration.[26] Several bacterial strains with the ability to convert synthetic polymers into CO_2, H_2O, CH_4 and biomass in methanogenic conditions have been identified; at the same time, other bacteria have exhibited production of H_2S, CO_2, and H_2O in sulfidogenic conditions. Metabolic pathways and degradation products can vary in response to environmental conditions during the process since these factors will determine the proliferation of certain microbial groups.[26]

Various factors can affect bacterial degradation efficiencies; a few of them are numbered below.

1. The type of plastic to be treated. It has been reported that nonhydrolyzable plastics (e.g., PVC, PS, PE, and PP), whose primary molecular chain is entirely made up of C-C bonds, have a lower degradation level than hydrolysable plastics such as phthalate and PET.[20]

2. Particle size of plastic. A lower particle size is generally related to a higher superficial area of the polymer; therefore, bacteria have access to a larger contact area. This enhances the biodegradation process as it favors a greater microbial population. High molecular weight polymers present a significantly lower degradation rate because of their little superficial area, which does not offer proper conditions for biofilm formation.[59]

3. The bacterial strain or consortium to be used in the process. Depending on the conditions of the procedure, the use of certain types of bacteria or the application of mix cultures or consortia will affect the efficiency of the process.[142] Some bacteria with plastic biodegradation features have been isolated from laboratory samples or directly from the environment. A recent review of the existing literature indicated that among the PE-degrading organisms, the following bacteria genera have been reported: *Acinetobacter, Achromobacter, Bacillus, Stenotrophomonas, Enterobacter, Lysinibacillus, Salinibacterium, Paenibacillus, Brevibacillus,*

Pseudomonas, as well as some actinomycetes namely *Streptomyces, Pseudonocardia, Actinoplanes, Sporichthya, Actinomadura*, and *Nocardiopsis*.[142] Likewise, unspecified *Streptomyces,* and bacterial strains *Ideonella sakaiensis (I. sakaiensis), Brevibacillus borstelensis*, and *Pseudomonas aestusnigri* were identified as PET consumers. Furthermore, bacteria belonging to *Pseudomonas* and *Bacillus* genera have also been listed as capable organisms for PVC biodegradation. *In vitro* degradation efficiency of some of these bacteria on several types of plastic has also been addressed. For instance, *Exiguobacterium* sp. YT2 lowered PS mass in 12.4% after 30 days of incubation at 30°C. When incubated with PP, *Pseudomonas* sp. provoked a mass loss of 60% of the polymer over a 175-day period at 30°C. A mixed cultured of *Pseudomonas citronellolis* and *Bacillus flexus* produced a 19% reduction of PS weight after 30 days at 30°C. *I. sakaiensis* almost completely degraded PET in 42 days of incubation at 30°C. Lastly, after 12 weeks of incubation at room temperature (25°C), *Pseudomonas aeruginosa* and *Corynebacterium* sp. achieved 9.3 ± 1.32% and 15.77 ± 3.5% mass loss of polyurethane (PU), respectively.[77]

An increase in efficiency is usually achieved when pretreatments are applied in bacterial biodegradation essays. Recently, a novel denitrifying bacteria isolated from compost, *Achromobacter denitrificans* Ebl13, was able to degrade PVC and LDPE after the polymers went through a thermal-oxidative pretreatment. Degradation efficiencies of Ebl13 strain were reported as mass loss, resulting in 12.3% for PVC and 6.5% for PE.[74]

Bacterial strains with important activity in the biodegradation of plastics have been isolated from aquatic environments as well. During a study of mangrove sediments, two microorganisms, identified as *Rhodococcus sp.* strain 36 and *Bacillus sp.* strain 27 were obtained. These strains diminished PP weight in 6.4% and 4% respectively, after 40 days' incubation in liquid synthetic medium.[10]

8.7 FUNGAL TREATMENT

Fungi are recognized as some of the most efficient biodegrading organisms in the planet because of the large variety of enzymes they possess. This enzymatic arsenal gives fungi an exceptional capacity to degrade sturdy compounds. Some authors consider fungi to be better at degrading plastics than bacteria.[14,85] Fungal mycelium secretes a large number of molecules that allow fungi to adhere on hydrophobic plastic surfaces. Fungi can use bioplastics and synthetic polymers as carbon sources. As noticed, fungi have a high potential for being applied in bioremediation processes; however, the slow growing rates and high costs associated with these processes have limited their application at large scale. Therefore, applied research focused on enhancing fungal bioremediation is still needed.[85] Some examples of enhancing procedures for fungal biodegradation of plastics include: application of pretreatments (sunlight, heat or acids, ozonation), utilization of fungi consortia, genetic modification of fungi, mixing with easily assimilated carbon sources, and addition of surfactants to the medium.

Filamentous fungi are able to colonize and adhere to plastic surfaces thanks to their ability to produce hydrophobins,[104] which are a group of small cysteine-rich proteins known for their capacity to form water repellent coatings in the surface of numerous materials.[116] When colonization of plastic takes place, fungal mycelia increase tension and pressure on the surface, generating cracks and holes; this will break down the material into smaller particles and will allow extracellular enzymes to have a larger superficial area, transforming these particles into more biodegradable compounds. Another important characteristic of fungi is their notable resilience and endurance since many fungal strains are able to grow in adverse environments such as extreme temperatures, basic or acid mediums, and hydric stress conditions.[14]

Wood-decay fungi—also known as ligninolytic fungi—have exhibited an important capacity to degrade bioplastics and synthetic polymers. It is believed that this capacity comes from the high amount of unspecific oxidoreductase enzymes they are able to secrete to the surrounding medium.[31,96,108] This family of fungal enzymes possess high redox potentials, for instance, laccases and peroxidases can participate in the biological degradation of PVC and PE. Hydrolases are another group of fungal enzymes with broad participation in the degradation process of synthetic polymers. Cutinase and lipase are a couple of hydrolytic enzymes which are actively involved in the biodegradation of PVC and PU.[120] Further research on these and other enzymes still needs to be made to accelerate a possible application of these catalytic technologies on a larger scale.

Several environmental factors such as pH, humidity, temperature, salinity, oxygen availability, and sunlight have an influence in the biodegradation effectiveness as these parameters affect microorganism's distribution and composition and the general enzymatic activity on the plastic surface. Recent analysis of 35 studies specifically targeting conventional plastic materials visibly demonstrated that environmental variables, rather than polymer type, appear as key factors affecting the prokaryotic composition of the plastisphere (ecosystem that has adapted to the environment contaminated with plastics). Nevertheless, chemical and physical properties of the materials may alter their degradability. For instance, a higher molecular weight (Mw), higher melting point, or the presence of additives, among others, will likely reduce the biodegradability of plastic polymers.[54,57]

Ascomycetes *Penicillium oxalicum* (*P. oxalicum*) and *Penicillium chrysogenum* (*P. chrysogenum*) were isolated from a soil with a high concentration of PE. Laboratory studies demonstrated that *P. oxalicum* and *P. chrysogenum* were able to degrade LDPE and HDPE and use them as a carbon source, achieving a degradation efficiency of 50% for both materials in 90 days. These results were attributed to their capacity to produce a biofilm on the polymer's hydrophobic surface and the secretion of diverse hydrolytic and oxidative enzymes.[84] The use of microorganisms' consortia has generated a great biotechnological interest in the last few years. It has been opined that the effective communication between bacteria and microscopic fungi (quorum sensing) allows these organisms to elaborate functional biofilms which will create a proper environment where enzymes will be secreted; this will ultimately permit an efficient biodegradation of recalcitrant materials.[103] Some bacterial–fungal associations discovered within the gut of several insects have shown a notable capability to degrade PS, for example, *Trichoderma* sp., *Penicillium* sp., *Hyphodermella* sp., and *Aspergillus* sp.[32]

LDPE can be used as sole carbon during the growth of a fungal consortium conformed by *Aspergillus niger*, *Aspergillus flavus*, and *Aspergillus oryzae*. This consortium produced an LDPE mass loss of 26.15% under laboratory-controlled conditions in 55 days. Structural changes in PE surface were detected using electronic microscopy analyses.[37]

Aspergillus terreus and *Aspergillus sydowii* are also able to degrade PE. During the degradation process, several metabolites were identified, and the lack of toxicity from those metabolites was proved.[105] The absence of negative effects from the PE degradation products on animals and plants postulates fungal biodegradation as one of the most environmentally friendly options for plastic waste treatment.

PS can also be degraded by fungal strains. *Cephalosporium* sp. and *Mucor* sp. have shown an interesting yet low potential for PS degradation. *Cephalosporium* sp. was able to diminish PS mass by 2.17% while *Mucor sp* decreased the mass of PS by only 1.81% in an 8-week period of time.[27]

In another work, PS was able to be used as a sole carbon source by an association of *Aspergillus sp.* with unidentified bacterial strains from cattle's gut. Moreover, it was detected that *Aspergillus sp.* effectively metabolized toxic additives from PS namely O-xylene, cis-9 hexadecenal, and tetradecanoic acid.[4]

Efforts to mitigate plastic pollution have led to the development and design of new materials, being bioplastic one of the best alternatives to conventional plastic nowadays. Bioplastics or biodegradable plastics can be rapidly broken under environmental conditions. It has been observed that thermophilic fungus *Thermomyces lanuginosus* can efficiently degrade bioplastics in compost or soil and under either controlled or environmental conditions. This indicates that native microbiota is very likely to degrade this type of plastics in landfills or waste deposits. Bioplastics are still under development to achieve a reduction of production costs. Thus, it is very important to keep studying different manufacturing techniques and the use of novel raw materials like starch, cellulose, molasses, among others, with the intention not only of reducing their cost but also of improving their biodegradability rates.

8.8 ENZYMATIC TREATMENT

Enzymes are biological catalysts that can accelerate several reactions such as oxidation, reduction, hydrolysis, esterification, synthesis, and molecular interconversions.[111] Enzymes can act on a broad range of substrates with different degrees of specification. Thus, several enzymes with noticeable activity have been observed during plastic biorecycling. Some of these enzymes are secreted by macro-organisms and many more are produced by microorganisms. During their catalytic cycle, plastic-degrading enzymes break down the hydrocarbon backbone of synthetic polymers generating products that are readily available for being absorbed or furtherly metabolized within the cells.[77]

It can be said that biotic oxidation of plastic is mediated directly or indirectly by enzymes; the action of this protein machinery results in splitting of polymeric chains into small oligomers or monomers.[26,77] As an illustrative example, it has been found that bacterial extracellular polymerase catalyzes primary fragmentation of synthetic polymers by either oxidizing or hydrolyzing the hydrocarbon chain; this exposes polar functional groups, increasing the solubility of the plastic. Also, this enzymatic catalysis can take place in aerobic and anaerobic conditions.[59]

PE, PET, and PP biodegradation can be achieved through the action of several types of procaryotic and eucaryotic enzymes like cutinases, esterases, lipases, lacases, peroxidases, proteases, and ureases.[20] Esterases are proteins that catalyze the hydrolytic rupture of the ester bonds of many polyesters and other synthetic polymers. There are a large variety of esterases which differ from each other in structure, substrate specificity, and biological functions.[77] An esterase obtained from *Nocardia sp.* can catalyze the breakdown of thin layers of translucent PET, while other esterase isolated from *Thermomonospora fusca* KW3b was able to hydrolyze ester bonds in PET fibers, exposing a hydroxyl group on the fiber's surface.[59]

Cutinases are an important subclass of esterases. Interestingly, cutinases have shown a strong capacity to hydrolyze high molecular weight polyesters.[28] Some cutinases have displayed α/β hydrolase activity while other cutinases have exhibited carboxilic ester hidrolase activity. A cutinase from *Thermobifida fusca* was found to be able to hydrolase aromatic polyesters. Other reports demonstrated that cutinases obtained from *Humicola insolens, Pseudomonas mendocina*, and *Fusarium solani* were able to degrade low crystallinity PET films, producing superficial modifications on the material.[59] Few cutinases have also been obtained from thermophile Actinomycetes, that is, *Thermobifida alba*, *Thermobifida cellulosilytica*, and *Thermomonospora curvata.*

Some cutinases possess a notorious versatility, being able to hydrolase a good number of different polyesters such as polycaprolactone (PCL), co-polyesters (like PET), and polytrimethylene terephthalate (PTT).[77] The examples of these cutinases are PETase and MHETase which have been isolated from *I. sakaiensis* 201-F. Both the enzymes can catalyze a series of reactions in PET. PETase catalyzes the initial reaction on PET to produce mono-(2-hydroxyethyl) terephthalic acid (MHET), while MHETase can convert MHET into ethylene glycol and terephthalic acid.[59] Lipase has also been used for the hydrolysis of PET. Biodegradation of PET using lipase from *Candida cylindracea* and *Pseudomonas sp.* was achieved by using nanoparticles of the plastic.[73]

An active search for novel plastic-degrading enzymes is currently taking place in the scientific community. Many studies look for a deeper comprehension of enzymatic mechanisms within many types of different organisms, which would ultimately lead to the detection of the enzymes involved in the degradation pathways. The long-term goal of plastic biocatalysis researchers is to produce efficient, stable, and economic enzymes that can be applied in the waste management industry. In order to do so, a lot of work has been done with extremophile microorganisms, since most of the plastic contaminated sites tend to have extreme conditions such as low or elevated temperatures, acidic or alkaline pH, high salt concentrations, or high osmotic pressure.[20]

In the case of halophilic and thermophilic enzymes, these frequently present a longer life cycle, which can be helpful for storage and transportation and would prevent a significant loss of enzymatic activity.[86]

The use of enzymes for plastic degradation is a promising approach. However, there are still some aspects that must be completely addressed if an effective application is desired. For instance, it is imperative to have a complete understanding of the exact degradation pathways and mechanisms of reaction. Additionally, there are still a lot of opportunities for biocatalysis optimization through bioinformatics and synthetic biology. This would allow us to reduce operational costs by enhancing enzymatic activity and increasing the reuse of these biological catalysts on a large, commercial scale.

8.9 CONCLUSIONS

Plastic accumulation in the environment is a pressing concern for modern societies that has been caused by the excessive use, the high recalcitrance, and the deficient management of synthetic polymers. The existing plastic treatment techniques such as landfill disposal, recycling, or incineration have proven to be inefficient and highly contaminating. Alternatively, plastic biodegradation seems to be the only option with nearly zero environmental impact, fitting into the current circular economy models thanks to the possibility of obtaining valuable products from it. However, biological treatment of plastic is still a slow and difficult process that requires more development and research in order to be implemented in vast contaminated sites. The big challenge today is to accomplish effective and inexpensive biorecycling technologies through the improvement of bioplastics, the use of efficient biodegrading insects, the standardization of process conditions in microbial cultures, and the obtention of highly active, stable enzymes for biocatalytic applications.

KEYWORDS

- **plastic**
- **waste management**
- **bacteria**
- **fungi**
- **arthropods**

REFERENCES

1. Aboelkheir, M. G.; Visconte, L. Y.; Oliveira, G. E.; Toledo Filho, R. D.; Souza, F. G. The Biodegradative Effect of *Tenebrio molitor* Linnaeus Larvae on Vulcanized SBR and Tire Crumb. *Sci. Total Environ.* **2019,** 649, 1075–1082.
2. Abraham, J.; Ghosh, E.; Mukherjee, P.; Gajendiran, A. Microbial Degradation of Low Density Polyethylene. *Environ. Prog. Sustain. Energy* **2017,** *36*, 147–154.
3. Ajith, N.; Arumugam, S.; Parthasarathy, S.; Manupoori, S.; Janakiraman, S. Global Distribution of Microplastics and its Impact on Marine Environment-A Review. *Environ. Sci. Pollut. Res. Int.* **2020,** *27*, 25970–25986.
4. Akerele, O. S.; Buraimoh, O. M.; Humphrey, I.; Ilori, M. O. Degradation of Polystyrene by Tropical Bacterial and Fungal Isolates. *J. Appl. Sci. Environ. Manage.* **2021,** *25*, 1721–1727.
5. Al Hosni, A. S.; Pittman, J. K.; Robson, G. D. Microbial Degradation of Four Biodegradable Polymers in Soil and Compost Demonstrating Polycaprolactone as an Ideal Compostable Plastic. *Waste Manage.* **2019,** *97*, 105–114.
6. Alhazmi, H.; Almansour, F. H.; Aldhafeeri, Z. Plastic Waste Management: A Review of Existing Life Cycle Assessment Studies. *Sustainability* **2021,** *13*, 5340.
7. Al-Salem, S. M.; Al-Hazza'a, A.; Karam, H. J.; Al-Wadi, M. H.; Al-Dhafeeri, A. T.; Al-Rowaih, A. Insights into the Evaluation of the Abiotic and Biotic Degradation Rate of Commercial Pro-Oxidant Filled Polyethylene (PE) Thin Films. *J. Environ. Manage.* **2019,** *250*, 109475.
8. Alshehrei, F. Biodegradation of Synthetic and Natural Plastic by Microorganisms. *J. Appl. Environ. Microbiol.* **2017,** *5*, 8–19.
9. Arthur, C.; Baker J.; Bamford, H. Eds; Proceedings of the International Research Workshop on the Occurrence, Effects and Fate of Microplastic Marine Debris. NOAA Technical Memorandum NOS-ORandR-30, 2009.
10. Auta, H. S.; Emenikea, C. U.; Jayanthia, B.; Fauziaha, S. H. Growth Kinetics and Biodeterioration of Polypropylene Microplastics by *Bacillus* sp. and *Rhodococcus* sp. Isolated from Mangrove Sediment. *Mar. Pollut. Bull.* **2018,** *127*, 15–21.
11. Babaremu, K. O.; Okoya, S. A.; Hughes, E.; Tijani, B.; Teidi, D.; Akpan, A.; Igwe, J.; Karera, S.; Oyinlola, M.; Akinlabi, E. T. Sustainable Plastic Waste Management in a Circular Economy. *Heliyon* **2022,** *8*, e09984.
12. Bahl, S.; Dolma, J.; Singh, J.; Sehgal, S. Biodegradation of Plastics: A State of the Art Review. *Mater. Today Proc.* **2021,** *39*, 31–34.
13. Bertocchini, F.; Arias, C. F. Why have we not yet Solved the Challenge of Plastic Degradation by Biological Means? *PLoS Biol.* **2023,** *21*, e3001979.
14. Bhavsar, P.; Bhave, M.; Webb, H. K. Solving the Plastic Dilemma: The Fungal and Bacterial Biodegradability of Polyurethanes. *World J. Microbiol. Biotechnol.* **2023,** *39*, 1–11.
15. Billen, P.; Khalifa, L.; Van Gerven, F.; Tavernier, S.; Spatari, S. Technological Application Potential of Polyethylene and Polystyrene Biodegradation by Macro-Organisms such as Mealworms and Wax Moth Larvae. *Sci. Total Environ.* **2020,** *735*, 139521.
16. Bombelli, P.; Howe, C. J.; Bertocchini, F. Polyethylene Bio-degradation by Caterpillars of the Wax Moth *Galleria mellonella. Curr. Biol.* **2017,** *27*, R292–R293.
17. Bonhomme, S.; Cuer, A.; Delort A. M.; Lemaire, J.; Sancelme, M.; Scott, G. Environmental Biodegradation of Polyethylene. *Polym. Degrad. Stab.* **2003,** *81*, 441–52.

18. Brandon, A. M.; Gao, S. H.; Tian, R.; Ning, D.; Yang, S. S.; Zhou, J.; Wu, W. M.; Criddle, C. S. Biodegradation of Polyethylene and Plastic Mixtures in Mealworms (Larvae of *Tenebrio molitor*) and Effects on the Gut Microbiome. *Environ. Sci. Technol.* **2018,** *52*, 6526–6533.
19. Briassoulis D. Mechanical Behaviour of Biodegradable Agricultural Films under Real Field Conditions. *Polym. Degrad. Stab.* **2006,** *91*, 1256–1272.
20. Cai, Z.; Li, M.; Zhu, Z.; Wang, X.; Huang, Y.; Li, T.; Gong, H.; Yan, M. Biological Degradation of Plastics and Microplastics: A Recent Perspective on Associated Mechanisms and Influencing Factors. *Microorganisms* **2023,** *11*, 1–17.
21. Campanale, C.; Stock, F.; Massarelli, C.; Kochleus, C.; Bagnuolo, G.; Reifferscheid, G.; Uricchio, V. Microplastics and their Possible Sources: The Example of Ofanto River in Southeast Italy. *Environ. Pollut.* **2019,** *258*, 113284.
22. Capitain, C.; Ross-Jones, J.; Möhring, S.; Tippkötter, N. Differential Scanning Calorimetry for Quantification of Polymer Biodegradability in Compost. *Int. Biodeterior. Biodegradation* **2020,** *149*, 104914.
23. Carmen, S. Microbial Capability for the Degradation of Chemical Additives Present in Petroleum-Based Plastic Products: A Review on Current Status and Perspectives. *J. Hazard. Mater.* **2021,** *402*, 123534.
24. Cassone, B. J.; Grove, H. C.; Elebute, O.; Villanueva, S. M. P.; LeMoine, C. M. R. Role of the Intestinal Microbiome in Low-Density Polyethylene Degradation by Caterpillar Larvae of the Greater Wax Moth, *Galleria mellonella*. *Proc. Biol. Sci.* **2020,** *287*, 20200112.
25. Castro-Aguirre, E.; Auras, R.; Selke, S.; Rubino, M.; Marsh, T. Insights on the Aerobic Biodegradation of Polymers by Analysis of Evolved Carbon Dioxide in Simulated Composting Conditions. *Polym. Degrad. Stab.* **2017,** *137*, 251–271.
26. Arela, M. C.; Masco, F. S.; Churac, R. A. C.; Rivera, J. E. V.; Calla, J. C. Biodegradation of plastic polymers by *Pseudomonas*. *Revista de Investigación: Ciencia, Tecnología y Desarrollo* **2020,** *6*, 46–59.
27. Chaudhary, A. K.; Vijayakumar, R. P. Studies on Biological Degradation of Polystyrene by Pure Fungal Cultures. *Environ. Dev. Sustain.* **2020,** *22*, 4495–4508.
28. Chen, S.; Su, L.; Chen, J.; Wu, J. Cutinase: Characteristics, Preparation and Application. *Biotechnol. Adv.* **2013,** *31*, 1754–1767.
29. Cole, M.; Lindeque, P.; Halsband, C.; Galloway, T. S. Microplastics as Contaminants in the Marine Environment: A Review. *Marine Pollut. Bull.* **2011,** *62*, 2588–2597.
30. Cook, T. How are Microplastics Transported to Polar Regions? *EOS* **2019,** 100.
31. Cowan, A. R.; Costanzo, C. M.; Benham, R.; Loveridge, E. J.; Moody, S. C. Fungal Bioremediation of Polyethylene: Challenges and Perspectives. *J. Appl. Microbiol.* **2022,** *132*, 78–89.
32. Cucini, C.; Leo, C.; Vitale, M.; Frati, F.; Carapelli, A.; Nardi, F. Bacterial and Fungal Diversity in the Gut of Polystyrene-Fed *Alphitobius diaperinus* (Insecta: Coleoptera). *Animal Gene* **2020,** *17*, 200109.
33. De Santana, F. S.; Gracioso, L. H.; Karolski, B.; dos Passo, G. B.; Mendes, M. A.; do Nascimento, C. A.; Perpetuo, E. A. Isolation of Bisphenol A-Tolerating/Degrading *Shewanella haliotis* Strain MH137742 from an Estuarine Environment. *Appl. Biochem. Biotechnol.* **2019,** *189*, 103–115.

34. De Souza-Machado, A. A.; Kloas, W.; Zarfl, C.; Hempel, S.; and Rillig, M. C. Microplastics as an Emerging Threat to Terrestrial Ecosystems. *Glob. Chang. Biol.* **2018,** *24*, 1405–1416.
35. Dowd, P. F.; Shen, S. K. The Contribution of Symbiotic Yeast to Toxin Resistance of the Cigarette Beetle (*Lasioderma serricorne*). *Entomologia Experimentalis et Applicata* **1990,** *56*, 241–248.
36. Drzyzga, O.; Prieto, A. Plastic Waste Management, a Matter for the 'Community'. *Microbial. Biotechnol.* **2019,** *12*, 66–68.
37. DSouza, G. C.; Sheriff, R. S.; Ullanat, V.; Shrikrishna, A.; Joshi, A. V.; Hiremath, L.; Entoori, K. Fungal Biodegradation of Low-Density Polyethylene using Consortium of Aspergillus Species under Controlled Conditions. *Heliyon* **2021,** *7*, e07008.
38. Elijah, I.; Ohimain, E. I.; Bawo, D. D.; Ere, D.; Peters, E. N. Biodeterioration of Plastics by Weevils; An Environmental and Stored Food Product Perspective. *Int. J. Agric. Environ. Sci.* **2015,** *3*, 51–55.
39. Frias, J.; Nash, R. Microplastics: Finding a Consensus on the Definition. *Mar. Pollut. Bull.* **2018,** *138*, 145–147.
40. Geyer, R.; Jambeck, J. R.; Law, K. L. Production, Use, and Fate of all Plastics Ever Made. *Sci. Adv.* **2017,** *3*, e1700782.
41. Ghatge, S.; Yang, Y.; Ahn, J. H.; Hur, H. G. Biodegradation of Polyethylene: A Brief Review. *Appl. Biol. Chem.* **2020,** *63*, 1–14.
42. Hakkarainen, K.; Ilomäki, L.; Lipponen, L.; Muukkonen, H.; Rahikainen, M.; Tuominen, T.; Lakkala, M.; Lehtinen, E. Students' Skills and Practices of using ICT: Results of a National Assessment in Finland. *Comput. Educat.* **2000,** *34* (2), 103–117.
43. Harrison, J. P.; Boardman, C.; O'Callaghan, K.; Delort, A. M.; Song, J. Biodegradability Standards for Carrier Bags and Plastic Films in Aquatic Environments: A Critical Review. *Royal Soc. Open Sci.* **2018,** *5*, 171792.
44. Helm, L. T.; Murphy, E. L.; McGivern, A.; Borrelle, S. B. Impacts of Plastic Waste Management Strategies. *Environ. Rev.* **2022,** *31*, 45–65.
45. Hirota, B.; Okude, G.; Anbutsu, H.; Futahashi, R.; Moriyama, M.; Meng, X. Y.; Nikoh, N.; Koga, R.; Fukatsu, T. A Novel, Extremely Elongated, and Endocellular Bacterial Symbiont Supports Cuticle Formation of a Grain Pest Beetle. *mBio* **2017,** *8*, e01482-17.
46. Hu, L.; Fu, J.; Zheng, P.; Dai, M.; Zeng, G.; Pan, X. Accumulation of Microplastics in Tadpoles from Different Functional Zones in Hangzhou Great Bay Area, China: Relation to Growth Stage and Feeding Habits. *J. Hazard. Mater.* **2022,** *424*, 127665.
47. Huang, C. Y.; Roan, M. L.; Kuo, M. C.; Lu, W. L. Effect of Compatibiliser on the Biodegradation and Mechanical Properties of High-Content Starch/Low-Density Polyethylene Blends. *Polym. Degrad. Stab.* **2005,** *90*, 95–105.
48. Huang, S.; Wang, H.; Ahmad, W.; Ahmad, A.; Vatin, N. I.; Mohamed, A. M.; Mehmood, I. Plastic Waste Management Strategies and their Environmental Aspects: A Scientometric Analysis and Comprehensive Review. *Int. J. Environ. Res. Public Health* **2022,** *19*, 4556.
49. Idumah, C. I.; Nwuzor, I. C. Novel Trends in Plastic Waste Management. *SN Appl. Sci.* **2019,** *1*, 1–14.
50. Ijgosse, J. Incineración y población recicladora: Una guía técnica sobre aprovechamiento energético. *Women In Informal Employment Globalizing and Organizing.* WIEGO. Nota técnica 11, 2019.

51. Jayan, N.; Skariyachan, S.; Sebastián, D. The Escalated Potential of the Novel Isolate *Bacillus cereus* NJD1 for Effective Biodegradation of LDPE Films without Pre-treatment. *J. Hazard. Mater.* **2023,** *455*, 131623.
52. Jeon, H. J.; Kim, M. N. Functional Analysis of Alkane Hydroxylase System Derived from *Pseudomonas aeruginosa* E7 for Low Molecular Weight Polyethylene Biodegradation. *Int. Biodeterior. Biodegradation* **2015,** *103*, 141–146.
53. Jiang, S.; Su, T.; Zhao, J.; Wang, Z. Isolation, Identification, and Characterization of Polystyrene-Degrading Bacteria from the Gut of *Galleria mellonella* (Lepidoptera: Pyralidae) Larvae. *Front. Bioeng. Biotechnol.* **2021,** *9*, 736062.
54. Kale, S. K.; Deshmukh, A. G.; Dudhare, M. S.; Patil, V. B. Microbial Degradation of Plastic: A Review. *J. Biochem. Technol.* **2015,** *6*, 952–961.
55. Kesti, S. S. K.; Thimmappa, S. C. T. First Report on Biodegradation of Low Density Polyethylene by Rice Moth Larvae, *Corcyra cephalonica* (Stainton). *Holistic Approach Environ.* **2019,** *9*, 79–83.
56. Khan, S.; Dong, Y.; Nadir, S.; Schaefer, D. A.; Mortimer, P. E.; Xu, J.; Ye, L.; Gui, H.; Wanasinghe, D. N.; Dossa, G. G. O.; Yu, M.; Sheng, J. Valorizing Plastic Waste by Insect Consumption. *Circ. Agric. Syst.* **2021,** *1*, 1–9.
57. Khruengsai, S.; Sripahco, T.; Pripdeevech, P. Low-Density Polyethylene Film Biodegradation Potential by Fungal Species from Thailand. *J. Fungi* **2021,** *7*, 1–18.
58. Kim, H. R.; Lee, H. M.; Yu, H. C.; Jeon, E.; Lee, S.; Li, J.; Kim, D. H. Biodegradation of Polystyrene by *Pseudomonas* sp. Isolated from the Gut of Superworms (Larvae of *Zophobas atratus*). *Environ. Sci. Technol.* **2020,** *54*, 6987–6996.
59. Koshti, R.; Mehta, L.; Samarth, N. Biological Recycling of Polyethylene Terephthalate: A Mini-Review. *J. Polym. Environ.* **2018,** *26*, 3520–3529.
60. Kotova, I. B.; Taktarova, Y. V.; Tsavkelova, A. E.; Egorova, M. A.; Bubnov, I. A.; Malakhova, D. V.; Shirinkina, L. I.; Sokolova, T. G.; Bonch-Osmolovskaya, A. E. Microbial Degradation of Plastics and Approaches to Make It More Efficient. *Microbiology* **2021,** *90*, 671–701.
61. Kundungal, H.; Gangarapu, M.; Sarangapani, S.; Patchaiyappan, A.; Devipriya, S. P. Efficient Biodegradation of Polyethylene (HDPE) Waste by the Plastic-Eating Lesser Waxworm (*Achroia grisella*). *Environ. Sci. Pollut. Res.* **2019,** *26*, 18509–18519.
62. Kundungal, H.; Synshiang, K.; Devipriya, S. P. Biodegradation of Polystyrene Wastes by a Newly Reported Honey Bee Pest *Uloma* sp. larvae: An Insight to the Ability of Polystyrene-Fed Larvae to Complete its Life Cycle. *Environ. Chall.* **2021,** *4*, 100083.
63. Kwadha, C. A.; Ong'amo, G. O.; Ndegwa, P. N.; Raina, S. K.; Fombong, A. T. The Biology and Control of the Greater Wax Moth, *Galleria mellonella. Insects* **2017,** *8* (61), 1–17.
64. Kyaw, B. M.; Champakalakshmi, R.; Sakharkar, M. K.; Lim, C. S.; Sakharkar, K. R. Biodegradation of Low Density Polythene (LDPE) by *Pseudomonas* Apecies. *Indian J. Microbiol.* **2012,** *52*, 411–419.
65. Kyrikou, I.; Briassoulis, D. Biodegradation of Agricultural Plastic Films: A Critical Review. *J. Polym. Environ.* **2007,** *15*, 125–150.
66. Law, K. L.; Narayan, R. Reducing Environmental Plastic Pollution by Designing Polymer Materials for Managed End-of-Life. *Nat. Rev. Mater.* **2021,** *7*, 104–116.
67. Lear, G.; Kingsbury, J. M.; Franchini, S.; Gambarini, V.; Maday, S. D. M.; Wallbank, J. A.; Weaver, L.; Pantos, O. Plastics and the Microbiome: Impacts and Solutions. *Environ. Microbiome.* **2021,** *16*, 1–19.

68. Li, W. C.; Tse, H. F.; Fok, L. Plastic Waste in the Marine Environment: A Review of Sources, Occurrence and Effects. *Sci. Total. Environ.* **2016,** 566–567, 333–349.
69. Lou, Y.; Ekaterina, P.; Yang, S. S.; Lu, B.; Liu, B.; Ren, N.; Corvini, P. F. X.; Xing, D. Biodegradation of Polyethylene and Polystyrene by Greater Wax Moth Larvae (*Galleria mellonella* L.) and the Effect of Co-diet Supplementation on the Core Gut Microbiome. *Environ. Sci. Technol.* **2020,** *54*, 2821–2831.
70. Lou, Y.; Li, Y.; Lu, B.; Liu, Q.; Yang, S. S.; Liu, B.; Ren, N.; Wu, W. M.; Xing, D. Response of the Yellow Mealworm (*Tenebrio molitor*) Gut Microbiome to Diet Shifts During Polystyrene and Polyethylene Biodegradation. *J. Hazard. Mater.* **2021,** *416*, 126222.
71. Lucas, N.; Bienaime, C.; Belloy, C.; Queneudec, M.; Silvestre, F.; Nava-Saucedo, J. E. Polymer Biodegradation: Mechanisms and Estimation Techniques – A Review. *Chemosphere* **2008,** *73*, 429–442.
72. Luo, L.; Wang, Y.; Guo, H.; Yang, Y.; Qi, N.; Zhao, X. Biodegradation of Foam Plastics by *Zophobas atratus* Larvae (Coleoptera: Tenebrionidae) Associated with Changes of Gut Digestive Enzymes Activities and Microbiome. *Chemosphere* **2021,** *282*, 131006.
73. Ma, M.; Wang, L.; Zhu, H. Enzymatic Degradation of Polyester-Nanoparticles by Lipases and Adsorption of Lipases on the Polyester-Nanoparticles. *Adv. Mater. Res.* **2012,** 418–420, 2302–2307.
74. Maleki Rad, M.; Moghimi, H.; Azin, E. Biodegradation of Thermo-Oxidative Pretreated Low-Density Polyethylene (LDPE) and Polyvinyl Chloride (PVC) Microplastics by *Achromobacter denitrificans* Ebl13. *Marine Pollut. Bull.* **2022,** *181*,113830.
75. Matyja, K.; Rybak, J.; Hanus-Lorenz, B.; Wróbel, M.; Rutkowski, R. Effects of Polystyrene Diet on *Tenebrio molitor* Larval Growth, Development and Survival: Dynamic Energy Budget (DEB) Model Analysis. *Environ. Pollut.* **2020,** *264*, 114740.
76. Mohan, K. Microbial Deterioration and Degradation of Polymeric Materials. *J. Biochem. Technol.* **2011,** *2*, 210–215.
77. Mohanan, N.; Montazer, Z.; Sharma P. K.; Levin D. V. Microbial and Enzymatic Degradation of Synthetic Plastics. *Front. Microbiol.* **2020,** *11* (580709), 1–22.
78. Moharir, R. V.; Kumar, S. Challenges Associated with Plastic Waste Disposal and Allied Microbial Routes for its Effective Degradation: A Comprehensive Review. *J. Clean. Prod.* **2019,** *208*, 65–76.
79. Montazer, Z.; Habibi- Najafi, M. B.; Levin, D. B. Microbial Degradation of Low-Density Polyethylene and Synthesis of Polyhydroxyalkanoate Polymers. *Can J Microbiol.* **2019,** *65*, 224–234.
80. Morse, R. A.; Flottum, K. In *Honey Bee Pests, Predators and Diseases*; 3rd Ed.; Root Company: Ohio, 1997.
81. Müller, R. J. Biodegradability of Polymers: Regulations and Methods for Testing. In *Biopolymers Online: Biology• Chemistry• Biotechnology• Applications*, 2002; vol *10*.
82. Müller, R. J.; Schrader, H.; Profe, J.; Dresler, K.; Deckwer, W. D. Enzymatic Degradation of Poly(ethylene terephthalate): Rapid Hydrolysis using a Hydrolase from *T. fusca*. *Macromol. Rapid Commun.* **2005,** *26*, 1400–1405.
83. Narancic, T.; O'Connor, K. E. Plastic Waste as a Global Challenge: Are Biodegradable Plastics the Answer to the Plastic Waste Problem?. *Microbiology* **2019,** *165*, 129–137.
84. Ojha, N.; Pradhan, N.; Singh, S.; Barla, A.; Shrivastava, A.; Khatua, P.; Ray, V.; Bose, S. Evaluation of HDPE and LDPE Degradation by Fungus, Implemented by Statistical Optimization. *Sci. Rep.* **2017,** *7*, 39515.

85. Okal, E. J.; Heng, G.; Magige, E. A.; Khan, S.; Wu, S.; Ge, Z.; Zhang, T.; Mortimer, P. E.; Xu, J. Insights into the Mechanisms Involved in the Fungal Degradation of Plastics. *Ecotoxicol. Environ. Saf.* **2023,** *262*, 115202.
86. Orellana, R.; Macaya, C.; Bravo, G.; Dorochesi, F.; Cumsille, A.; Valencia, R.; Rojas, C.; Seeger, M. Living at the Frontiers of Life: Extremophiles in Chile and Their Potential for Bioremediation. *Front. Microbiol.* **2018,** *9*, 2309.
87. Organisation for Economic Co-operation and Development (OECD). Global Plastics Outlook [Online] 2023. https://www.oecd.org/environment/plastics/
88. Pan, D.; Su, F.; Liu, C.; Guo, Z. Research Progress for Plastic Waste Management and Manufacture of Value-Added Products. *Adv. Compos. Hybrid Mater.* **2020,** *3*, 443–461.
89. Park, S. Y.; Kim, C. G. Biodegradation of Micro-Polyethylene Particles by Bacterial Colonization of a Mixed Microbial Consortium Isolated from a Landfill Site. *Chemosphere* **2019,** *222*, 527–533.
90. Peng, B. Y.; Li, Y.; Fan, R.; Chen, Z.; Chen, J.; Brandon, A. M.; Criddle, C. S.; Zhang, Y.; Wu, W. M. Biodegradation of Low-Density Polyethylene and Polystyrene in Superworms, Larvae of *Zophobas atratus* (Coleoptera: Tenebrionidae): Broad and Limited Extent Depolymerization. *Environ. Pollut.* **2020,** *266*, 115206.
91. Peng, B.Y.; Su, Y.; Chen, Z.; Chen, J.; Zhou, X.; Benbow, M. E.; Criddle, C. S.; Wu, W. M.; Zhang, Y. Biodegradation of Polystyrene by Dark (*Tenebrio obscurus*) and Yellow (*Tenebrio molitor*) Mealworms (Coleoptera: Tenebrionidae). *Environ. Sci. Technol.* **2019,** *53*, 5256–5265.
92. Pivokonsky, M.; Cermakova, L.; Novotna, K.; Peer, P.; Cajthaml, T.; Janda, V. Occurrence of Microplastics in Raw and Treated Drinking Water. *Sci. Total Environ.* **2018,** *643*, 1644–1651.
93. Plastic Waste Makers Index, 2023. Copyright © 2023. The Minderoo Foundation Pty Ltd. [Online] 2023. https://cdn.minderoo.org/content/uploads/2023/02/04205527/Plastic-Waste-Makers-Index-2023.pdf
94. Prasad, R. K.; Chatterjee, S.; Mazumder, P. B.; Sharma, S.; Datta, S.; Vairale, M. G. Study on Cellulase (B-1, 4-endoglucanase) Activity of Gut Bacteria of *Sitophilus oryzae* in Cellulosic Waste Biodegradation. *Bioresour. Technol. Rep.* **2019,** *7*,100274.
95. Prata, J. C. Airborne Microplastics: Consequences to Human Health? *Environ. Pollut.* **2018,** *234*, 115–126.
96. Ratuchne, A.; Lonardoni, E. A.; Bueno, C. E.; Reis, G. F.; Rezende, M. I.; Urbano, A.; Biz, G.; Couto de Almeida, R. S.; Panagio, L. A. *Pleurotus ostreatus* and a Novel Fungal Composite: Development and Bioremediation of Plastic Wastes. *Resour. Conserv. Recycl. Adv.* **2023,** 200167.
97. Ren, L.; Men, L.; Zhang, Z.; Guan, F.; Tian, J.; Wang, B.; Wang, J.; Zhang, Y.; Zhang, W. Biodegradation of Polyethylene by *Enterobacter* sp. D1 from the Guts of Wax Moth *Galleria mellonella*. *Int. J. Environ. Res. Public Health* **2019,** *16*, 11.
98. Rezania, S.; Park, J.; Din, M. F. M.; Taib, S. M.; Talaiekhozani, A.; Yadav, K. K.; Kamyab, H. Microplastics Pollution in Different Aquatic Environments and Biota: A Review of Recent Studies. *Mar. Pollut. Bull.* **2018,** *133*, 191–208.
99. Rigamonti, L.; Grosso, M.; Møller, J.; Sanchez, V. M.; Magnani, S.; Christensen, T. H. Environmental Evaluation of Plastic Waste Management Scenarios. *Resour. Conserv. Recy.* **2014,** *85*, 42–53.
100. Rillig, M. C.; Ingraffia, R.; De Souza Machado, A. Microplastic Incorporation into Soil in Agroecosystems. *Front. Plant Sci.* **2017,** *8*, 1805.

101. Riudavets, J.; Salas, I.; Pons, M. J. Damage Characteristics Produced by Insect Pests in Packaging Film. *J. Stored Prod. Res*. **2007,** *43*, 564–570.
102. Ru, J.; Huo, Y.; Yang, Y. Microbial Degradation and Valorization of Plastic Wastes. *Front. Microbiol.* **2020,** *11*, 1–20.
103. Ruiz, A.; Herráez, M.; Costa-Gutierrez, S. B.; Molina-Henares, M. A.; Martínez, M. J.; Espinosa-Urgel, M.; Barriuso, J. The Architecture of a Mixed Fungal–Bacterial Biofilm is Modulated by Quorum-Sensing Signals. *Environ. Microbiol.* **2021,** *23*, 2433–2447.
104. Sánchez, C. Fungal Potential for the Degradation of Petroleum-Based Polymers: An Overview of Macro-and Microplastics Biodegradation. *Biotechnol. Adv.* **2020,** *40*, 107501.
105. Sangale, M. K.; Shahnawaz, M.; Ade, A. B. Gas Chromatography-Mass Spectra Analysis and Deleterious Potential of Fungal Based Polythene-Degradation Products. *Sci. Rep.* **2019,** *9*, 1599.
106. Sanluis-Verdes, A.; Colomer-Vidal, P.; Rodriguez-Ventura, F.; Bello-Villarino, M.; Spinola-Amilibia, M.; Ruiz-Lopez, E.; Illanes-Vicioso, R.; Castroviejo, P.; Aiese Cigliano, R.; Montoya, M.; Falabella, P.; Pesquera, C.; Gonzalez-Legarreta, L.; Arias-Palomo, E.; Solà, M.; Torroba, T.; Arias, C. F.; Bertocchini, F. Wax Worm Saliva and the Enzymes Therein are the Key to Polyethylene Degradation by *Galleria mellonella. Nat. Commun.*. **2022,** *13*, 5568.
107. Sarmah, P.; Rout, J. Algal Colonization on Polythene Carry Bags in a Domestic Solid Waste Dumping Site of Silchar Town in Assam. *Phykos* **2018,** *48*, e77.
108. Šašek, V.; Vitásek, J.; Chromcová, D.; Prokopová, I.; Brožek, J.; Náhlík, J. Biodegradation of Synthetic Polymers by Composting and Fungal Treatment. *Folia Microbiol.* **2006,** *51*, 425–430.
109. Shilpa B. N.; Meena, S. S. Microbial Biodegradation of Plastics: Challenges, Opportunities, and a Critical Perspective. *Front. Environ. Sci. Eng.* **2022,** *16*, 161.
110. Siddiqui, J.; Pandey, G. A Review of Plastic Waste Management Strategies. *Int. Res. J. Environ. Sci.* **2013,** *2*, 84.
111. Singh, B.; Sharma, N. Mechanistic Implications of Plastic Degradation. *Polym. Degrad. Stab.* **2008,** *93* (3), 561–584.
112. Singh, P.; Sharma, V. P. Integrated Plastic Waste Management: Environmental and Improved Health Approaches. *Procedia Environ. Sci.* **2016,** *35*, 692–700.
113. Skariyachan, S.; Manjunatha, V.; Sultana, S.; Jois, C.; Bai, V.; Vasist, K. S. Novel Bacterial Consortia Isolated from Plastic Garbage Processing Areas Demonstrated Enhanced Degradation for Low Density Polyethylene. *Environ. Sci. Pollut. Res. Int.* **2016,** *23*, 18307–18319.
114. So, M. W. K.; Vorsatz, L. D.; Cannicci, S.; Not C. Fate of Plastic in the Environment: From Macro to Nano by Macrofauna. *Environ. Pollut.* **2022,** *300*, 118920.
115. Sonnendecker, C.; Oeser, J.; Richter, P. K.; Hille, P.; Zhao, Z.; Fischer, C.; Lippold, H.; Blázquez-Sánchez, P.; Engelberger, F.; Ramírez-Sarmiento, C. A.; Oeser, T.; Lihanova, Y.; Frank, R.; Jahnke, H. G.; Billig, S.; Abel, B.; Sträter, N.; Matysik, J.; Zimmermann, W. Low Carbon Footprint Recycling of Post-Consumer PET Plastic with a Metagenomic Polyester Hydrolase. *Chem. Sus. Chem.* **2022,** *15*, e202101062.
116. Sunde, M.; Kwan, A. H.; Templeton, M. D.; Beever, R. E.; Mackay, J. P. Structural Analysis of Hydrophobins. *Micron* **2008,** *39*, 773–84.

117. Suresh, B.; Maruthamuthu, S.; Kannan, M.; Chandramohan, A. Mechanical and Surface Properties of Low-Density Polyethylene Film Modified by Photo-Oxidation. *Polym. J.* **2011,** *43*, 398–406.
118. Tang, Z. L.; Kuo, T. A.; Liu, H. H. The Study of the Microbes Degraded Polystyrene. *Adv. Technol. Innov.* **2017,** *2*, 13–17.
119. Taniguchi, I.; Yoshida, S.; Hiraga, K.; Miyamoto, K.; Kimura, Y.; Oda, K. Biodegradation of PET: Current Status and Application Aspects. *ACS Catal.* **2019,** *9*, 4089–4105.
120. Temporiti, M. E. E.; Nicola, L.; Nielsen, E.; Tosi, S. Fungal Enzymes Involved in Plastics Biodegradation. *Microorganisms* **2022,** *10*, 1180.
121. Tournier, V.; Topham, C. M.; Gilles, A.; David, B.; Folgoas, C.; Moya-Leclair, E.; Kamionka, E.; Desrousseaux, M. L.; Texier, H.; Gavalda, S.; Cot, M.; Guémard, E.; Dalibey, M.; Nomme, J.; Cioci, G.; Barbe. S.; Chateau, M.; André, I.; Duquesne, S.; Marty, A. An Engineered PET Depolymerase to Break Down and Recycle Plastic Bottles. *Nature* **2020,** *580*, 216–219
122. Tourova, T. P.; Sokolova, D. S.; Nazina, T. N.; Gruzdev, D. S.; Laptev, A. B. Phylogenetic Diversity of Microbial Communities from the Surface of Polyethylene Terephthalate Materials Exposed to Different Water Environments. *Microbiology* **2020,** *89*, 96–106.
123. Tschan, M. J. L.; Brulé, E.; Haquette, P.; Thomas, C. M. Synthesis of Biodegradable Polymers from Renewable Resources. *Polym. Chem.* **2012,** *3*, 836–851.
124. Urbanek, A. K.; Mirończuk, A. M.; García-Martín, A.; Saborido, A.; de la Mata, I.; Arroyo, M. Biochemical Properties and Biotechnological Applications of Microbial Enzymes Involved in the Degradation of Polyester-Type Plastics. *Biochim. Biophys. Acta Proteins Proteomics* **2020a,** *1868*, 140315.
125. Urbanek, A. K.; Rybak, J.; Wróbel, M.; Leluk, K.; Mirończuk, A. M. A Comprehensive Assessment of Microbiome Diversity in *Tenebrio molitor* Fed with Polystyrene Waste. *Environ. Pollut.* **2020b,** *262*, 114281.
126. Vanapalli, K. R.; Sharma, H. B.; Ranjan, V. P.; Samal, B.; Bhattacharya, J.; Dubey, B. K.; Goel, S. Challenges and Strategies for Effective Plastic Waste Management During and Post COVID-19 Pandemic. *Sci. Total Environ.* **2021,** *750*, 141514.
127. Vega, F. E.; Dowd P. F. The Role of Yeasts as Insect Endosymbionts. In *Insect-Fungal Associations: Ecology and Evolution*; Oxford University Press: New York, 2005; pp 211–243.
128. Viel, T.; Manfra, L.; Zupo, V.; Libralato, G.; Cocca, M.; Costantini, M. Biodegradation of Plastics Induced by Marine Organisms: Future Perspectives for Bioremediation Approaches. *Polymers* **2013,** *15*, 2673.
129. Wang, H. M.; He, Y. and Sen, B. Research and Management of Plastic Pollution in Coastal Environments of China. *Environ. Pollut.* **2019,** *248*, 898–905.
130. Wang, Z.; Xin, X.; Shi, X.; Zhang, Y. A Polystyrene Degrading *Acinetobacter* Bacterium Isolated from the Larvae of *Tribolium castaneum. Sci. Total Environ.* **2020,** *726*, 138564.
131. Waring, R. H.; Harrisa, R. M.; Mitchell, S. C. Plastic Contamination of the Food Chain: A Threat to Human Health? *Maturitas* **2018,** *115*, 64–68.
132. Weber, C.; Pusch, S.; Opatz, T. Polyethylene Bio-degradation by Caterpillars? *Curr. Biol.* **2017,** *27*, 744–R745.
133. Wei, R.; Zimmermann, W. Microbial Enzymes for the Recycling of Recalcitrant Petroleum-Based Plastics: How Far are We?. *J. Microbial. Biotechnol.* **2017,** *10*, 1308–1322.

134. Wilkes, R. A.; Aristilde, L. Degradation and Metabolism of Synthetic Plastics and Associated Products by *Pseudomonas* sp: Capabilities and Challenges. *J. Appl. Microbiol.* **2017,** *123*, 582–593.
135. Woo, S.; Song, I.; Cha, H. J. Fast and Facile Biodegradation of Polystyrene by the Gut Microbial Flora of *Plesiophthalmus davidis* Larvae. *Appl. Environ. Microbiol.* **2020,** *86*.
136. Wu, Q.; Tao, H.; Wong, M. H. Feeding and Metabolism Effects of Three Common Microplastics on *Tenebrio molitor* L. *Environ. Geochem. Health* **2019,** *41*, 17–26.
137. Yang, C. Z.; Yaniger, S. I.; Craig-Jordan, V.; Klein D. J.; Bittne, G. D. Most Plastic Products Release Estrogenic Chemicals: A Potential Health Problem That Can Be Solved. *Environ. Health Perspect.* **2011,** *109*, 989–996.
138. Yang, J.; Yang, Y.; Wu, W. M.; Zhao, J.; Jiang, L. Evidence of Polyethylene Biodegradation by Bacterial Strains from the Guts of Plastic-Eating Waxworms. *Environ. Sci. Technol.* **2014,** *48*, 13776–13784.
139. Yang, S. S.; Wu, W. M. Biodegradation of Plastics in *Tenebrio* Genus (Mealworms). In *Microplastics in Terrestrial Environments. The Handbook of Environmental Chemistry*; He, D., Luo, Y., Eds.; Springer: Cham, 2020; p 95.
140. Yang, S. S.; Brandon, A. M.; Andrew Flanagan, J. C.; Yang, J.; Ning, D.; Cai, S. Y.; Fan, H. Q.; Wang, Z. Y.; Ren, J.; Benbow, E.; Ren, N. Q.; Waymouth, R. M.; Zhou, J.; Criddle, C. S.; Wu, W. M. Biodegradation of Polystyrene Wastes in Yellow Mealworms (larvae of *Tenebrio molitor* Linnaeus): Factors Affecting Biodegradation Rates and the Ability of Polystyrene-Fed Larvae to Complete their Life Cycle. *Chemosphere* **2018a,** *191*, 979–989.
141. Yang, S. S.; Wu, W. M.; Brandon, A. M.; Fan, H. Q.; Receveur, J. P.; Li, Y.; Wang, Z. Y.; Fan, R.; Mcclellan, R. L.; Gao, S. H.; Ning, D.; Phillips, D. H.; Peng, B. Y.; Wang, H.; Cai, S. Y.; Li, P.; Cai, W. W.; Ding, L. Y.; Yang, J.; Zheng, M.; Ren, J., Zhang, Y. L.; Gao, J.; Xing, D.; Ren, N. Q.; Waymouth, R. M.; Zhou, J.; Tao, H. C.; Picard, C. J.; Benbow, M. E.; Criddle, C. S. Ubiquity of Polystyrene Digestion and Biodegradation within Yellow Mealworms, Larvae of *Tenebrio molitor* Linnaeus (Coleoptera: Tenebrionidae). *Chemosphere* **2018b,** *212*, 262–271.
142. Yang, X. G.; Wen, P. P.; Yang, Y. F.; Jia, P. P.; Li, W. G.; Pei, D. S. Plastic Biodegradation by *in vitro* Environmental Microorganisms and *in vivo* Gut Microorganisms of Insects. *Front. Microbiol.* **2023,** *13*, 1001750.
143. Yang, Y.; Wang, J.; Xia, M. Biodegradation and Mineralization of Polystyrene by Plastic-Eating Superworms *Zophobas atratus*. *Sci. Total Environ.* **2020,** *708*, 135233.
144. Yang, Y.; Yang, J.; Wu, W. M.; Zhao, J.; Song, Y.; Gao, L.; Yang, R.; Jiang, L. Biodegradation and Mineralization of Polystyrene by Plastic-Eating Mealworms: Part 1. Chemical and Physical Characterization and Isotopic Tests. *Environ. Sci. Technol.* **2015a,** *49*, 12080–12086.
145. Yang, Y.; Yang, J.; Wu, W. M.; Zhao, J.; Song, Y.; Gao, L.; Yang, R.; Jiang, L. Biodegradation and Mineralization of Polystyrene by Plastic-Eating Mealworms: Part 2. Role of Gut Microorganisms. *Environ. Sci. Technol.* **2015b,** *49*, 12087–12093.
146. Zhang, J.; Gao, D.; Li, Q.; Zhao, Y.; Li, L.; Lin, H.; Bi, Q.; Zhao, Y. Biodegradation of Polyethylene Microplastic Particles by the Fungus *Aspergillus flavus* from the Guts of Wax Moth *Galleria mellonella*. *Sci. Total Environ.* **2020,** *704*, 135931.

CHAPTER 9

Ultrasound-Assisted Techniques for Managing and Reducing Wastes: Challenges, Potentialities, and Perspectives

ALMA ROSA TORRES-RODRÍGUEZ, WENDY Y. VILLASTRIGO-LÓPEZ, JUANITA D. FLORES-VALDEZ, ADALI CASTAÑEDA-FACIO, ROSA IDALIA NARRO-CESPEDES, RAÚL RODRÍGUEZ-HERRERA, and AIDÉ SAENZ-GALINDO*

Universidad Autónoma de Coahuila, Facultad de Ciencias Químicas, Saltillo, Coahuila, México

*Corresponding author.

ABSTRACT

This chapter presents the fundamentals of ultrasound as an alternative, clean, and green energy, as well as an overview of the different current applications of energy in various areas of chemistry, highlighted in the different syntheses of polymeric materials, particularly in the fields of medical, pharmaceutical, automotive, textile, and food among others. Ultrasonic energy also has an impact on the extraction of bioactive organic compounds of pharmaceutical interest, such as from a renewable natural source or agroindustrial waste, proving that ultrasonic energy is a viable nonconventional alternative, presenting acceptable and improved extraction yields when compared with the conventional extraction methods, likewise the reduction of extraction times, adding the use of friendly solvents, such as water, ethanol, among others. Nanotechnology is a vital field where ultrasonic energy has a significant impact on the surface modification of carbon nanostructures, such as graphene or carbon nanostructures, such as graphene or carbon nanotubes

Agroindustrial Waste Management and Natural Resources Conservation: Environmental Challenges and Transformation into Value-Added Products. José Juan Buenrostro-Figueroa, Juan Manuel Tirado-Gallegos, Mónica L. Chávez-González, A. K. Haghi, and Shaheemath Suhara K. K. (Eds.)

(CNTs), as well as the dispersion of metal nanoparticles, metal oxides, polymers, and carbon nanostructures, in different matrices where polymeric matrices synthetic and natural.

9.1 INTRODUCTION

Currently, green and/or clean energies are excellent viable alternative energies to continue developing research, innovations, or improving processes or products in different areas such as medicine, polymeric materials, metallurgy, ceramics, automotive, textile, pharmacy, nanotechnology, food, packaging, among others.

Green chemistry or sustainable chemistry is a philosophy that aims to prevent environmental pollution, helping to reduce the production of waste in the chemical processes involved, in addition to designing environmentally friendly processes involving friendly and/or green energies; it is based on 12 principles that highlight microwave, wind, solar, hydraulic, and ultrasonic energies, among others. Ultrasonic or sonochemical energy is a type of energy based on cavitation, a physical phenomenon where bubbles collide and reach very high temperatures and pressures, which promotes the generation of mechanical energy that is transformed into thermal energy, increasing the temperature; in the phenomenon of cavitation, there are different factors, such as frequency, intensity, temperature, chemical solvent, nature and chemical of the sample, power, and pressure. This type of energy is considered economical, and the instrumentation used for this type of energy is easily accessible. The production of ultrasonic energy can be generated in a bath or with an ultrasonic tip; the latter can vary the power or frequency of the energy.

Unlike other types of energy, ultrasound energy is considered the most environmentally friendly. Currently, ultrasound energy is used in different areas, such as food, nanotechnology, textile, automotive, medicine, and pharmacy, among others. An area of great interest is the obtaining of organic compounds of pharmaceutical interest from natural sources such as leaves, seeds, stems, flowers, or waste, assisting the extraction with ultrasonic energy, using solvents, green or friendly, where solvents stand out, polar extractors, such as water, ethanol methanol or mixtures of them, to date it has been reported that this type of methodology ensures good yields. Another novel area is nanotechnology, and ultrasound energy has an important role; for example, it is useful for the dispersion of nanoparticles or nanostructures in polymers to obtain natural and/or synthetic polymeric nanocomposites, dispersion at the nanometric level is highly viable, assisted by ultrasonic energy, this type of polymeric nanocomposites have essential applications in different areas.

The objective of this chapter is to provide a general overview of ultrasound energy as an environmentally friendly green energy that has an impact on green chemistry; the foundation and its current and novel applications are generally presented, as well as the advantages and disadvantages.

9.2 GREEN CHEMISTRY

Chemistry is an area that has enormously contributed to the development of human beings, from obtaining essential products, such as food, medicines, textiles, and homes, among others. All the scientific advances that have been made throughout history have brought significant advantages to the way of life of human beings. However, at an environmental level, they have set the tone for severe environmental problems.[1]

In 2021, Bolaños Guerra published about Nobel Prize Winner Mario Molina, who carried out his postdoctoral stay with Dr. Rowland at the University of California in 1973. Molina Carried out a study on the effect of chlorofluorocarbons (CFCs) on the ozone, which was produced mainly in the refrigeration industry. The results obtained by Molina were not favorable since CFCs persist in the atmosphere for decades before dissociating in the stratosphere, which would bring about the formation of atomic chlorine, thus causing the decrease in the ozone layer until its total disappearance, generating human health and environmental problems due to the increases in ultraviolet radiation.[2]

The concept of green chemistry became present after the approval of the pollution prevention law issued in 1990, defined as a greener or more environmentally friendly chemical process or product, to prevent contamination before it occurs, minimizing the use of hazardous chemicals or materials.

Green chemistry is based on 12 fundamental principles and is depicted in Figure 9.1 and described in the following.

1. Prevention. It is better to prevent the generation of polluting waste than to treat it after it is generated.
2. Atomic economy. Synthesis methods must be designed in such a way as to maximize the incorporation of all the materials used in the process to obtain the final product.
3. Less dangerous chemical synthesis. Synthesis methods must be designed to use and generate substances with little or no toxicity to human health and the environment.

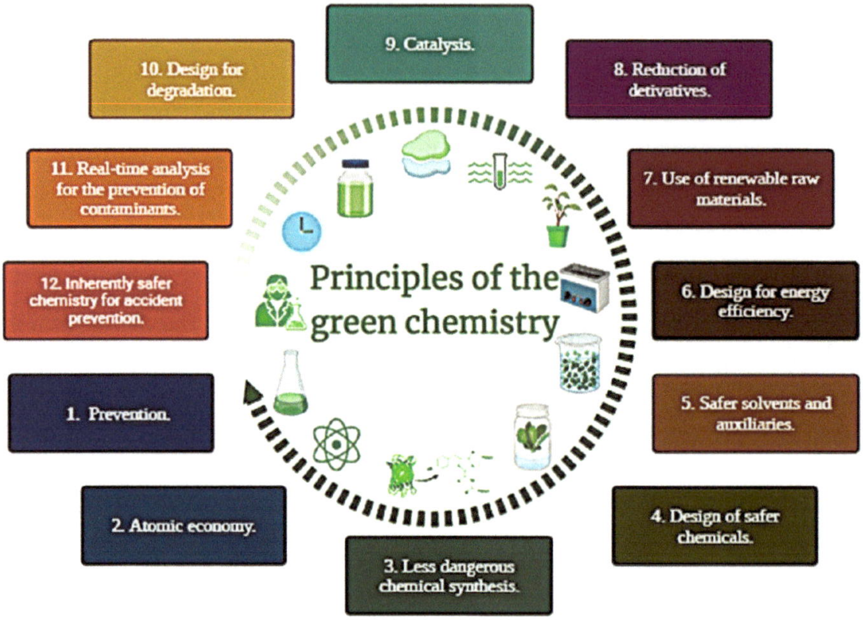

FIGURE 9.1 Principles of green chemistry.

4. Design of safer chemicals. Chemical products must be designed in such a way that effectiveness is maintained while toxicity is reduced.
5. Safer solvents and auxiliaries. Reduce the use of auxiliary substances.
6. Design for energy efficiency. Energy consumption must be analyzed based on its economic and environmental impact to minimize it.
7. Use of renewable raw materials. Use renewable raw materials as long as it is technically and economically viable.
8. Reduction of derivatives. Avoid unnecessary referrals to obtain products.
9. Catalysis. Use selective and reusable catalysts to avoid the use of stoichiometric reagents.
10. Design for degradation. Generate chemical products whose functions do not ultimately persist in the environment, but rather harmless degradation products are generated.
11. Real-time analysis for pollution prevention. Methodologies must be developed to be able to be monitored and controlled in real time prior to the formation of dangerous substances.
12. Inherently safer chemistry for accident prevention. Choose chemical substances that do not generate any risk of chemical accidents.

Today one of the most important topics is using environmentally friendly energies to obtain products. As described by Green Chemistry Principle 6 on Design for Energy Efficiency, this refers to the fact that the energy of chemical processes must be recognized by their environmental and economic impacts, which must be minimized. Furthermore, it describes that, if possible, these processes should be carried out at ambient temperature and pressure.[3] The use of fossil fuels, as well as electrical energy, has a significant environmental impact, which is why their consumption must be minimized. An alternative that has generated good results is the use of biological processes based on microorganisms, in which toxic components are not used, and temperatures are moderately controlled. Another alternative is the use of alternative energies such as ultrasound, microwaves, and plasma, which allow reactions to be accelerated and energy consumption to be reduced.[4]

The use of technologies such as ultrasound originated in 1927. It was mainly used to activate reactions since the effect generated by sound waves on the physical, chemical, and biological properties of various systems was discovered. In the 1950s, it was possible to explain its mechanism, as well as its application in synthesis processes, classifying it within sustainable alternative energies.[5] In 2022, Zhu et al.[6] conducted a study on the application of ultrasound combined with nanomaterials for cancer therapy. They determined that ultrasound is a safe external stimulation technique for biomedical applications since it is noninvasive, nontoxic, and nonpolluting. On the contrary, it treats cancerous tissues more deeply, causing the membranes and organelles to rupture, thus leading to the death of this type of cells without causing harmful effects on healthy surrounding tissues. Within biomedicine, ultrasound would be useful as a treatment for different types of cancer cells and/or increase the permeability of membranes for the administration of drugs more efficiently and directly.[6] Kutlu et al., in 2022,[7] carried out an exhaustive bibliographic review on ultrasound technology and its effect on color, as well as its mechanism of action and factors that influence the preservation of food. They determined that nonthermal methods such as ultrasound, focused on sustainable green chemistry, allow the physical and chemical properties of food to be preserved to the maximum, in addition to shorter processing times, little loss of nutrients, as well as higher yield. On the contrary, they defined that sonic frequencies that range between 20 and 100 kHz allow food processing due to their ability to provide high energy to the fluid medium, and frequencies greater than 100 kHz achieve lower intensities, which allows monitoring food quality to identify changes that may affect food properties from the food production line.[7]

9.3 FUNDAMENTALS OF ULTRASOUND

Ultrasound, also called ultrasonic waves, are acoustic waves that belong to the category of mechanical waves, whose frequency range varies from 20 to 100 MHz, which is why they are beyond the frequency of human hearing, which is why they are not perceived.[8,9] There are two types of ultrasound waves depending on the frequency:

(a) High-frequency ultrasound (0.1–10 MHz). This type of ultrasound does not modify the composition of the medium.
(b) Powerful ultrasound (16–100 MHz). It produces high-energy waves, which generate physical and chemical changes.

Mechanically, ultrasound is described as sound vibrations that are generated by an acoustic emitter, which is backed by piezoelectrics. That is, it consists of two ceramic elements in a sandwich configuration and an amplifying acoustic horn that changes its size precisely and reproducibly in the face of an electromagnetic field, in addition to having a power source, an emitter, and an amplifier signal as shown in Figure 9.2.

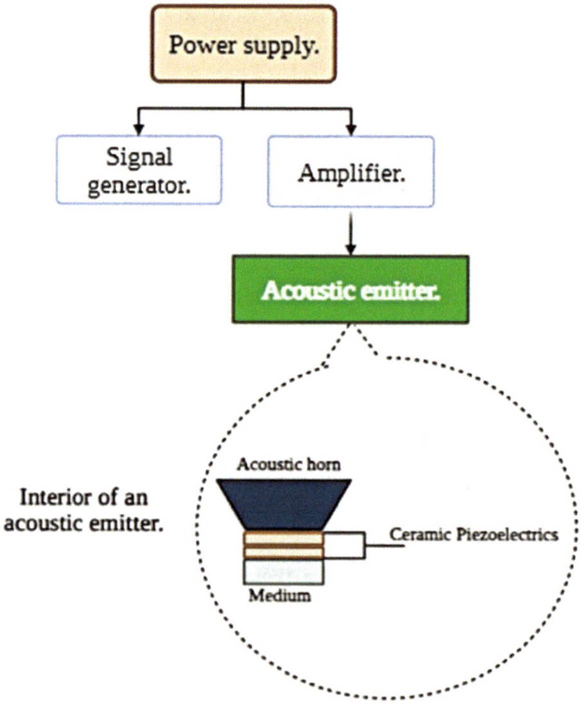

FIGURE 9.2 Composition of high-intensity ultrasound.

The frequency of the generated sound wave is defined by the frequency of the electromagnetic field. According to the frequency generated, ultrasound and acoustic waves are used in various applications, as shown in Figure 9.3.

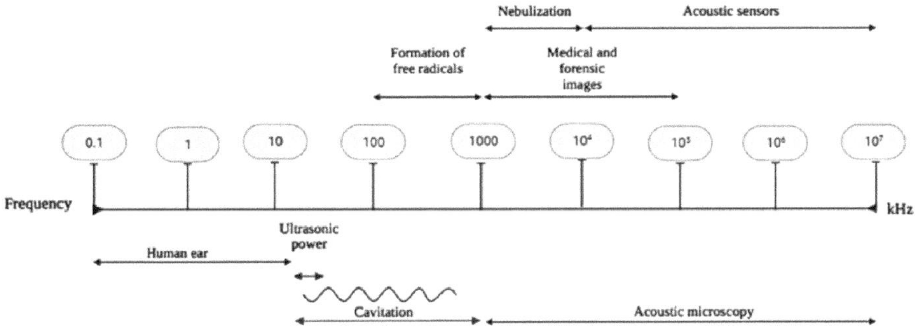

FIGURE 9.3 Frequency of ultrasound and acoustic waves.

At frequencies ranging between 20 and 40 kHz, it is regularly used for food processing. At frequencies greater than 1 MHz, it is used for obtaining images, as well as at frequencies between 300 and 600 kHz, it is used in specific applications in sonochemistry since it is the area where there is greater production of free radicals.[11] Ultrasound causes changes according to the propagation medium used, so it begins with a heating process, which is due to the adsorption of acoustic energy, better known as acoustic energy, or better known as acoustic transmission and is based on the cavitation phenomenon, which is shown in Figure 9.3, this phenomenon refers to the formation growth and destruction by the implosion of small bubbles as a result of pressure.[12] The phenomenon occurs in the transition of the acoustic wave from the negative half cycle to the positive half cycle, better known as expansion and compression. Furthermore, the appearance of this phenomenon depends on the physical and chemical properties of the liquids, as well as their viscosity.

In the transmission of an acoustic wave in a liquid medium, the total pressure of the system is defined as

$$P = P_h \pm P_a$$

where P_h is the hydrostatic pressure and P_a the acoustic pressure; because there are pressure variations, we have the following:

- Positive or compression half cycle: It is one in which the average distance between the molecules decreases, so the pressure will be given by

$$P = P_h \pm P_a$$

- Negative or compression half cycle: The distance between the molecules will increase, so the pressure will be given as

$$P = P_h \pm P_a$$

If the expansion and compression cycles continue, at some point, the average distance between the molecules will exceed the critical molecular distance necessary to maintain the liquid without any chemical alteration, generating cavitation bubbles that will subsequently increase in size until they implode as shown in Figure 9.3.[13]

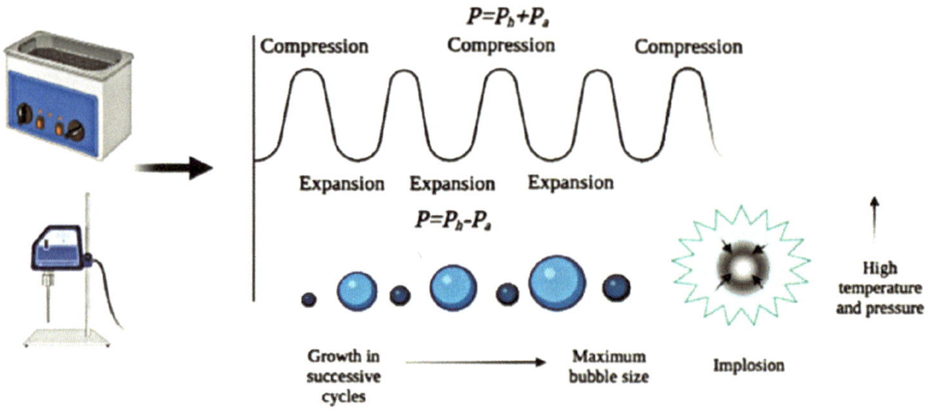

FIGURE 9.4 Cavitation phenomenon.

There are two types of states in cavitation:

- Stable state: Given the oscillation in the acoustic field, a bubble will be generated and subsequently broken up.
- Transient state: This state is generated when there are variations in pressure in short periods with respect to time, thus leading to the breakdown of molecular bonds in the fluid, causing a rapid increase in temperature.

When a fluid or liquid is under constant stress, the bubbles within it will increase in size until they reach a maximum volume, which is why the wave does the same when changing from potential energy to kinetic energy, thus giving way to implosion. Implosion between neighboring bubbles generates microcurrents that collide with the surface of the interface. These microcurrents have an average speed between 100 and 340 m/s, depending on the diameter and pressure of the bubbles.[10]

9.4 APPLICATIONS OF ULTRASOUND IN CHEMISTRY

The ecological processes of ultrasound make possible its connection with green chemistry, and thanks to this, it has been used in a great variety of applications in diverse areas such as organic chemistry, sonocatalysis, electrochemistry, polymer chemistry, biomass conversion, materials preparation, extractions, enzymatic catalysis, pharmacy, etc. This is due to the fact that ultrasound is a unique technique for matter–energy interaction as it enhances physical and chemical changes in a liquid medium by cavitation process.[14,15] The following sections describe the application of ultrasound in different areas of chemistry.

9.4.1 ORGANIC CHEMISTRY

Ultrasound is used specifically in organic synthesis because it provides a particular activation as a result of a high interaction intensity between the acoustic waves and the matter at the molecular and/or atomic level; in addition, it is considered a "green" energy because it helps to reduce reaction times, as well as the use of hazardous chemicals and solvents.[16] Table 9.1 shows some of the investigations where ultrasound has been used in organic synthesis.

9.4.2 SONOCATALYSIS

The combination of sonochemistry and catalysis is carried out by a series of chemical reactions, reducing reaction times and thus being better than conventional methods.[23] To degrade the contaminants, in sonocatalysis, diffusion, and sorption of the main components on the solid surface to be treated is the primary process to trigger a series of heterogeneous reactions at the active sites of the simple.[24] Thus, it has been employed to degrade diverse pollutants since this technique uses and takes advantage of simultaneous oxidation, thermolysis, shock waves, mass transfer, etc.[25] Table 9.2 shows some pollutants that have been successfully degraded using sonocatalysis.

TABLE 9.1 Ultrasound Applications in Organic Chemistry.

Reaction	Conditions	Description	References
Thiocyanation	Ultrasonic radiation, 25°C	The use of ultrasound at room temperature provides α-thiocyanocetones employing enamines and ammonium thiocyanate, and heating after radiation can lead to the selective synthesis of 2-amino thiazoles.	[17]
Aldolic condensation	High-intensity ultrasound, 95°C, 90 W for 90 min	The high pressure and temperatures caused by high-intensity ultrasound favored the generation of pyrazines in a glucose-glycine Maillard reaction system.	[18]
Condensation	Ultrasonic radiation	The synthesis of β-enamine by condensation of 1,3-diketones using ultrasound resulted in a highly efficient and eco-sustainable process, as excellent yields (90–90%) were obtained.	[19]
Indole	Ultrasonic radiation, 80°C, 75–120 min	Developed a simple and efficient protocol for the synthesis of functionalized 2-oxo-benzol[1,4]oxazine. They also reported an increase in yield of about 30% compared to a conventionally heated synthesis.	[20]
Isoquinolines	Ultrasonic radiation, 15 min	The reaction of phthalaldehyde, methyl-amine, activated acetylenic compounds, alkyl bromides, and triphenylphosphine in water, all under ultrasonic irradiation, produced and reduced the reaction time of pyrrido[2,1]isoquinoline with excellent anti-oxidant properties, in addition to obtaining high percentage yields.	[21]
Pyrrolpyridines	Ultrasonic radiation	Described the first ultrasound-assisted process to obtain 1,2-diaryl-substituted azaindol by sequential *N*-arylation followed by coupling-cyclization under Pd/C-Cu catalysis, resulting in a product with promising anticancer properties.	[22]

9.4.3 ELECTROCHEMISTRY

The sonoelectrochemistry is the combination of ultrasound and electro-chemistry; it provides different advantages in electrochemistry, such as in reaction rate, cleaning, surface activation, and increased mass transport in an electrode.[34] Therefore, in recent years, ultrasound–electrochemistry has been widely used in different areas such as medicine, food processing, luminescence, disinfection and sterilization, wastewater treatment, etc.[35]

TABLE 9.2 Application of Sonocatalysis in the Degradation of Aqueous Contaminants.

Technique	**Catalyst**	**Pollutant**	**Percentage of degradation**	**References**
Sonocatalysis	ZnO/bentonitE	Textile dye	---	[26]
Sonocatalysis	CuS/MoS_2	Rhodamine B y Cr	98 y 88.52	[27]
Sonocatalysis/ Ultraviolet (UV)	CuO nanoparticles	Victoria Dyes Blue y Direct Red	99	[29]
Sonocatalysis	SnO_2/multiwalled carbon nanotube (MWCNT)	Methylene blue	99	[29]
Sonocatalysis/UV	NiO nanoparticles	Methylene blue	>90	[30]
Sonocatalysis	MoS_2/CNTs	Hydroxychloroquine	70	[31]
Sonocatalysis	ZnFe/V	Pymetrozine	73	[32]
Sonocatalysis	$CoFe_2O_4/g\text{-}C_3N_4$	chlorhydrate	26.71	[33]

Xie et al.,[36] employed sonoelectrochemistry for the degradation of a pollutant, p-nitrophenol, obtaining 94.1% removal of p-nitrophenol in 2 h by an electrochemical process enhanced with double-pulse ultrasound. Moreover, Lin and Hourng[37] investigated the effect of ultrasound on hydrogen production by water electrolysis, where they employed, for the first time, electrochemical impedance spectroscopy to examine the polarization impedance phenomena in the sonoelectrolysis of water. They reported that ultrasound enhanced the activity and concentration impedances, with a cell voltage of 2 V. Accelerating hydrogen release, they further deduced that 3.5 kW was the appropriate power for sonoelectrolysis of water for hydrogen production, obtaining a storage efficiency of 15%.[37] Mansoorianfar et al.,[38] fabricated TiO_2 nanotubes, loaded with vancomycin for its controlled release, on implantable Ti 6-4, by means of a sonoelectrochemical process, obtaining the best results when using a voltage between 60 and 75 V, since the material showed inhibition against *S. aureus* bacteria. Ren et al.,[39] used sonoelectrochemistry to degrade chlorpyrifos, which has been one of the five most commercial pesticides, besides being an organophosphate currently used in agriculture. The optimum conditions were 20 V, 2 mg/L electrolyte, and 200 W ultrasonic power at 20°C, which led to 93.3% degradation.[39]

9.4.4 SYNTHESIS OF POLYMERIC MATERIALS

The effects of ultrasound in the synthesis of polymer materials can be of two types: physical and chemical. Sonic radiation of liquids can cause only physical changes; for example, some ultrasound-induced changes in polymeric systems are charge dispersion, encapsulation of inorganic particles with polymers, particle size modification in polymeric powders, welding and cutting of thermoplastics, etc.[40]

Basically, the effect of ultrasound application on polymers consists of three main fields:

- Polymer degradation (depolymerization).
- Synthesis of polymers.
- Synthesis of polymer nanocomposites.

9.4.5 *ULTRASONIC DEPOLYMERIZATION*

Depolymerization is an important process step depending on the application, especially for biomedical devices and drug delivery systems. This type of degradation differs from chemical degradation, as the latter refers to the destruction of a substance, while depolymerization is the reduction of the molecular weight of the polymer without causing significant changes in its chemical structure; specifically, it is well known that irradiating a polymer solution reduces the viscosity of the solution.[16,41] Some applications of ultrasonic depolymerization are presented in Table 9.3.

9.4.6 *POLYMER SYNTHESIS*

Most synthetic polymers are obtained from monomers containing reactive double bonds. The most common method of polymer synthesis is by a free radical initiation process. In this sense, ultrasound is capable of producing highly reactive radicals thanks to the cavitation process.[16]

Borai et al.,[48] synthesized an emulsion polymer for lanthanide sorption. The synthesis consisted of dissolving carboxymethyl cellulose in double distilled water with *N,N*-methylenediacrylamide with triacetic acid (dissolved in sodium hydroxide). Ferrous sulfate was dissolved in water and mixed with methyl acrylate and acrylic acid. The mixture was subjected to ultrasonic irradiation at a temperature of 5°C for 15 min. Finally, they added ammonium persulfate, subsequently left in a bath for 3 h at 70°C. They obtained an emulsion polymer, confirming that the cavitation process, coming from sonication, causes the excitation of the molecules, forming free radicals that bind as side branches, thus making the polymer more grafted.[48] On the contrary, Collins et al.,[49] reported, for the first time, sonochemically initiated RAFT polymerization in organic solvents. They reported that a frequency of 490 kHz ultrasound triggered pyrolysis of some molecules present in organic solvents, providing

TABLE 9.3 Ultrasonic Depolymerization Applications.

Material	Technique	Optimal conditions	Important results	References
Guar gum	Enzymatic depolymerization assisted with ultrasound	US: 60 W, 30 min	Ultrasound-assisted enzymatic depolymerization resulted in a polysaccharide with a low degree of polymerization, a viscosity and consistency index, without some change in the core chemical structure, which could make it useful for use in food products.	[42]
Carboxymethyl cellulose	Ultrasonic/UV radiation depolymerization	US: 60 W UV: 8 W, 18 min	They observed no change in the chemical structure using the combined radiation, in addition to obtaining a maximum reduction in viscosity.	[43]
Sodium alginate	Ultrasonic depolymerization	59 kHz, 35°C, 2–4 h	Ultrasonic depolymerization proved to be able to reduce the molecular mass of alginate in a controlled manner, which represents a fast, effective, and low-cost method to adjust the material properties depending on the desired application.	[44]
Egg yolk	High-intensity ultrasonic depolymerization	270 W	The egg yolk granules were depolymerized, obtaining an average particle size between 181.4 nm. In addition, the presence of ultrasound increases the dissolution of the components present in the yolk. Improving the emulsifying activity index of the yolk granules.	[45]
Silicate typo Li_2O-Na_2O-SiO_2	Ultrasonic depolymerization	1000–1800 W	It decreases the molecular weight of the chain due to the fact that the ultrasound breaks Si-O bond units, thus decreasing the melt viscosity as the ultrasound intensity increases. This has the potential to be used in metallurgy, as it can improve slag fluidity and ion diffusion.	[46]
Polyethylene	Enzymatic depolymerization assisted with ultrasound	-	They achieved the depolymerization of polyethylene, which is an emerging green technology for polyethylene recycling.	[47]

a source of initiator radicals. They used this advantage to synthesize acrylates and acrylamides.[49] Selvaraj et al.,[50] report the radical polymerization of acrylonitrile by ultrasound-assisted multisite phase transfer catalysis (300 W) in a two-phase chlorobenzene/water system at 60°C in an inert atmosphere, where from the results obtained, they deduced a suitable mechanism.[50] McKenzie et al.,[51] studied RAFT sonopolymerization in an aqueous medium of different polymers, demonstrating that the use of ultrasound determines a high efficiency since the high frequency produces very little shear force, which avoids potential depolymerization events due to chain breakage.

9.4.7 SYNTHESIS OF POLYMERIC NANOCOMPOSITES

The synthesis of polymer nanocomposites has aroused great interest thanks to the fact that the addition of different nanofillers, such as metal nanoparticles, metal oxides, CNTs, nanofibers, etc., improve the physical and chemical properties of raw polymers. However, to achieve this, it is required that the nanocomposite presents a uniform dispersion of the nanofillers and no agglomeration.[52,53] Therefore, the synthesis of these can be assisted with ultrasound because it has proven to be beneficial not only for achieving uniform distribution and homogeneity of the nanofiller with the polymer matrix but also for its high monomer conversion, high polymerization rate, and high molecular weight of the polymer.[54] Table 9.4 presents different investigations where they perform ultrasound-assisted synthesis of polymer nanocomposites.

9.4.8 PHARMACEUTICAL

Ultrasound is known as a green technology, which means that the process is environmentally friendly. Its basis is based on the transmission of sonic waves in a liquid medium, creating alternating cycles of compression, in which a negative pressure is formed that breaks the liquid to establish a vacuum and subsequently bubbles or cavities. The use of ultrasound has become relevant in recent years because it has been shown to be an effective, simple, energy-saving, and fast method. It stands out in comparison with conventional methods because of its low investment, higher reaction yield percentages are obtained, and shorter reaction and/or extraction times are used.[62]

The first commercial application of ultrasound was in 1917; since then, the use of this unconventional technique has expanded to a large number of applications. The most important thing involving ultrasound is in the area of

TABLE 9.4 Ultrasound Applications in the Synthesis of Polymer Nanocomposites.

Polymer	Nanofiller	Characteristics	References
Chitosan and polyvinyl alcohol	$BiOCl:Eu^{3+}$ nanoparticles	The compatibility of nanoparticles with polymers improved with ultrasonic radiation, additionally, this nanocomposite has potential application in forensic medicine.	[55]
Polystyrene, polyvinyl alcohol	SiO_2 nanoparticles	The material obtained presented a good homogeneity; in addition, they evaluated its flame retardancy, where they demonstrated that the incorporation of nanoparticles influences the improvement of this, compared to the polymeric matrices separately.	[56]
Polymethylmethacrylate	Cloisite 30B organomodified	They reported that the physical effect of ultrasound, the cavitation, generates intense microturbulence in the system, thus playing an important role in the formation of radicals.	[57]
Polyaniline	$LaNiSbWO_4$-G	Due to the small grain size, the intrinsic properties of polyaniline, and the synergistic effects of the composite, it results in a fast electron separation, so the potential use would be in photocatalytic mechanisms.	[58]
Poly[2-(dimethylamino) ethyl methacrylate)	Ag, Pd nanoparticles	Demonstrated that it is possible to obtain this composite using ultrasound as a reducing source, this type of material has potential catalytic applications.	[59]
Polyester	Cellulose-modified SiO_2 nanoparticles	The nanoparticles of the polymer matrix were successfully ultrasonically modified. This filler improves the thermal resistance and robustness of the polyester membranes.	[60]
Polymethylmethacrylate	Hydroxyapatite nanoparticles	The composite showed good dispersion of the filler, and the biological evaluation of the composite showed that it exhibited good cell adhesion.	[61]

processing, as it has been the most employed in the pharmaceutical and food industries.[63,64]

The pharmaceutical industry faces serious problems involving the environment because the synthesis of various drugs unfortunately requires solvents that are toxic; in addition, the processes consume large amounts of energy, are expensive, and waste is generated.[65] Therefore, the industry has focused on finding alternatives to this problem, and one of them is the use of green chemistry for the manufacture and development of drugs.[66] The principles of green chemistry can be implemented in drug synthesis by choosing green solvents (preferably water), and alternative reaction media and minimizing energy consumption by using more energy-efficient equipment such as ultrasound.[67]

Water has begun to be used in drug synthesis. Because it is an inexpensive, nonflammable, environmentally friendly, abundantly available solvent, it also favors cavitation, which makes it an ideal solvent in sonochemistry.[68] Long et al.[69] synthesized 2-sulfonylquinolines in ultrasonically accelerated water without using organic solvents. When compared with conventional heating conditions, the use of ultrasound not only improved the efficiency and increased the reaction rate but also minimized side reactions. Concluding that this synthesis process was economical, practical, environmentally friendly, and allowed for obtaining good yields and high energy efficiency.

Ultrasound-assisted, catalyst-free synthesis of biologically active heterocycles has also been reported. Heterocycles have a wide range of applications, especially when they are used for the manufacture of agrochemical and veterinary drugs.[70] Pyrazoles and their derivatives are biologically active moisture molecules found in drugs such as Celebrex[71] and Viagra.[72] Methods already exist to synthesize them, but they suffer from low yields, by-products are formed, require high reaction times and are toxic and expensive solvents.[73] Shabalala et al.[74] reported on a green and catalyst-free synthesis method under ultrasonic irradiation in a single vessel via multicomponent reaction of aromatic aldehydes, hydrazine monohydrate, ethyl acetoacetate and malonitrile/ammonium acetate in water. Under ultrasonic irradiation, the pyranopyrazoles were selectively obtained faster and obtained yields higher than 90%, concluding that ultrasound increases the reaction kinetics and yield.

On the contrary, stimuli-sensitive polymeric microcapsules synthesized by ultrasound have emerged to develop more effective drug delivery systems, and these are focused on finding better properties such as structure, size, stability, loading capacity, and drug release. Hence the use of ultrasound in the synthesis of these microcapsules and/or nanoparticles.[75]

Behbahani et al.[76] optimized and characterized the preparation of ultrasound-assisted synthesis of solid lipid nanoparticles loaded with curcumin. The synthesis was performed by encapsulating curcumin by microemulsion using ultrasound, stearic acid, tripalmitin as solid lipids, and tween 80 as surfactant. The relationship between particle size and entrapment efficiency of solid lipid nanoparticles was operated by a central composite design under a similar response surface method. The variables included were lipid ratio, surfactant ratio, drug/lipid ratio, sonication time (5 min), and homogenization time. The curcumin-loaded solid lipid nanoparticles presented a spherical shape with a size of 112 nm and an entrapment efficiency of 98.7%, concluding that with the smallest nanoparticle size, the maximum entrapment efficiency was obtained. The drug release profile was examined in aqueous media, and 30% of curcumin loaded on the solid lipid nanoparticles was gradually released over 49 h, which revealed prolonged and efficient drug release.

Geesi et al.[77] prepared titanium dioxide (TiO_2) nanoparticles doped with copper (Cu) as a nanocatalyst to synthesize pyridopyrimidines by sonochemistry. Tetrabutyl titanate was used as a precursor in the sol-gel process. Copper chloride and a sonication time of 30 min were used for the synthesis of Cu-doped TiO_2 nanoparticles. The catalytic efficiency of the prepared heterogeneous nanocatalysts was then evaluated against substituted 3-(aminomethyl)pyridin-2-amine pyridopyrimidines and different arylmethanethiols. In general, this Cu-doped TiO_2 nanoparticle-catalyzed ultrasound-assisted synthesis was found to produce pyridopyrimidine heterocycles with yields up to 94%.

Borah et al.[78] used water as a green solvent for the synthesis of coumarin-pyrazole, employing citric acid as a catalyst under the influence of ultrasonic irradiation. A comparison of this reaction with the conventional method was performed, finding that the ultrasound-assisted reaction was carried out at room temperature and was, therefore, more energetically efficient compared to the conventional method. They also observed that ultrasound increased the reaction rate and provided high yields. This methodology offers advantages such as environmental friendliness, a wide range of substrates, short reaction times, and mild synthesis conditions.

9.4.9 FOOD

In the food sector, the use of ultrasound has several advantages. It has been used to obtain extracts and any solvent for extraction according to the type of extract desired without altering its organoleptic properties, and the process is performed at room temperature, resulting in a reduction of energy and extraction time required.[79] It improves process speed and efficiency in the food industry. According to the intensity and frequency, ultrasonic waves are classified into low and high-intensity waves. Low-intensity waves have a frequency higher than 100 kHz; this type of wave is used to evaluate the shape and size of the food, determining the composition of some of them, such as meat, fish, and poultry. High-intensity waves are characterized by being disruptive, which means that they induce effects on the physical, biochemical, and mechanical properties of food. Their frequency varies from 20 to 100 kHz. They have a wide range of applications in defoaming, sono-crystallization, emulsification, freezing, drying, softening, and thawing.[80]

Fermentation processes involve enzymes and catalysts, generated by microorganisms, generally processed through primitive conditions, which is maturation (also known as aging). In addition, some foods freshly prepared by fermentation are not suitable for immediate consumption because of their stiff taste, odor, aroma, and color. However, there are disadvantages to ripening such as long times, large storage space, and the presence of microorganisms that lead to food contamination.[81,82] Therefore, it is necessary to include emerging technologies to accelerate ripening and fermentation. Ultrasound has gained ground in the food industry because it has been shown to stimulate the growth of microorganisms by promoting enzymatic activities, accelerating their activity, and improving the quality of fermented food with favorable texture, flavor, and color compared to those produced by conventional fermentation.[82,83]

Barukcic et al.[84] studied the influence of high-intensity ultrasound on the quality of reconstituted sweet whey to replace heat treatments, that is, pasteurization. They used it in the fermentation of pasteurized whey with yogurt culture and Lactobacillus acidophilus monoculture. They subjected the whey to different treatments using 400 and 600 W power for 5, 6, 8, and 10 min at a temperature of 55°C. The parameters analyzed were microbiological quality, protein content, acidity, and viscosity. They observed that whey fermentation by ultrasonic culture resulted in a higher viable cell count. They concluded that ultrasound is very useful because it allows homogenization of the whey, thus increasing stability, and when microbiological cultures for fermentation are pretreated with ultrasound, fermentation is faster. From an economic point of view, this process reduces costs because the fermentation time is shorter, and the same effect as pasteurization is achieved.

Daccache et al.[85] performed a low-intensity ultrasound-assisted fermentation of Hanseniaspora sp. to produce cider from Lebanese apples. They evaluated the effect of pulse duration and growth phase when applying ultrasound using a synthetic medium. For this, they first designed a treatment chamber that allows recycling of the growth medium. The optical duration of the ultrasound pulse on yeast growth rate was 0.5 s, followed by a rest period of 6 s for 6 h in the lag and log phase. They observed that these parameters led to a faster consumption of glucose in the medium during fermentation compared to the untreated culture. In the growth kinetics at 12 h of fermentation there was a growth due to the fact that the stationary phase was reached faster and the biomass was higher when compared to the untreated culture. In conclusion, ultrasound-assisted fermentation serves to stimulate biomass growth.

Ultrasound is also used in the food industry to extract bioactive compounds of agri-food origin.[86] It is a green technology because of the advantages it offers in terms of selectivity, better quality, less toxic chemicals, and energy efficiency. In addition to extraction, it has been reported to inactivate certain microorganisms, especially Gram-negative bacteria, which is a prerequisite for food safety.[87]

Da Rocha et al.[88] extracted bioactive compounds from grape pomace using microwaves and ultrasound. For the extraction, they prepared an acidic aqueous solution of grape pomace with 2% citric acid as solvent and powers of 250, 350, and 450 W at times of 5, 10, and 15 min. The results showed that the best extraction time was 10 min using microwaves, where 45% of anthocyanins were recovered, and the antioxidant activity by 2,2'-azino-bis(3-ethylbenzothiazoline-6-sulfonic acid (ABTS) and 2,2-diphenyl-1-picrylhydrazyl (DPPH) increased. Ultrasound and microwave promote solvent penetration into the food matrix by molecular motion of dipole rotation, and ionic conduction is adapted for the extraction of polar compounds.

Emulsions are widely used in the production of foods such as milk, butter, sausages, ice cream, and beverages. Most emulsions are a system of two immiscible liquids, usually water and oil. Ultrasound has been used as an emulsifying method in food processing for many years. It has advantages such as long-term stability of the emulsion, low energy consumption, and easy control.[89]

Lia et al.[90] employed high-intensity ultrasound to improve the emulsifying properties of the chicken myofibrillar protein, improve rheological properties, and stabilize the emulsion. They subjected myofibrillar protein suspensions to ultrasound with a frequency of 20 kHz, a power of 450 W, and a time of 0, 3, and 6 min. The treatment increased the stability index of the myofibrillar protein emulsion, improved the elasticity viscosity, significantly reduced the size of the myofibrillar protein, and promoted the formation of smaller and more uniform emulsion droplets. It also induced structural changes in the myofibrillar protein and the interfacial proteins around the oil droplets, which contributed to improving the emulsifying properties of the myofibrillar protein, the rheological properties, and the storage capacity of the oil–water emulsion.

9.5 CONCLUSION

Alternative, green, or environmentally friendly energies are a viable solution today; they represent an alternative for reducing environmental pollution in different processes at the laboratory, pilot plant or industrial level. However, it is important to review the disadvantages that these present in already established processes. Ultrasound or ultrasonic energy is considered the most viable in terms of costs, production, and implementation in some chemical or physical processes. It currently presents applications of great interest, where it has been shown that using this type of energy offers important advantages such as increases in process performance, reduction in time, selectivity and purity of products, as well as cost reduction, as access to instrumentation with economically profitable investments, simple and economical maintenance, for this reason, it is an excellent alternative to be used in different areas, it can be implemented at a laboratory or industrial level. However, one of the main disadvantages is the increase in temperature during prolonged use times, which can be controlled with power, time, or frequency. To date, it is one of the most used friendly energies in different processes, from the cleaning of material in a laboratory to the ideal dispersion of a colloidal suspension at the nanometric level, to mention an example.

KEYWORDS

- **ultrasound energy**
- **nanotechnology**
- **polymers**
- **polymeric composite materials**

REFERENCES

1. Török, B.; Dransfield, T. Green Chemistry. In *Green Chemistry*; Elsevier, 2018; pp 3–16. https://doi.org/10.1016/B978-0-12-809270-5.00001-7
2. Bolaños Guerra, B. Mario Molina: Pionero de la justicia ambiental. *Educación Química* **2021,** *32* (4), 43. https://doi.org/10.22201/fq.18708404e.2021.4.80331
3. Gonzalez, M. A. Introduction to Green and Sustainable Chemistry. In *Encyclopedia of Sustainable Technologies*; Elsevier, 2017; pp 487–495. https://doi.org/10.1016/B978-0-12-409548-9.10249-0
4. Pino, A. L.; Marcedes, E. Química verde: enfoque sistémico. Ediciones UNL, 2020. https://bibliotecavirtual.unl.edu.ar:8443/bitstream/handle/11185/5630/PINO-química%20verde_DIGITAL.pdf?sequence=1.
5. Ellstrom, C. J.; Török, B. Application of Sonochemical Activation in Green Synthesis. In *Green Chemistry*; Elsevier, 2018; pp 673–693. https://doi.org/10.1016/B978-0-12-809270-5.00024-8
6. Zhu, Y.; Liu, Y.; Khan, K.; Arkin, G.; Tareen, A. K.; Xie, Z.; He, T.; Su, L.; Guo, F.; Lai, X.; Xu, J.; Zhang, J. Ultrasound Combined with Nanomaterials for Cancer Therapy. *Mater. Today Adv.* **2023,** *17*, 100330. https://doi.org/10.1016/j.mtadv.2022.100330
7. Kutlu, N.; Pandiselvam, R.; Kamiloglu, A.; Saka, I.; Sruthi, N. U.; Kothakota, A.; Socol, C. T.; Maerescu, C. M. Impact of Ultrasonication Applications on Color Profile of Foods. *Ultrason. Sonochem.* **2022,** *89*, 106109. https://doi.org/10.1016/j.ultsonch.2022.106109
8. Servili, M.; Veneziani, G.; Taticchi, A.; Romaniello, R.; Tamborrino, A.; Leone, A. Low-Frequency, High-Power Ultrasound Treatment at Different Pressures for Olive Paste: Effects on Olive Oil Yield and Quality. *Ultrason. Sonochem.* **2019,** *59*, 104747. https://doi.org/10.1016/j.ultsonch.2019.104747
9. Khataee, A.; Rahim Pouran, S.; Hassani, A. Editorial Note-Special Issue on "Ultrasonic Nanotechnology: New Insights into Industrial and Environmental Applications". *Ultrason. Sonochem.* **2020,** *65*, 104878. https://doi.org/10.1016/j.ultsonch.2019.104878
10. Luisa Rentería-Monterrubio, A.; Chávez-Martínez, A.; Juárez-Moya, J.; Sánchez Vega, R.; Manuel Tirado, J.; Alberto Reyes-Villagrana, R. High-Intensity Ultrasound and Its Interaction with Foodstuff and Nanomaterials. In *Trends and Innovations in Food Science*; El-Samragy, Y., Ed.; IntechOpen, 2022. https://doi.org/10.5772/intechopen.104437
11. Kentish, S. E. Engineering Principles of Ultrasound Technology. In *Ultrasound: Advances for Food Processing and Preservation*; Elsevier, 2017; pp 1–13. https://doi.org/10.1016/B978-0-12-804581-7.00001-4
12. Vera, Y. C.; Ordoñez, V. M. G.; Daza, A. C. Efecto del tratamiento ultrasónico en la extracción de aceite crudo de palma (*Elaeis guineensis*). *Información Tecnológica* **2023,** *34* (1), 183–192. https://doi.org/10.4067/S0718-07642023000100183
13. Calderón, J.; Marroquin, A.; Luviano, L.; Maqueda, V.; Marín, E.; Calderón, A. Sonido, ultrasonido y cavitación. *Revista Latinoamericana de Educación Física* **2019**, *13* (4).
14. Chatel, G.; How Sonochemistry Contributes to Green Chemistry?. *Ultrason. Sonochem.* **2018,** *40*, 117–122.
15. Saurabh Puri, S.; Kaur, B.; Parmar, A.; Kumar, H. Applications of Ultrasound in Organic Synthesis - A Green Approach. *Curr. Org. Chem.* **2013,** *17*, 1970–1828.
16. Chen, D.; Sharma, S.; Mudhoo, A. In *Handbook on Applications of Ultrasound Sonochemistry for Sustainability*; CRC Press: Boca Raton, FL, 2012; pp 213–262.

17. Liu, Y.; Zhang, T.; Wan, J. Ultrasound-Promoted Synthesis of α-Thiocyanoketones via Enaminone C═C Bond Cleavage and Tunable One-Pot Access to 4-Aryl-2-aminothiazoles. *J. Organ. Chem.* **2022,** *87*, 8248–8255. https://doi.org/10.1021/acs.joc.2c00708
18. Zhang, R.; Zhang, Y.; Sun, Y.; Yu, H.; Yang, F.; Guo, Y.; Xie, Y.; Yao, W. High-Intensity Ultrasound Promoted the Aldol-type Condensation as an Alternative Mean of Synthesizing Pyrazines in a Maillard Reaction Model System of D-glucose-13C6 and L-glycine. *Ultrason. Sonochem.* **2022.** https://doi.org/10.1016/j.ultsonch.2022.105913
19. Redjemia, R.; Bouzina, A.; Bouone, Y.; Mansouri, A.; Bahadi, R.; Berredjem, M. Copper (I) Bromide (CuBr): A Highly Efficient Catalyst for the Synthesis of β-enaminone Derivatives using Ultrasound Irradiation under Solvent-Free Conditions. *Res. Chem. Intermed.* **2022,** 4947–4962. https://doi.org/10.1007/s11164-022-04853-z
20. Jaiswal, P.; Sharma, V.; Priohodko, J.; Mashevskaya, I.; Chaundhary, S. "On Water" Ultrasound-Assisted One Pot Efficient Synthesis of Functionalized 2-oxo-benzo[1,4] oxazines: First Application to the Synthesis of Anticancer Indole Alkaloid, Cephalandole A. *Tetrahedron Lett.* **2017,** *58*, 2077–2083. http://dx.doi.org/10.1016/j.tetlet.2017.03.048
21. Sharafian, S.; Hossaini, Z.; Rostami, F.; Khalilzadeh, M. Ultrasound-Promoted Green Synthesis of pyrido[2,1-a]isoquinoline Derivatives and Studies on their Antioxidant Activity. *Comb. Chem. High Throughput Screen.* **2021,** *24*, 119–128. http://dx.doi.org/10.2174/1386207323666200606212501
22. Venkateshwarlu, R.; Singh, S.; Siddaiah, V.; Ramamohan, H.; Dandela, R.; Pal, M. Ultrasound Assisted One-Pot Synthesis of 1,2-diaryl azaindoles via Pd/C-Cu Catalysis: Identification of Potential Cytotoxic Agents. *Tetrahedron Lett.* **2019**. https://doi.org/10.1016/j.tetlet.2019.151326
23. Gawande, M.; Bonifacio, V.; Luque, R.; Branco, P.; Varma, R. Solvent-Free and Catalystsfree Chemistry: A Benign Pathway to Sustainability. *Chemsuschem* **2014,** *7*, 24–44.
24. Kuna, E.; Behling, R.; Valange, S.; Chatel, G.; Colmenares, J. Sonocatalysis: A Potential Sustainable Pathway for the Valorization of Lignocellulosic Biomass and Derivatives. In *Chemistry and Chemical Technologies in Waste Valorization. Topics in Current Chemistry Collections*; Lin, C., Eds.; Springer: Cham, 2017. https://doi.org/10.1007/978-3-319-90653-9_1)
25. Gholami, P.; Khataee, A.; Chemshmeh, R.; Bhatnagar, A. A Review on Carbon-Based Materials for Heterogeneous Sonocatalysis: Fundamentals, Properties and Applications. *Ultrason. Sonochem.* **2019,** *58*, 104681. https://doi.org/10.1016/j.ultsonch.2019.104681
26. Cheshmeh, R.; Jorfi, S.; Safari, M.; Sadegh, M. Enhanced Sonocatalysis of Textile Wastewater using Bentonitesupported ZnO Nanoparticles: Response Surface Methodological Approach. *J. Environ. Manage.* **2016,** *179*, 47–57. http://dx.doi.org/10.1016/j.jenvman.2016.05.001
27. Dharman, R.; Shejale, K.; Kim, S. Efficient Sonocatalytic Degradation of Heavy Metal and Organic Pollutants using CuS/MoS2 Nanocomposites. *Chemosphere* **2022**, *305* (7), 135415. https://doi.org/10.1016/j.chemosphere.2022.135415
28. Chauhan, M.; Kaur, N.; Bansal, P.; Kumar, R.; Srinivasa, S.; Chaudhary, G. Proficient Photocatalytic and Sonocatalytic Degradation of Organic Pollutants Using CuO Nanoparticles. *J. Nanomater.* **2020,** 1–15. https://doi.org/10.1155/2020/6123178
29. Zaman, S.; Zhangc, K.; Karim, A.; Xinb, J.; Sol, T.; Gongc, J. Sonocatalytic Degradation of Organic Pollutant by SnO_2/MWCNT Nanocomposite. *Diam. Relat. Mater.* **2017,** *76*, 177–183.

30. Darbandi, M.; Eynollahi, M.; Badri, N.; Mohajer, M.; Li, Z. NiO Nanoparticles with Superior Sonophotocatalytic Performance in Organic Pollutant Degradation. *J. Alloys and Compd.* **2021**. https://doi.org/10.1016/j.jallcom.2021.161706
31. Dastborhan, M.; Khataee, A.; Oskoui, S.; Yoon, Y. Synthesis of Flower-Like MoS2/ CNTs Nanocomposite as an Efficient Catalyst for the Sonocatalytic Degradation of Hydroxychloroquine. *Ultrason. Sonochem.* **2022,** *87* (14), 106058. https://doi.org/10.1016/j.ultsonch.2022.106058
32. Keyikoglu, R.; Khataee, A.; Lin, H.; Orooji, Y. Vanadium (V)-Doped ZnFe Layered Double Hydroxide for Enhanced Sonocatalytic Degradation of Pymetrozine. *Chem. Eng. J.* **2022**, 134730. https://doi.org/10.1016/j.cej.2022.134730
33. Zhang, J.; Zhao, Y.; Zhang, K.; Zada, A.; Qi, K. Sonocatalytic Degradation of Tetracycline Hydrochloride with CoFe2O4/g-C3N4 Composite. *Ultrason. Chem.* **2023,** *94*, 106325. https://doi.org/10.1016/j.ultsonch.2023.106325
34. Foroughi, F.; Lamb, J.; Burheim, O.; Pollet, B. Sonochemical and Sonoelectrochemical Production of Energy Materials. *Catalysts* **2021,** *11* (2), 284. https://doi.org/10.3390/catal11020284
35. Zhang, M.; Zhang, Z.; Liu, S.; Peng, Y.; Chen, J.; Ki, S. Ultrasound-Assisted Electrochemical Treatment for Phenolic Wastewater. *Ultrason. Sonochem.* **2020**, *65*, 105058. https://doi.org/10.1016/j.ultsonch.2020.105058
36. Xie, F.; Xu, Y.; Xia, K.; Jia, C.; Zhang, P. Alternate Pulses of Ultrasound and Electricity Enhanced Electrochemical Process for p-nitrophenol Degradation. *Ultrason. Sonochem.* **2016,** *28*, 199–206. http://dx.doi.org/10.1016/j.ultsonch.2015.07.011
37. Lin, M.; Hourng, L. Ultrasonic Wave Field Effects on Hydrogen Production by Water Electrolysis. *J. Chinese Inst. Eng.* **2014**, *37* (8). https://doi.org/10.1080/02533839.2014.912787
38. Mansoorianfar, M.; Khataee, A.; Riahi, Z.; Shahin, K.; Asadnia, M.; Razmjou, A.; Najafabadi, A.; Mei, C.; Orooji, Y.; Li, D. Scalable Fabrication of Tunable Titanium Nanotubes via Snoelectrochemical Process for Biomedical Applications. *Ultrason. Sonochem.* **2020**, *64*, 104783. https://doi.org/10.1016/j.ultsonch.2019.104783
39. Ren, Q.; Yin, C.; Cheng, Z.; Re, Y.; Xie, X.; Li, Y.; Zhao, X.; Xu, L.; Yang, H.; Li, W. Efficient Sonoelectrochemical Decomposition of Chlorpyrifos in Aqueous Solution. *Microchem. J.* **2019,** *145*, 146–153. https://doi.org/10.1016/j.microc.2018.10.032
40. Suslick, K.; Price, G. Applications of Ultrasound to Materials Chemistry. *Annu. Rev. Mater. Sci.* **1999,** *29* (1), 295–326.
41. Gogate, P.; Prajapat, A. Depolymerization Using Sonochemical Reactors: A Critical Review. *Ultrason. Sonochem.* **2015,** *27*, 480–494. http://dx.doi.org/10.1016/j.ultsonch.2015.06.019
42. Prajapat, A.; Subhedar, P.; Gogate, P. Ultrasound Assisted Enzymatic Depolymerization of Aqueous Guar Gum Solution. *Ultrason. Sonochem.* **2016,** *29*, 84–92. http://dx.doi.org/10.1016/j.ultsonch.2015.09.009
43. Prajapat, A.; Das, P.; Gogate, P. A Novel Approach for Intensification of Enzymatic Depolymerization of Carboxymethyl Cellulose using Ultrasonic and Ultraviolet Irradiations. *Chem. Eng. J.* **2016,** *290*, 391–399. https://doi.org/10.1016/j.cej.2016.01.074
44. Dodero, A.; Vicini, S.; Castellano, M. Depolymerization of Sodium Alginate in Saline Solutions via Ultrasonic Treatments: A Rheological Characterization. *Food Hydrocoll.* **2020**, *109*, 106128. https://doi.org/10.1016/j.foodhyd.2020.106128

45. Geng, F.; Xie, Y.; Wang, Y.; Wang, J. Depolymerization of Chicken Egg Yolk Granules Induced by High-Intensity Ultrasound. *Food Chem.* **2021,** *354*, 129580. https://doi.org/10.1016/j.foodchem.2021.129580
46. Min, Y.; Shi, Z.; Liu, C. Ultrasonic Depolymerization of Li_2O-Na_2O-SiO_2 Type Silicate Melt. *Ultrason. Sonochem.* **2017,** *39*, 727–732. https://doi.org/10.1016/j.ultsonch.2017.05.035
47. Brackman, R.; Veloso, C.; Machado, A.; Antunes, M. Enzymatic Post-Consumer Poly(ethylene terephthalate) (PET) Depolymerization using Commercial Enzymes. *3 Biotech* **2023**, *13* (5). https://doi.org/10.1007/s13205-023-03555-6
48. Borai, E.; Hamed, M.; El-kamash, A.; Abo, M. Sonochemical Synthesis and Characterization of Emulsion Polymer for Sorption of Lanthanides. *J. Mol. Liq.* **2018,** *255*, 556–561. https://doi.org/10.1016/j.molliq.2018.01.151
49. Collins, J.; McKenzie, T.; Notthling, M.; Logan, S.; Ashokkumar, M.; Qiao, G. Sonochemically Initiated RAFT Polymerization in Organic Solvents. *Macromolecules* **2019,** *52*, 185–195.
50. Selvaraj, V.; Sakthivel, P.; Rajendran, V. Effect of Ultrasound in the Free Radical Polymerization of Acrylonitrile under a New Multi-Site Phase-Transfer Catalyst – A Kinetic Study. *Ultrason. Sonochem.* **2015,** *22*, 265–271. https://doi.org/10.1016/j.ultsonch.2014.05.002
51. McKenzie, T.; Colombo, E.; Fu, Q.; Ashokkumar, M.; Qiao, G. Sono-RAFT Polymerization in Aqueous Medium. *J. Ges. Dtsch. Chem.* **2017**. http://dx.doi.org/10.1002/anie.201706771
52. Tsai, T.; Lin, M.; Chuang, Y.; Chou, P. Effects of Modified Clay on the Morphology and Thermal Stability of PMMA/Clay Nanocomposites. *Mater. Res. Forum* **2013,** *138*, 230–237. https://doi.org/10.1016/j.matchemphys.2012.11.051
53. Layek, R.; Samanta, S.; Chatterjee, D.; Nandi, A. Physical and Mechanical Properties of Poly(methyl methacrylate)-Functionalized Graphene/Poly(vinylidine fluoride) Nanocomposites: Piezoelectric β Polymorph Formation. *Polymer* **2010,** *51*, 5846–5856. https://doi.org/10.1016/j.polymer.2010.09.067
54. Poddar, M.; Moholkar, V.; Chakma, S. Sonochemical Shyntesis of Polymer Nanocomposites. *Mater. Res. Forum* **2022,** *129*, 233–281. https://doi.org/10.21741/9781644902035-10
55. Kalaburgi, B.; Prasadb, D.; Lavanya, D.; Darshan, G.; Veeranna, V.; Hanumantharaju, N.; Venkatesulu, A.; Taj, M.; Sharma, S.; Nagabhushanaa, H. Nanocomposites of PVA/Chitosan Blend with BiOCl:Eu^{3+} Prepared by Sonochemical Route: Forensic and Optoelectronics Applications. *Coll. Surfaces A Physicochem. Eng. Aspects* **2022,** *657*, 130446. https://doi.org/10.1016/j.colsurfa.2022.130446
56. Masjedi, M.; Ghanbari, D.; Salvarti, M.; Bagheri, S. Sonochemical Synthesis of Spherical Silica Nanoparticles and Polymeric Nanocomposites. *J. Clust. Sci.* **2016,** *27*, 39–53. https://doi.org/10.1007/s10876-015-0897-3
57. Kumar, M.; Sharma, S.; Moholkar, V. Sonochemical Synthesis of PMMA/Cloisite 30BNanocomposites: A Mechanistic Investigation. *Macromol. Symp.* **2016,** *361*, 82–100. https://doi.org/10.1002/masy.201500009
58. Oh, W.; Nahar, K.; Liu, Y.; Sung, C.; Youn, K.; Jung, C.; Dowla, M. Sonochemical Synthesis of Quaternary LaNiSbWO4-G-PANI Polymer Nanocomposite for Photocatalytic Degradation of Safranin-O and Gallic Acid Under Visible Light Irradiation. *J. Photochem. Photobiol. A Chem.* **2020**, *394* (12), 112484. https://doi.org/10.1016/j.jphotochem.2020.112484

59. Wang, J.; Fan, B.; Thang, S. Sonochemical Preparation of Polymer–Metal Nanocomposites with Catalytic and Plasmonic Properties. *Nanoescale Adv.* **2021,** *3,* 3306–3315.
60. Ahmadizadegan, H.; Esmaielzadeh, S.; Ranjbar, M.; Marzban, Z.; Ghavas, F. Synthesis and Characterization of Polyester Bionanocomposite Membrane with Ultrasonic Irradiation Process for Gas Permeation and Antibacterial Activity. *Ultrason. Sonochem.* **2018,** *41,* 538–550. https://doi.org/10.1016/j.ultsonch.2017.10.020
61. Parra, C.; González, G.; Albano, C. Synthesis and Characterization of Composite Materials HDPE/HA and PMMA/HA Prepared by Sonochemistry. *Macromol. Symp.* **2009,** *286,* 60–69. https://doi.org/10.1002/masy.200951208
62. Hernández, J. A.; Reyes, A.; Hernández, I.; Peláez, A.; Campos, R. G. Sustainbale Technologies (Microwaves, Ultrasound and Supercritical CO_2 for the Extraction of Bioactive Compounds. *ICAP Agric. Sci. Bull.* **2023,** *9,* 20–25.
63. Masson, T. J.; Sonochemistry: Current Use and Future Prospects in the Chemical and Processing Industries. *Philos. Trans. Royal Soc. A Math. Phys. Eng. Sci.* **1999,** *357,* 355–369.
64. Azuola, R.; Vargas, P. Ultrasound-Assisted Extraction. *Technol. Move* **2007,** *20,* 30–40.
65. Tucker, J. L.; Faul, M. M. Industrial Research: Drug Companies must Adopt Green Chemistry. *Nature* **2016,** *534,* 27–29.
66. Chatel, G.; Leclerc, L.; Naffrechoux, E.; Bas, C.; Kardos, N.; Goux-Henry, C.; Andrioletti, B.; Draye, M. Ultrasonic Properties of Hydrophobic bis(trifluoromethylsulfonyl)imide-Based Ionic Liquids. *J. Chem. Eng.* **2012,** *57,* 3385–3390.
67. Stefani, H. A.; Cella, R. Reactions Under Ultrasound. In *Encyclopedia of Physical Organic Chemistry*; Wang, Z., Wille, U., Juaristi, E., Eds.; John Wiley & Sons, Inc.; New York, USA, 2017; pp 1–25.
68. Butler, R. N.; Coyne, A. G. Organic Synthesis Reactions On-Water at the Organic-Liquid Water Interface. *Org. Biomol. Chem.* **2016,** *14,* 9945–9960.
69. Long, Y. X.; Yong, J. L.; Qu, J.; Duan, Y.; Hu, J.; Jian, L. K.; Cao, Z.; Wei-Min, H. A Base-Free, Ultrasound Accelerated One-Post Synthesis of 2-sulfonylquinolines in Water. *Green Chem.* **2017,** *23,* 5642–5646.
70. Arora, P.; Arora, V.; Lamba, S. A.; Wadhwa, D. Importance of Heterocyclic Chemistry: A Review. *Int. J. Pharm. Res. Sci.* **2017,** *3,* 2947–2955.
71. Penning, T. D.; Talley, J. J.; Bertenshaw, S. R.; Carter, J. S.; Collinds, P. W.; Docter, S.; Graneto, M. J.; Lee, L. F.; Malecha, J. W.; Miyashiro, J. M.; Roland, S. R.; Rogier, D. J.; Stella, S. Y.; Gary, D. A.; Burton, E. G.; Cogburn, J. N.; Gregory, S. A.; Koboldt, C. M.; Perkins, W. E.; Seibert, K.; Veenhuizen, A. W.; Zhang, Y. Y.; Isakson, P. C. Synthesis and Biological Evaluation of the 1,5-diarylpyrazole Class of Cyclooxygenase-2 Inhibitoris: Identification of 4-[5-(4-methylphenyl)-3-(trifluoromethyl)-1-H-pyrazol-1-yl] Benzenesulfonamide (SC-58635, celecoxib). *J. Med. Chem.* **1997,** *40,* 1347–1365.
72. Terrett, N. K.; Bell, A. S.; Brown, D.; Ellis, P.; Sildenafil (VIAGRATM), A Potent and Selective Inhibitor of Type 5 cGMP Phosphodiesterase with Utility for the Treatment of Male Erectile Dysfunction. *Bioorg. Med. Chem. Lett.* **1996,** *6,* 1819–1824.
73. Draye, M.; Chatel, G.; Duwald, R. Ultrasound for Drug Synthesis: A Green Approach. *Phamaceuticals* **2020,** *13,* 23.
74. Shabalala, N. G.; Pagadala, R.; Jonnalagadda, S. B. Ultrasonic-Accelerated Rapid Protocol for the Improved Synthesis of Pyrazoles. *Ultrason. Sonochem.* **2015,** *27,* 423–429.

75. Li, Z.; Wang, Z.; Du, X.; Shi, C.; Cui, X. Sonochemistry-Assembled Stimuli-Responsive Polymer Microcapsules for Drug Delivery. *Adv. Healthcare Mater.* **2018,** *7*, 1–22.
76. Behbahani, E. S.; Ghaedi, M.; Rostamizadeh, M. A. Optimization and Characterization of Ultrasound Assisted Preparation of Curcumin-Loaded Solid Lipid Nanoparticles: Application of Central Composite Design, Thermal Analysis and X-ray Diffraction Techniques. *Ultrason. Sonochem.* **2017,** *38*, 271–280.
77. Geesi, M. H.; Ouerghi, O.; Elsanousi, A.; Kaiba, A.; Riadi, Y. Ultrasound-Assisted Preparation of Cu-doped TiO_2 Nanoparticles as a Nanocatalyst for Sonochemical Synthesis of Pyridopyrimidines. *Polycyclic Aromatic Compounds* **2020,** *42*, 1–11.
78. Borah, B.; Dwivedi, K. D.; Kumar, B.; Chowhan, L. R. Recent Advances in the Microwave- and Ultrasound-Assisted Green Synthesis of Coumarin-Heterocycles. *Arab. J. Chem.* **2022,** *15*, 103654.
79. Gallo, M.; Ferrara, L.; Naviglio, D. Application of Ultrasound in Food Science and Technology: A Perspective. *Foods* **2018,** *7*, 164.
80. Bhargavaa, N.; Mora, R. S.; Kumarb, K.; Sharanagat, V. S. Advances in Application of Ultrasound in Food Processing: A Review. *Ultrason. Sonochem.* **2021,** *70*, 105293.
81. Ojha, K. S.; Mason, T. J.; O´Donnell, C. P.; Kerry, J. P.; Tiwari, B. K. Ultrasound Technology for Food Fermentation Applications. *Ultrason. Sonochem.* **2017,** *34*, 410–417.
82. Pawar, S. V.; Rathod, V. K.; Role of Ultrasound in Assisted Fermentation Technologies for Process Enhancements. *Prep. Biochem. Biotechnol.* **2020,** *50*, 627–634.
83. Yu, Z.; Su, Y.; Zhang, Y.; Zhu, P.; Mei, Z.; Zhou, X.; Yu, C. Potential Use of Ultrasound to Promote Fermentation, Maturation and Properties of Fermented Foods: A Review. *Food Chem.* **2021,** *30*, 129805.
84. Barukcic, I.; Jakopovic, K. L.; Herceg, Z.; Karlovic, S.; Bozanic, R. Influence of High Intensity Ultrasound on Microbial Reduction, Physico-Chemical Characteristics and Fermentation of Sweet Whey. *Innov. Food Sci. Emerg. Technol.* **2015,** *27,* 94–101.
85. Daccache, M. A.; Koubaac, M.; Salameh, D.; Marounb, R. G.; Louka, M. N.; Vorobiev, E. Ultrasound-Assisted Fermentation for Cider Production from Lebanese Apples. *Ultrason. Sonochem.* **2020,** *63*, 104952.
86. Singla, M.; Sit, N. Application of Ultrasound in Combination with Other Technologies in Food Processing: A Review. *Ultrason. Sonochem.* **2021,** *73*, 105506.
87. Perera, C. O.; Alzahrani, M. A. Ultrasound as a Pre-Treatment for Extraction of Bioactive Compounds and Food Safety: A Review. *LWT* **2021,** *142*, 111114.
88. Da Rocha, C. B.; Noreña, C. P.; Microwave-Assisted Extraction and Ultrasound-Assisted Extraction of Bioactive Compounds from Grape Pomace. *Int. J. Food Eng.* **2020,** *16*, 20190191.
89. Zhou, L.; Zhang, J.; Xing, L.; Zhang, W.; Applications and Effects of Ultrasound Assisted Emulsification in the Production of Food Emulsions: A Review. *Trends Food Sci. Technol.* **2021,** *110*, 493–512.
90. Li, K.; Fu, L.; Zhao, Y. Y.; Xue, S. W.; Wang, P.; Xu, X. L.; Bai, Y. H. Use of High-Intensity Ultrasound to Improve Emulsifying Properties of Chicken Myofibrillar Protein and Enhance the Rheological Properties and Stability of the Emulsion. *Food Hydrocoll.* **2020,** *98*, 105275.

CHAPTER 10

Application of AI-Enabled Computer Vision Technology for Segregation of Industrial Plastic Wastes

TANUSHREE BHATTACHARJEE and SOHAM BHAVE

Department of Chemical Engineering, Vishwakarma Institute of Technology, Bibwewadi, Pune, India

*Corresponding author.

ABSTRACT

Plastic waste generation is growing alarmingly at the global level. Since a major chunk of it is nonbiodegradable, it is detrimental to the health of the ecology. Mixed plastic waste hampers the recycling initiatives and imposes a grave and threatening challenge to sustainability efforts. Commercial plastics differ in chemical composition, and disposing of them makes recycling cumbersome and results in low-quality plastic with poor market potential and economic viability. Further, recycling mixed plastic waste is an energy-intensive process that enhances operational costs and exacerbates carbon footprint. Mostly, the mixed plastic waste is either landfilled or incinerated, emanates greenhouse gasses, pollutes the environment, and inflicts health-hazardous effects to both humans and ecology. Unsound management of mixed plastic waste leads to microplastic generation and the leaching of toxic chemicals into the environment. Therefore, it is imperative to deal with mixed plastic waste efficiently by proper handling, segregating, and processing for a sustainable future. Different methods, such as manual sorting, sensor-based sorting, magnetic density separation, electrostatic separators, near-infrared (NIR) spectroscopy-based methods, robotics, artificial intelligence (AI), etc., have come to the fore as emerging technologies for effective segregation of plastics. Researchers in the field of AI are aiming to use AI

Agroindustrial Waste Management and Natural Resources Conservation: Environmental Challenges and Transformation into Value-Added Products. José Juan Buenrostro-Figueroa, Juan Manuel Tirado-Gallegos, Mónica L. Chávez-González, A. K. Haghi, and Shaheemath Suhara K. K. (Eds.)

successfully and effectively for sorting mixed plastic waste. Advancements in AI, convolutional neural networks (CNNs), and deep neural networks are emerging as powerful tools for the segregation of mixed plastic waste. Thus, computer vision employing image-based classification is of great help in segregating the plastics into poly(ethylene terephthalate) (PET), high-density poly(ethylene) (HDPE), poly(vinyl chloride) (PVC), low-density poly(ethylene) (LDPE), poly(propylene) (PP), poly(styrene) (PS), and others based on their resin identification code. This book chapter demonstrates the application of computer vision technology-driven, image-based classification for segregating mixed plastic waste, enabling its amenability to recycling processes and, in turn, facilitating plastic waste management.

10.1 INTRODUCTION

Belgian chemist Leo H. Baekeland produced the plastic Bakelite in 1907.[78] Since then, plastic has been extensively made and used because of its impressive merits such as longevity, lightweight, light, chemical,[2] and corrosion resistance,[122] as well as with additional benefits of low manufacturing cost, flexibility in shape, color, and touch, plastic has been used to substitute materials such as woods, metals, ceramics, etc.[128,135] Because of the reasons mentioned above, plastic production increased tremendously since 1950, rising from 2 million tons (Mt).[87] At this rate, the requirement for plastic by 2050 is expected to be quadrupled,[137] and nearly 12,000 Mt of plastic will have to be landfilled.[41] As per the latest study,[87] 390.7 Mt of plastic comprises different types of plastic, broadly characterized as fossil-based and circular plastic. Fossil-based plastic contains poly(ethylene terephthalate) (PET), high-density poly(ethylene) (HDPE), poly(vinyl chloride) (PVC), low-density poly(ethylene) (LDPE), poly(propylene) (PP), poly(styrene) (PS), poly(urethane) (PUR), other fossil-based thermoplastics, and thermosetting plastic. In contrast, circular plastic comprises recycled and bio-based plastic.[87] As per the latest report,[125] nearly half of the plastic produced is used only once and thrown away. As per the latest statistics,[3] out of 353 Mt of plastic waste produced, only 9% is recycled, 50% is landfilled, 19% is incinerated, and 22% is abandoned in the environment or unrestrained sites. Thus, this plastic waste produced in such huge quantities is called mixed plastic waste (MPW), a complex blend of PET, PS, PP, PVC, HDPE, etc., as reported by Plug and Play Tech Center in 2020.[89] However, the MPW can be permutations and combinations of these different types of plastic.

MPW can lead to detrimental effects on human and animal life and the environment. MPW's low recycling rate and harmful environmental impact make it of utmost importance to treat the MPW. According to a study,[43] nearly 8 Mt of plastic waste goes into the ocean yearly. Ingestion of plastic waste has impacted marine life. It has resulted in the deaths of millions of birds and mammals,[124] which in turn has affected 267 different species of birds and animals.[43] Plastic exposure to humans through ingestion, inhalation, and adsorption can lead to respiratory problems, cancer, skin irritations, etc.[28] Plastic waste in soil alters the soil's physical and chemical properties and nutritional content, ultimately affecting the surrounding ecosystem.[24,62] Also, it is found that plastic in the form of nanoparticles can travel through the food chain; for example, as reported by Karin Mattsson et al.,[76] nanoparticles present in zooplanktons can travel through the food chain and can end up in a fish which changes the fish behavioral patterns. Also, Bhattacharya et al.[11] reported that plastic nanoparticles in algae can hinder photosynthesis in algae, affecting the food chain. According to the latest research,[27] nano- and microplastics are also found in apples and carrots.

However, effective management of MPW can solve the above problems. As aptly mentioned by Wang et al.,[133] the 3R policy, that is, reduce, reuse, and recycle, as well as energy recovery from plastic waste, can reduce the impact of plastic on the environment. Management of MPW can be classified as land plastic waste management (LPWM) and ocean plastic waste management (OPWM). Different strategies have been implemented for effective management of LPWM. As reported by Kibria et al.,[59] strategies such as a landfill, other recycling techniques such as Mechanical (closed and open loop), chemical recycling, biological recycling, and energy conversion processes, such as incineration, pyrolysis, gasification, etc., have been implemented for LPWM—also, Kibria et al.[59] reported various techniques such as "floating sea bin," "ocean clean up array," and "interceptors" for effective OPWM.

Waste pretreatment and sorting plastic waste play an essential role in waste management and recovery.[26,86] A recent study[80] shows that the recycling industry operates at 40% capacity because of a lack of differentiated collection and separation at the source, affecting raw materials supply. The latest article has aptly said that properly sorting and separating waste is not only crucial for compliance but also integral to creating an effective waste management plan for businesses, as it enhances recycling quality, reduces impurities, and prevents materials from ending up in a singular waste stream bound for landfill, thus helping achieve environmental goals.[112]

Hence, different types of sorting techniques have been developed, such as gravity separation, electrostatic separation, magnetic density separation, floatation,[104] laser-induced breakdown spectroscopy, X-ray fluorescence, Fourier transformed infrared technique, near-infrared spectroscopy (NIR),[80] triboelectric separation, ultrasound,[90] and robotics, and AI.[72,100] However, sorting plastic waste is also a challenge. As reported by Cesar Lubongo and Paschalis Alexandridis,[72] material recovery facilities (MRFs) dependent on manual sorting have outlined a low yield of materials.

In contrast, MRFs that depend on automated sorting have reported tangling of plastic films, cords, and rope chains around the equipment that affects the performance of the equipment. They also noted the challenge of black plastic sorting. However, advanced sorting technologies that combine the usage of deep learning, NIR, and visual infrared spectroscopy (VIS) can help to tackle the challenges of sorting plastic. These methods sort the plastic based on their color and type. According to the latest research,[7] computer vision and sensor fusion have been used to identify the plastic in municipal waste, thereby improving the sorting efficiency of the plant. Collaborative robots are also used to assist humans in the recycling process. Thus, AI and robotics have been found to enhance the conventional approach and make them more efficient. One of the typical deep learning algorithms used for this process is convolutional neural network (CNN).[1,114]

10.2 METHODS USED FOR PLASTIC SORTING

10.2.1 MANUAL SORTING

Material sorting is one of the most common and more accessible methods of sorting mixed plastic waste. Plastic waste collected from industries and households contains other materials such as glass, metals such as aluminum, etc. Hence, plastic waste must be separated from other materials for efficient recycling. Manual sorting is used when materials other than plastic are present in the plastic stream and are easily visible. Manual sorting is the preliminary process used initially in recycling operations to separate glass, cardboard, bulky items, metals, films, etc.[103] Plant operators usually carry out Manual Sorting on conveyor belts as the plastic waste stream is carried forward to the other processing sections through the conveyor belt. However, manual sorting is challenging and time-consuming. It is difficult for the operators to sort a variety of plastic. Hence, automated sorting technologies, such as air classification, eddy currents, optical methods, robotics, and AI, have been developed to overcome the drawbacks of manual sorting.[90]

10.2.2 GRAVITY SEPARATION

10.2.2.1 DRY

10.2.2.1.1 Air Classifier

An air classifier sorts dry plastic waste based on gravity. In air classifiers, the separation of lighter and heavier materials takes place with the help of air. A constant air current enters the column as the waste stream does. The lighter waste stream particles are thrown upward as they come into contact with the air circulation, while the heavier particles are dumped down. Air classifiers primarily remove light pollutants from the primary plastic stream, such as paper, glass powder, dust, tiny foam particles, polymer foils, etc. Aspirators and wind shifters of air-cyclone-based methods are used to achieve the separation. However, it is observed that different behavior of particles influences the separation when put through a stream of air. It is important to note that air-based classification and sorting of plastic need to consider several parameters such as morphological and morphometrical characteristics, particle density, etc. These variables affect the separation efficiency and associated operational setup by interacting with a single polymer terminal velocity.[33] Air classifiers are employed at the beginning of the process or after gravity separation to separate complicated plastic-rich parts from end-of-life electronics.

10.2.2.1.2 Ballistic Separator

According to a study,[25] the underlying idea behind ballistic separation is the distinctive motion displayed by particles of different shapes, sizes, and weights classified within spatial bounds as either two-dimensional (2D) or two-dimensional (3D) forms. Ballistic separation is utilized for mixed garbage comprising plastics and plastic waste streams. Products that are flat and low in weight, such as cardboard, paper, textiles, and fibrous materials, can be categorized as 2D structures. Conversely, a 3D construction may be constructed with rolling and heavy items such as bottles, stones, and plastic containers. The ballistic separator or Ballistic Screen is used for effective separation for both 2D and 3D formations. The ballistic separator is a machine that typically consists of several perforated plates or screening paddles. Depending on the physical features and input rate of the waste materials, the quantity, size, and form profile of the screening paddles or perforated plates may change. Materials are sorted using circular, slanted screening paddles or plates (10–20°) according to 2D/3D forms, weight, and structural characteristics. While lighter 2D materials flow higher, heavier 3D materials flow lower. By allowing smaller particles to pass between the screening screens, continuous shaking produces a third material stream.

10.2.2.2 WET

10.2.2.2.1 Sink–Float Separation

Processes for sink–float separation rely on different material densities. In a tank with a particular density, less dense objects float on the surface of a liquid, while more dense materials sink to the bottom of the tank. The density differential between the materials being separated is crucial to efficiency.[104] Hence, in contrast to close-density polymers such as polyolefins (PP, LDPE, and HDPE), this approach is successful in separating plastics from heavy metals or plastics with different densities (e.g., PET from PP/PE or ABS from PS). Its efficiency may be compromised by impurities such as additions, fillers, air bubbles, or changes.

10.2.2.2.2 Jigging

Hori et al.[46] reported that Jigging is one of the oldest gravity concentration methods. A technique known as "improved gravity-based separation" is jigging. Pistons push the water up and down through a bed material during this process. The materials are separated based on their densities. The systematic and repetitive pulsation applied to the method plays a crucial role. The frequency and amplitude of these pulsations are closely related to the separated materials' physical, morphological, and morphometrical attributes. This process is quite efficient for plastic waste segregation as effective polymer separation occurs because of their low-density differences. The repetitive actions help particles arrange themselves in the layers within the bed based on their densities. Heavier particles tend to settle in the lower layers, while the lighter particles are found in the uppermost layers.

10.2.2.2.3 Hydrocycloning

According to a study,[18] hydrocloning technology is used for sorting materials based on their densities. This technology relies on the centrifugal and centripetal forces produced because of the spinning fluid and the varying fluid resistance experienced by particles with different characteristics. A slurry with a specified solid–liquid ratio and suitable operating pressure is introduced into the cyclone, where the fluid pressure induces a rotational motion, separating materials with varying densities, such as polymers and contaminants or different types of polymers. The cyclone separator will carry the lighter fraction to the top and the heavier fraction to the bottom. Before implementing the hydrocyclone-based technique, materials should be milled. The vortex is aided in formation by rotating movement caused by the tangential introduction of the slurry at the intake. While lighter materials accumulate at the center vortex and depart through the outlet, heavier materials travel outward and downward from the barrel to the cone, exiting at the apex under the influence of centrifugal force. While selecting the hydrocyclones, several factors must be considered to ensure their optimum and efficient use. These factors include the requirement for relatively intricate fluid dynamic circuits, which may involve pumps, storage bins, pipes, valves, and more. Additionally, there is a need for optimal control over the feed characteristics, such as maintaining a consistent water-to-solids ratio and ensuring the power of the polymer's particle morphology, morphometrics, and surface conditions.

10.2.3 ELECTROSTATIC SEPARATION

Effective handling of dielectric particles involves electrostatic separation. Dielectric particles can easily be separated according to their polarity when they are electrostatically charged.[92] Electrostatic separation involves two electrodes, one positive and one negative, for separating the plastic particles. The triboelectric effect is used for charging the plastic particles. In simple words, the triboelectric effect means when two particles are rubbed against each other, and during this process, one particle gets more electrons while the other loses some electrons. This creates an electric charge between them. The triboelectric effect is the basis for generating electricity through processes such as friction or rubbing. When dissimilar plastic particles rub together, they exchange electric charges, affecting their surfaces. In an electric field with electrodes, these charged particles are separated as they respond to attractive or repulsive forces, a method applicable to various polymers as per a study.[33]

(+) ABS - PP - PC - PET - PS - PE - PVC - PTFE (-).

The plastic closest to the positive end of any two plastics in the above sequence will be positively charged. In contrast, the plastic closest to the opposing end will be negatively charged. For instance, if PET and PE are rubbed against one other, PET will have a positive and a negative charge. However, when PET rubs against PP, PET will have a negative charge, whereas PP will have a positive charge.

However, the electrostatic separator can only be used under some specified conditions. However, this can be disadvantageous. Some of the requirements are as follows:

1. **Operating condition:** The mixed plastic waste stream has to be dry.
2. **Shape and size:** Particle size and surface characteristics should be such that they should not affect the chargeability attributes of the particles.
3. Additives or fillers.[103]
4. There should not be any dirt on the particle surface as it can hinder the particle surface charging.

10.2.4 MAGNETIC DENSITY SEPARATION

A study[8] shows that MDS separates particles based on their densities using a magnetic fluid. This magnetic fluid usually consists of water and magnetic particles made up of iron oxide, varying in size from 10 to 20 nm, suspended or evenly distributed in the liquid. The magnetic field creates a magnetic force, a form of artificial gravity.[93] The magnetic field's force on the particles varies significantly in the vertical direction. The vertical variation in the liquid density of magnetic particles is accompanied by an exponential pattern in the force. Plastic particles float when their effective density is equal to their own density. Particles suspend at various heights depending on their densities.[48] This method enables the separation of plastic debris from building and demolition trash that has comparable densities, such as PP and PE,[107] and PVC and rubber.[73] However, correctly implementing the above methodology requires accurately fulfilling the underlying steps -

1. Making the polyolefin surface hydrophilic, also called wetting.[47]
2. To ensure a smooth flow of particles within the magnetic field and to prevent any disruption caused by turbulence, it is essential to properly manage the process of feeding, separating, and collecting both before, during, and after the separation.

These methods aid in the production of secondary raw materials of high quality, which consist of a single polymer stream with little to no contaminants.

10.2.5 FLOTATION

According to Wang et al.,[129] the flotation process relies on varying surface wettability characteristics of materials. Floatation works similarly to the sink–float process, where separation is fundamentally rooted in the difference between the densities of materials and the surrounding medium. Generally, centrifugal force is utilized to enhance the separation process. However, the floatation process differs from the sink–float process, where the floatation process occurs in the water. Here, a carrier is introduced, which consists of air bubbles. The air bubbles help lift the hydrophobic particles, which stick to the bubbles and rise, while hydrophilic particles tend to sink. As per a study,[39] plastic surface characteristics (hydrophobic and hydrophilic) help separate the plastic particles. Conditioners and collectors,[110] as well as air flow rate and agitation (also called flotation cell operative conditions),[109] can help to increase or decrease the hydrophobic and hydrophilic properties of plastic surfaces. Takoungsakdakun and Pongstabodee[117] reported that the flotation process for plastics is carried out in an alkaline medium. Once the process starts, the hydrophobic polymers are recovered from the while hydrophilic sunk to the bottom. As per Buchan and Yarar,[20] this technique is quite popular and has some underlying disadvantages:

1. The floatation process is wet; hence, water must be recovered and processed before utilization as it may contain some reagents and contaminants.
2. The presence of contaminants or pollutants and chemical or physical changes on the plastic surface can change the plastic's hydrophobic or hydrophilic surface properties.
3. Different compositions of plastic waste, that is, the wide variation in plastic waste.

Floatation helps separate plastics, such as PS, PVC, PET, MPO, PC, etc.

10.2.6 SENSOR-BASED SORTING

Sorting of plastic from different materials or polymers is carried out when the specific polymers or materials exhibit unique physical properties, allowing efficient separation. The separation process often involves designing and arranging the equipment and handling system. These systems are often created to influence how the particles respond to the chosen physical property that drives the separation. Typically, the confluence of the particle behavior is done by altering the trajectory of the particles as they leave the separating device or by concentrating them in different parts of the apparatus. When these various streams are mechanically separated, concentrates, waste products, and occasionally one or more intermediate product classes known as "middlings" are produced. "The physical properties of the materials play a crucial role in separating the waste."

The use of sensors for sorting uses online algorithms and robotic units to carry out the sorting process. It consists of four main parts, which are as follows:

1. **Detection:** This is the first step. In this step, the main aim is to detect the materials in the stream. Sensors are used to detect the various materials in the stream.
2. **Identification:** Once detected, the materials need to be identified. The materials are determined based on their characteristics.
3. **Topological Assessment:** After identification, materials are assessed topologically. The materials are evaluated based on spatial arrangement and physical properties within the stream.
4. **Sorting:** Automated devices are sorted once the materials are detected, identified, and assessed. These devices use information gathered by sensors and analytical algorithms to separate materials based on their characteristics.

Implementing the aforementioned strategy includes two essential components. The first is the sensing principle, which tries to perceive the materials. This means selecting the method or technology used to recognize and distinguish different materials in the waste stream. The second one is the actuator logic or architecture. In this aspect, the logic and mechanism (actuators) are required to collect the desired material from the waste flow stream. Once the materials have been identified, physical separation and sorting are also required. Two further categories for sorting methods using sensors are sensing and collecting. There are three critical elements in each of these situations. The first is the conveyor belt, which feeds materials into the sorting system. Second, a sensor is connected to the computer. The sensor collects the data from the materials as they move along the conveyor belt, and the computer analyses this data to identify the materials. The third and last one is the pneumatic system, which separates the material based on the information gathered by the sensors. Sensors are nondestructive and do not require direct physical contact with textiles.

10.2.6.1 VISIBLE SPECTROSCOPY

Two methods are used when the materials are sorted in the visible range: the spectroscopic analytical technique or digital imaging. The spectroscopic analytical technique is mainly used in the 400–700 nm wavelength range. However, it has been found that both of the approaches mentioned above could be more efficient for plastic recycling. Materials are identified through visual properties, such as visible spectra or digital imaging ([RGB] and hue, saturation, and luminance [HSY]), making spectroscopic and imaging methods effective for waste polymer separation, primarily used at the process outset for large, easily recognizable polymer products.[52,150] However, these two techniques may separate plastic materials from mono-materials after recycling.

10.2.6.2 NEAR-INFRARED SPECTROSCOPY

One of the most popular technologies for recycling plastic is NIR. As per a study,[10] this technique collects reflected spectra from polymers after being illuminated by the light source. Typically, NIR spectroscopy operates within the 1000–7000 nm wavelength range. It can sometimes be extended to cover the short-wave infrared region (SWIR), which spans 1000–2500 nm. NIR spectroscopy is valuable for analyzing plastics by their unique near-infrared spectral responses, providing composition and property details to assist in sorting and recycling. The majority of an NIR sorting system consists of a conveyor belt, an illumination system, an optical sensor, and a separating unit with compressed nozzles. Different polymers, such as PP, PS, PVS, PET, PE, etc., can be separated with the help of near-infrared sensors, as these plastics can be distinguished by their unique spectral signature in the wavelength range. NIR spectroscopy is not only used for separating plastics from mixed plastic streams but also for separating plastics from other waste materials such as paper, glass, wood, stone, etc.; NIR spectroscopy is widely used for the following reasons:

1. Direct contact with the target object is not needed.
2. NIR spectroscopy has a high detection speed.
3. Multiple detections of the object are possible.
4. There is no interference of the color.

5. NIR spectroscopy uses fiber optics and flexible structures to light and record the spectral response of plastics. This enables the investigation and identification of these materials by comparing their infrared absorption bands to reference spectral libraries.

However, the flakes' size and position affect the sensor's ability to detect the plastic. Hence, the flakes' size and position on the conveyor belt are crucial factors that impact the sensor's ability to see the plastic accurately. The physical dimensions of the sensing probe affect the area they investigate, consequently affecting the spatial resolution of each sensing unit. The low surface reflectance of black or highly dark polymers makes identification difficult.

10.2.6.3 HYPERSPECTRAL IMAGING

A cutting-edge, quick, nondestructive technology called hyperspectral imaging (HSI) may quickly and accurately record things' spectral and spatial characteristics. This information is compiled into a dataset called "hypercube," which includes both spatial information (the *x*- and *y*-coordinates of each pixel) and spectral information (the *z*-axis showing the spectrum associated with each pixel). Depending on the needs of the application, the sensor of the device determines the precise spectral range studied, ranging from visible (VIS) to infrared (NIR), SWIR, or mid-wave infrared (MWIR). The utilization of the methodology has increased in many sectors over the years. HSI is used in the food, chemical, pharmaceutical, and agricultural industries. The application of HSI can also be found in recycling technologies such as recycling glass,[15] analyzing various plastic waste types,[48,106,107,124] creating recycled aggregates from concrete,[17,18,103] and ensuring the quality of compost through different activities or processes,[32] etc.

The popularity and extensive usage of the HSI device are because of some peculiar features,[103] which are as follows:

1. Continuous monitoring of sizeable plastic waste streams on the conveyor belt can be achieved using the scan line camera architecture.
2. The shape and structure of the material can be easily identified, which can help to sort the materials effectively.
3. Different sampling techniques corres pond to different time frames, particularly when developing specialized product-oriented control or quality actions is required.

4. Due to HSI detectors' characteristics, fast and reliable recognition procedures may be implemented practically. These detectors can process pictures, spectra, or both.
5. No environmental impacts or safety concerns are associated with using the HSI device.

The device is affordable, with a relatively low cost.

As far as polymer recycling is concerned, HSI is a highly effective and efficient tool, as reported by Jansen et al.[54] Finding spectral regions, such as NIR or SWIR (1000–1700 or 1000–2500 nm), where polymer molecules absorb light as a result of overtone or combination vibrations is necessary for using online sorting and quality control approaches.[137] Each polymer may now be identified more easily thanks to high-speed MWIR spectral cameras that can produce distinctive spectral fingerprints.[16] These cutting-edge cameras can help to sort the black polymers, which is impossible by sensors in the 400–2500-nm range. The challenge with black polymers is their high light absorption, resulting in low reflectance, as reported by Rozenstein et al.[96]

HSI-based sorting system consists of several vital components. They are:

1. **Conveyor belt:** Used to carry materials to be sorted.[104]
2. **Sensing unit:** The sensing device continually and steadily gathers spectral data. After that, these spectra are processed by a classification engine that was previously built using a reference library of material spectra. Based on their spectral properties, this engine recognizes and locates certain things.
3. **Air nozzles:** At the end of the conveyor belt, a set of compressed air nozzles are used to separate the recognized items. These nozzles discharge air bursts, which aid in separating the identified components.[86,120]

10.2.6.4 X-RAY FLUORESCENCE

In this technique, materials are sorted by detecting the wavelengths and energy released by the particular materials once exposed to X-rays. When the atoms in the fabric absorb the X-ray energy, they emit the X-ray fluorescence radiation as they release energy. The specific materials in the material influence the wavelength and power of the emitted radiation. As per a study,[13] the extensive usage of the X-ray fluorescence (XRF) technique is in the recycling sector, especially for wood recycling. XRF technique helps evaluate harmful elements such as arsenic, chromium, copper, and others in materials to be sorted. XRF is primarily used for sorting mixed plastic waste, for example, separating PVC from PET, as reported by Brunner et al.[20] Due to its ability to extract brominated plastic from the stream of shredded plastic, this approach is anticipated to be widely employed. Generally speaking, bromine is employed as a flame retardant, particularly in the field of electronics.

However, advancements in the XRF method have also been reported. Energy dispersive X-ray fluorescence (EDXRF) has been proposed. This method adds tracers to the polymer matrix.[11] These advancements enhance the capabilities of XRF in plastic sorting and recycling.

XRF and EDXRF sorting settings are equivalent. An X-ray beam is pointed toward the waste flow stream (particles traveling down a conveyor belt) in EDXRF's operating phase for analysis and sorting. This X-ray beam is focused, travels through the material, and finally arrives at the detector, where signal processing is done on the information gathered by the sensor. During this process, the tracers in the materials are identified. Their amount is evaluated, and based on their predefined criteria, the corresponding plastic items are selected by blast of air.

The XRF approach can be done without sample preparation or collecting beforehand. It is effective in finding impurities on the surface of polymers, specific products, and black or very dark polymers. However, this technique has one disadvantage: it cannot differentiate between different polymers. Also, safety concerns are associated with using X-ray sources in the process.

10.2.6.5 LASER-INDUCED BREAKDOWN SPECTROSCOPY

Laser-induced disintegration, a powerful laser pulse, is used in the analytical technique of spectroscopy to vaporize the substance and produce plasma fumes.[31] After that, charge-coupled device (CCD) technology (a spectrometric instrument) is used to analyze the radiation that the material's ablated portion of the material emits.[43] As per a study,[102] it is possible to identify the atomic emission lines by analyzing the waste material's constituent elements. LIBS can be used for polymers to determine the presence of elements such as carbon and hydrogen and calculate their ratio (C/H).[6] Recently, a polymer particle recognition system was developed, which involved a compact and reliable architecture. This system uses statistical analysis based on the information gathered through laser-induced plasma spectroscopy (LIPS). The information obtained is then further processed using the linear and rank collections within the 200–800 nm wavelength range.[5] Alternatively, the different energetic materials may be distinguished using an instant ratio analysis of molten bands.[4] Polymers, such as PVC, PP, PS, PET, HDPE, LDPE, etc., can easily be identified using the latter approach. The LIPS-based detection controls the mechanical components utilized in the sorting architecture, which helps separate the polymers into the proper bins.

10.2.7 AUXILIARY SEPARATION TECHNOLOGIES

Auxiliary separation technology is used in plastic recycling to purge the plastic waste stream of items having traits or origins unrelated to plastics. This procedure is often applied at the process's start (scalping) or the end (refining). Ferrous materials (low-grade stainless steel, nickel alloys, and similar materials) and nonferrous materials are the materials that are frequently removed. Magnetic separation[135] and eddy current separation[144] are commonly used to achieve the separation.

10.2.7.1 MAGNETIC SEPARATION

Magnetic separation is a common technique in plastic recycling. It involves the use of various types of magnetic separators.[117] The magnetic separators are as follows:

1. **Belt magnets:** These magnets are often positioned above the stream of plastic. A moving belt crosses the magnetic field at roughly a 90-degree angle to the material flow. As the belt exits the magnetic field, ferrous metal particles are drawn to the magnet, separated from plastics, and released.
2. **Magnetic pulleys:** A conveyor belt's end is usually where magnetic head pulleys are mounted below the belt. The magnet keeps Ferrous metal particles on the belt, while plastic particles go ahead and may be recovered separately.
3. **Drum magnets:** Drum magnets are frequently positioned within feeder chutes or in the space between chutes and conveyors. While plastic garbage flows, the revolving drum traps ferrous metals until a separator releases them for disposal.

Magnets are typically placed at the start of plastic recycling plants to eliminate larger magnetic contaminants from the material stream. At the same time, high-intensity permanent magnets, as recommended by Svoboda and Fujita,[116] are employed at the end for more comprehensive refining and quality control of the final products.

10.2.7.2 EDDY CURRENT SEPARATION

The auxiliary separation method utilizes a high-speed magnetic rotor system to eliminate nonferrous materials from the waste plastic stream. Rapid rotor rotation causes conducting metals to experience an eddy and electric current. Dry recyclables, plastics, and glass are separated from the repelled metals as they travel over the rotor by the magnetic field produced by the conductive metals, which opposes the magnetic field of the rotor and draws the generated electric current.

The recycling process begins with the use of eddy current separators. The effectiveness of the most recent eddy separation, which is dependent on feed particle size and in accordance with its separation principle, is the main deciding factor in this decision, as explained by Van der Walk et al.,[127] wherein particle morphological characteristics, mass, specific conductivity, magnetic field strength, and dispersion collectively affect the strength of repulsive forces.

10.2.8 ROBOTICS AND AI

Robots guided by AI have been used extensively to tackle challenges in plastic sorting, which can replace or complement traditional optical sorters, enhancing sorting efficiency and learning capabilities over time by mimicking human brain processes. Several companies offer AI-based sorters for plastic sorting, with some material recovery facilities (MRFs) already integrating them into their processing lines, employing methods such as deep learning, visible imaging spectroscopy (VIS), and sometimes NIR for plastic identification and sorting.[72] However, sorting plastic according to their resin identification number is also challenging. CNN, a deep learning method, has proven to be highly effective in sorting plastic materials based on their resin identification numbers. Bobulski and Kubanek,[14] Subashini and Vignesh[114] have successfully used CNN to sort mixed plastic waste according to their resin identification number. Also, novel techniques have been developed to separate plastic waste from other types, such as wood, glass, and metals. A recent study[33] shows that machine learning (ML) and mel-frequency cepstral coefficients (MFCC) feature extraction have been utilized for plastic detection. In this unique audio feature-based technology, a sensing unit identifies the plastic from other materials such as glass, wood, metal, paper, etc.

Then, a mechanical setup controlled by microcontrollers automatically separates the plastics. Thus, this innovative technology has transformed the conventional waste management system into an intelligent automatic plastic segregation system. According to the latest research,[6] integrating computer vision and sensor fusion and utilizing the data collected from municipal waste can enhance the efficiency of automated sorting plants. They also create collaborative robots to assist the humans at the sorting stations. According to the latest study,[89] High-tech systems comprising cameras and trained computer algorithms guide robotic arms equipped with sensors to efficiently identify and extract recyclable items such as cans, glass, and plastic containers from mixed waste, significantly outperforming humans with double the speed. These advanced recycling robots can discern various attributes such as colors, textures, shapes, and sizes of plastic materials, enhancing waste sorting accuracy and potentially doubling the resale value of recovered materials, which is crucial as quality standards become stricter. In pursuit of reaching zero waste, San Francisco is at the forefront of this technology, with companies such as Recology investing millions in state-of-the-art optical sorter robotics to improve recycling plant efficiency. According to the latest research,[38] Artificial intelligence (AI) and robotics optimize waste logistics, cutting transportation distance by 36.8%, reducing costs by 13.35%, and saving time by 28.22%, while also enhancing waste identification, sorting accuracy, energy conversion, pyrolysis, and carbon emissions estimation through chemical analysis integration.

10.3 PROPOSED AI SYSTEMS

AI is found to help segregate mixed plastic waste by separating plastics from other materials such as wood, cardboard, glass, metal, etc., or by separating plastics from different types of plastic based on resin identification numbers with the help of advanced technologies such as CNN, NIR, and VIS. These advanced technologies have been found to enhance efficiency and reduce transportation distance, costs, and time while improving waste accuracy to meet stricter quality standards and sustainable waste management practices.

However, while using a particular AI system, some facts related to the algorithms that would be implemented should be known. It helps to pick the right kind of algorithm. Some of these facts are as follows -

1. **Explainability:** Some ML algorithms are often referred to as "black boxes," these algorithms. However, only a few studies have created these models with high precision and low errors, but they are difficult to understand and explain, for example, neural networks.[97] On the contrary, other ML algorithms, such as k-nearest neighbors (KNN) or decision trees, are simple to predict but not always precise.[98]
2. **Several functions and examples:** Neural Networks can handle millions of dataset functions and many examples, while SVM has a very modest capacity.[114]
3. **Qualitative and quantitative attributes:** Certain ML algorithms cannot handle qualitative properties in the datasets. Hence, they require specialized techniques such as unique coding to convert qualitative to quantitative properties.[36,71]
4. **Drive speed:** The algorithm's time to build the model. This time can vary depending on the algorithm; for example, neural networks train slowly, while decision trees are much faster.[128] Some specialized libraries offer highly efficient implementations of specific algorithms, and certain algorithms take advantage of multiple processor cores, substantially reducing modeling time on machines with multiple cores.[131]
5. **Predicting speed:** Predictive Speed refers to the rate at which the algorithm can generate predictions. SVM and some neural networks are extremely fast when predicting,[61,145] while KNN, deep, or recursive neural networks (RNN) are relatively slow.[112]

Also, along with these facts, some advantages and disadvantages of these algorithms should be known—Table 10.1 highlights some of the pros and cons.

Considering these different ML algorithms and their advantages and disadvantages, Tables 10.2 and 10.3 use these different ML algorithms to segregate plastic among plastic wastes and other wastes.

However, while using these methods, a few things may be improved. The drawbacks of AI models are described in Figure 10.1. Compared to actual garbage, the plastic pictures used to train the models are far less complicated. The model's performance may be impacted by this.[141] Additionally,

TABLE 10.1 Mertis and Demerits of ML Algorithms.

Algorithm	Advantage	Disadvantage	References
DNN	Outperforms other solutions across domains, reducing the need for extensive feature engineering while remaining adaptable to new challenges.	Demands substantial data, incurs high computational training costs, lacks a robust theoretical foundation, and may yield complex, challenging-to-understand findings.	[57]
ANN	Artificial neural networks (ANNs) exhibit nonlinear adaptability, fault tolerance, and the ability to model complex, non-direct relationships, enabling generalization and uncovering hidden connections.	Artificial neural networks (ANNs) can be challenging to interpret due to their black-box nature and susceptibility to overfitting.	[51]
CNN	CNNs can extract features from images and classify detected objects, streamlining visual recognition tasks.	The performance of CNN relies on the quality of the image and the presence of distinguishable features within the image.	[70]
R-CNN	R-CNN, which is faster, excels in detecting small objects.	It takes a long time to train the network. The selective search algorithm is a fixed algorithm. Thus, no learning occurs at that stage. This could lead to lousy candidate region proposals.	[41]
SVM	Support vector machines (SVM) excel in avoiding local optima and the curse of dimensionality. They are less prone to overfitting, particularly effective when class division is clear, and perform well in high-dimensional spaces, especially when there are more features than samples. Additionally, SVMs are relatively memory efficient.	The prediction accuracy of support vector machines (SVM) heavily depends on the chosen kernel function and is sensitive to missing values. SVMs could be more effective in training with large-scale data, making them unsuitable for large datasets. Furthermore, they could perform better when the dataset contains significant noise.	[41]
KNN	They are easy to make predictions.	These models give low accuracy and are computationally slower during the prediction phase.	[58]
Decision tree	They learn quickly and often require minimal or no preprocessing of the input data.	Decision trees are prone to overfitting and cannot handle nonlinear data. They could be more precise and more complex to predict.	[37,46]

TABLE 10.2 ML Algorithms for Spectroscopic-Based Methods.

Spectroscopic method	Plastic samples	ML algorithm	Result	Shortcomings	References
NIR	PET, PE, PVC, PP, PS	SVM	Low success rate		[62]
	Composite wood	PCA	High purity	Resource-intensive and controlled environment requirements	[78]
	PE, PP	PCA	Ambiguous result		[105]
	PET, PE, PVC, PP, PS, ABS	PCA	Perfect accuracy		[148]
	PET, PLA	PLS-DA	High detection efficiency		[124]
	PET, PE, PVC, PP, PS	PCA + PLS-DA	Near-perfect classification		[92]
	PP, PS, ABS	Levenberg Marquardt	High accuracy for PP, PS, and ABS	Misclassification of PS and ABS	[68]
	PET, HDPE, PVC, LDPE, PP, PS, PC	PCA-SVM/LDA, SAM, PLS-DA	Almost perfect accuracy	Confusion between HDPE/LDPE and PET/PC	[35]
Raman	PET, PE, PVC, PP, PS, PMMA, POM	SVM	Perfect accuracy		[22]
	PP, PS, ABS	SVM	High accuracy at 7.5* noise, good accuracy at 15* noise	Noise sensitivity in data	[80]
	Black plastics consisting of PET, PP, PS	PCA-NN, c-mean clustering	The high test set classification rate (>93.6%)	Poor PS classification efficiency	[95]
	Black plastics consisting of PET, PP, PS	PCA-FRBFNNs classifier	93% accuracy for PET at 70 fps, nearly perfect accuracy for PP and PS.	PS and PET misclassified	[8]
LIBS	PET, HDPE, LDPE, PP, PS	PCA-ANN	Perfect accuracy, except for LDPE and HDPE		[56]
	PTFE, HDPE, PVC, PE, PP, PS, ABS, PC, PMMA, POM, PU	VI-PLS-DA	Accuracy of 89.55% without visible inspection (VI) and 98.18% with VI		[69]
	PE, PS, PP, PC, PA, ABS	KNN, SIMCA	Very good accuracy for KNN and SIMCA	High PC false alarms	[30]
	PVC, ABS, PA, PMMA	CNN	perfect accuracy	LDPE and HDPE are misclassified	[85]

TABLE 10.3 ML Algorithms for Segregating Plastic from Other Wastes.

ML algorithm	Solid waste material	mAP (%)	Recall (%)	Detection speed	References
Mask R-CNN	Organic waste in municipal solid waste (MSW)	81			[50]
	Pipes and cables	>98	88.9		[132]
	Bottles, paper, cups, batteries, cans, paper, milk bottles	93		59 fps	[39]
R-CNN	Concrete, bricks, metal, plastic, rubber, wood	98.7			[64]
Faster R-CNN	E-waste (mobile, laptop, TV, etc.)	90–96.7	>90.9		[84]
	Construction fasteners, including nails and screws	89.1			[133]
Fast R-CNN	Plastic water bottles	88.0	96.7	4.5 fps	[149]
YOLO v5	PP, PS, ABS	98.5			[119]
	Different domestic wastes	94.5		140 fps	[140]
YOLO v4	Various types of E-waste	97.8	97.1		[99]
	Plastic, fabric, paper, metal, glass	93.1	92.2	24% faster than Faster R-CNN	[142]
YOLO v3	Metal, glass, plastic, paper, cardboard, trash	81.4 92.12	86 91	80 fps	[76]
	Flow of water within residual waste materials	87.5%		41 fps	[146]
	Organic waste, metal, paper, cardboard, glass, plastic	95.0	87.0	3.0 fps	[65]
	Extracted chemicals from waste PCBs	>95.0		12.5 fps	[72]
SSD	Standard beverage containers: glass bottles, plastic bottles, and cans	87.7		15.9 fps	[122]
	Nonbiodegradable and biodegradable wastes	96.5		Real-time	[58]
	Plastic bottles and aluminum foils for packaging in beverages	95.0		160 fps	[66]
FSSD	Plastic, glass, cardboard, paper, metal	83.48		40 fps	[75]
DSSD	Plastic, paper, shell, ample bone, cigarette	85		2 fps	[67]

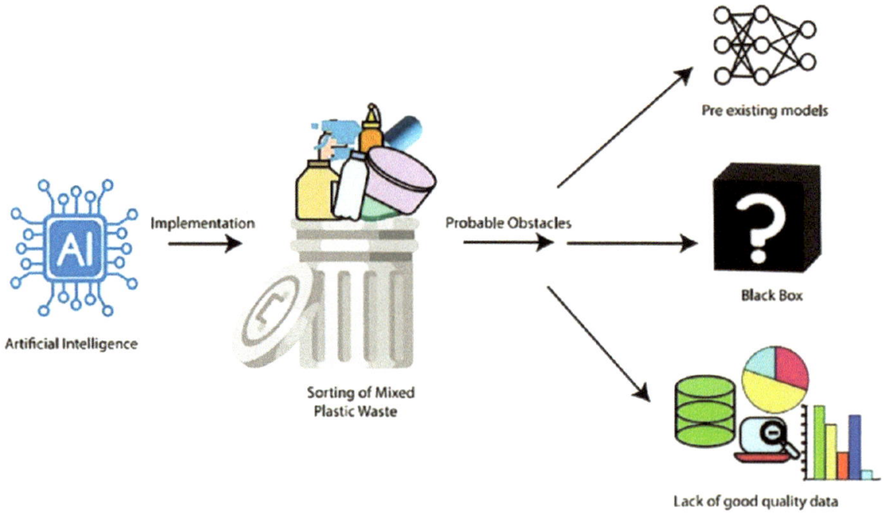

FIGURE 10.1 Main thremplementing AI in plastic sorting. Pre-existing moe challenges in idels, black box models, and lack of good-quality data are the three problems faced while implementing an AI model. Here, the Black box refers to the AI model's intricateness, making it strenuous for researchers to comprehend its mechanism. The lack of good quality data alludes to scarcity and noisy data, which makes it tough to train the AI models. Pre-existing models emphasize that AI models for sorting plastic waste are pre-defined and not custom-made.

the model's performance should not be harmed by the size of the data used to train it. Systems for waste detection and identification need expensive technology and use a lot of energy.[152] People find it challenging to understand these AI models since they are black boxes.[46] The complexity of these models has only recently been adequately explained by a small number of methods.

10.4 CONVOLUTIONAL NEURAL NETWORK

The AI models and their implementations, along with other methods to segregate the mixed plastic waste, are schematically represented in Figure 10.2.

As seen in Figure 10.2, the CNNs are used with Spectroscopic and Computer vision methods. CNN is used in Spectroscopic methods to segregate the mixed plastic waste, while CNN is used in the computer vision-based method to segregate the plastic from other wastes. The difference between both methods is highlighted in Table 10.4.

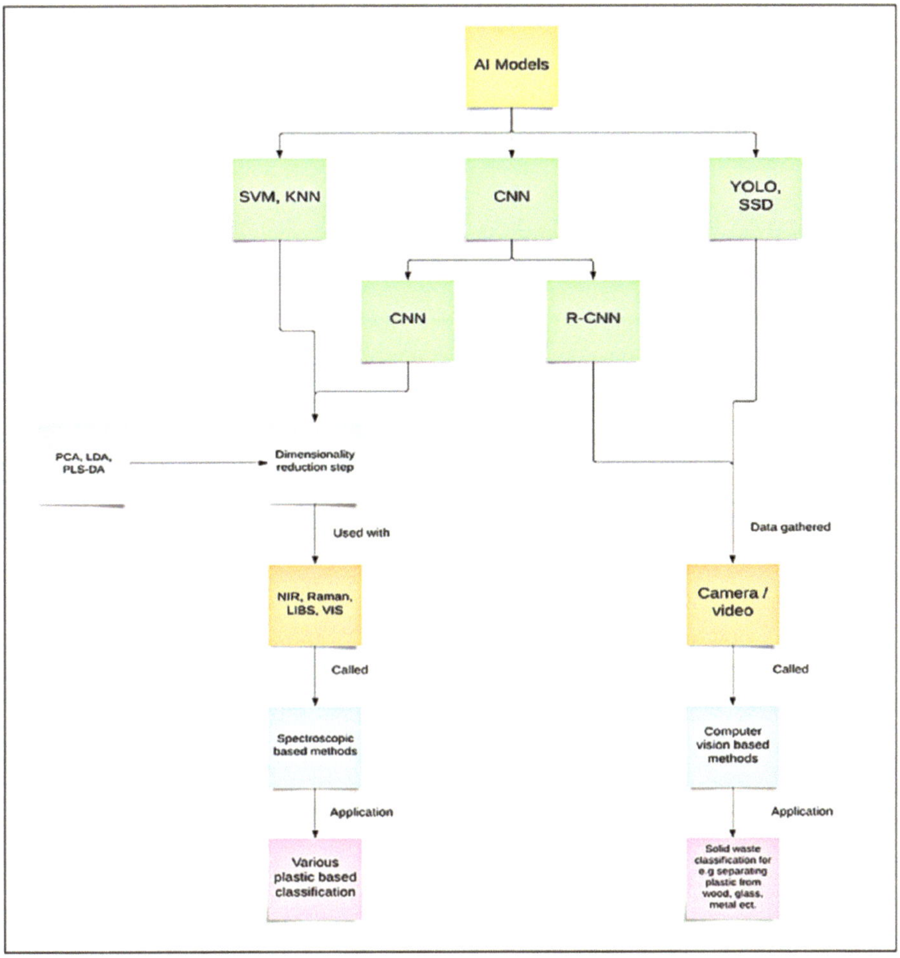

FIGURE 10.2 AI models and their implementations along with Spectroscopic based methods and Computer Vision based methods for sorting plastic waste

10.4.1 SPECTROSCOPIC-BASED METHOD

Spectroscopic methods classify materials based on their chemical structures and compositions of different elements within materials. Hence, spectroscopic methods classify materials into single categories,[147] for example, Identifying the PET from PS, PP, ABS, etc. However, the combination of computer vision-based methods such as CNN (Fig. 10.2) and spectroscopic-based methods has caught the eye of many researchers

TABLE 10.4 Differentiation Between Spectroscopic and Computer Vision-Based Methods.

Attributes	Spectroscopic based methods	Computer vision-based methods
Object recognition is based on	Employing diverse electromagnetic spectra to irradiate matter, the microscopic structures can absorb, reflect, or scatter light, yielding distinct spectral information.	The information captured through the camera consists of color, pixel distribution, and brightness, converting these into digital signals for effective processing.
Attributes procured	Various chemical structures, composition of different elements in the material.	Different characters, shapes, patterns, colors, labels, etc.
Forms of data	Data is collected when the intensity of light emitted interacts with the materials in the form of absorption, reflection, and scattering.	Red, green, and blue (RGB) channel images.
Large dataset	Not required.	In most cases, it is recommended to utilize a big dataset to reduce overfitting, and the distribution of the training set should closely match that of real-world problems.
Dimensionality reduction tools	LDA, PLS-DA, PCA	—
ML algorithms	SVM, KNN, CNN	R-CNN, YOLO, and SSD series.
Detection speed	NIR: Very fast. Raman: Also very fast, with speeds reaching up to 70 fps. LIBS: Fast, speeds reach up to 10 fps.	R-CNN, YOLO, and SSD series help to achieve real-time detection.
Accuracy / Precision	High	It leans on both the peculiar task on hand and the training dataset.
Material positioning and stature assessment	No	Algorithms help to locate the different materials, while cameras can help to measure the distance.
Implementation	Classification of mixed plastic waste.	Detecting all kinds of waste, for example, separating plastic from metals, wood, etc., or segregating different types of E-waste.

worldwide. Numerous examples have been used in a combination of CNN and spectroscopic-based methods such as NIR, LIBS, MIR, etc.[83] Neo et al.[82] combined CNN and FTIR, Raman spectroscopic techniques. The model was used to identify the polymers from Raman spectroscopic and FTIR datasets of over 20 polymers. The proposed model effectively classified the plastic with an accuracy of 0.949 and 0.967 for Raman spectroscopic and FTIR datasets.

Also, the proposed model was used to differentiate between weathered and unaged plastic samples with an astounding accuracy of 0.954 for HDPE and 0.906 for PET. Chen et al.[23] used CNN and NIR spectroscopy to address the challenge of shrink sleeves hindering the mechanical recycling of plastic bottles, particularly in NIR-based sorting processes. The research outcome shows that specific sleeve designs, including black coloring and large-area printing with high shrinkage, and incorporating sleeved bottle data into training datasets, along with selecting appropriate hyperparameters, can significantly improve the efficiency of the NIR-based sorting, thus improving the prospects for their mechanical recycling. According to the latest research,[45,139] the problem of black plastic sorting has been overcome by combining CNN with NIR spectroscopy and fluorescence spectrometer. According to a recent study,[151] a combination of CNN and MIR has been implemented to classify black plastic materials in mixed plastic waste. Dark plastics pose challenges due to their low signal-to-noise ratio. However, this novel approach achieves an impressive 100% classification accuracy for different types of dark plastics and can identify materials in multi-plastic blends with the same accuracy. MIR can collect spectral data up to 400 Hz, while the CNN model can make predictions with a speed of 8200 Hz. This state-of-the-art approach can enable real-time, highly accurate characterization of MPW. According to Ji et al.,[55] combining the RGB channels and limited NIR data creates a cost-effective sorting solution for low-value recyclable waste, achieving high material and color recognition accuracy.

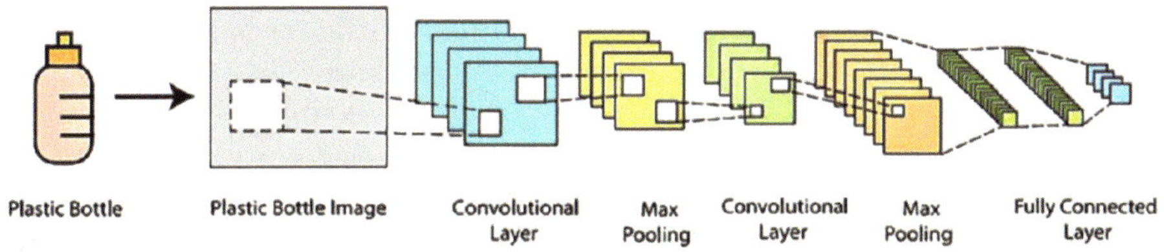

FIGURE 10.3 Convolutional Neural Network

10.4.2 COMPUTER VISION-BASED METHOD

Computer vision-based methods classify materials based on their visible macroscopic properties. Hence, computer vision-based methods classify materials early,[147] for example, indentifying the plastic from metal, wood, cardboard, etc. CNN has proven to be one of the effective computer vision-based methods for sorting plastic waste.[143] According to the latest research,[14,114] CNN is applied to segregate mixed plastic waste according to their resin identification code (Fig. 10.4). According to Santos et al.,[100] improper handling of the CRD waste can lead to detrimental effects on the environment. Thus, the use of various technologies, including the CNN can lead to the effective handling and management of CRD plastic waste.

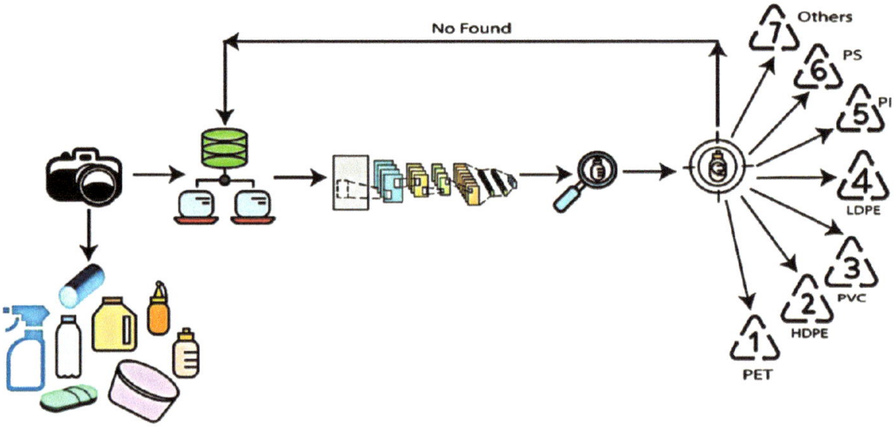

FIGURE 10.4 Convolutional Neural Network using Camera images to identify the plastics based on their resin identification code

Tessarini and Fileti[121] proposed using an audio signal and CNN to segregate the mixed plastic waste. With the growing population and its needs, plastic consumption will also increase, making recycling plastics a high priority. Manual sorting is efficient as it lacks resin identification symbols. However, this method is different. Audio signals differ among resins. Plastic samples were crushed, and their audio signals were recorded to create a database. Mel-Frequency Cepstral Coefficients (MFCC) were used for feature extraction, while CNN and LSTM were employed for classification. A recent study[53] has used a unique transfer learning approach and mask R-CNN for plastic bottle segmentation and detection. This model showed a mean average precision of around 59.4, demonstrating the potential of deep learning in plastic waste detection. Mask R-CNN has been used to segregate residual pipes and cables, bottles, papers, batteries, cans, cups, etc.[39,132] Faster R-CNN can segregate E-waste, waste bottles, etc.[84,149]

10.5 FUTURE OF AI IN PLASTIC SORTING

Segregation of solid waste, based on spectroscopic and computer vision-based methods, is an ingenious but arduous task. Most of the research conducted till now is laboratory and experimental based. Still, during implementation on an industrial scale, things get more complicated and trickier as real-life scenarios are to be dealt with. Hence, the following realms need more investigation to improvise the research from an industrial perspective.

1. Spectroscopic-based methods.
    1.1 Similar plastics such as HDPE/LDPE, PS/ABS, and black and dark-colored plastics are categorized as similar plastics when evaluating the effectiveness of different neural networks such as ANN, CNN, and others.
    1.2 Impact of different datasets on the behavior of the model and its outcomes and results.
2. Computer vision-based methods.
    2.1 Performance evaluation of models such as YOLO, FSSD, and R-CNN in a real-world waste detection task.
    2.2 Using tools such as R-FCN within the YOLO series to detect small waste particles.
    2.3 Building suitable quality datasets that will represent industrial scenarios.
    2.4 Assessment of the potential of semi-supervised, one-shot, or few-shot learning and finding applications for implementing these methodologies.
3. Sensor combination.
    3.1 Creating effective multitask sorting systems by exploring the possibility of integrating multiple sensors.
4. Data acquisition and transmission.
    4.1 The reliability of the categorization and position information supplied and collected from sensors to industrial robots should be examined as they carry out the sorting procedure on a large scale.

By concentrating on these areas, we may overcome major problems and create solid waste sorting systems that perform as well as humans in real-time categorization.

However, not only can AI be applied to the sorting process in plastic waste management, but it can also be applied to the pyrolysis process[24] and the recycling process.[38]

10.6 CONCLUSION

Plastic sorting is a critical process in recycling and waste management, and it involves various methods, both manual and automated. Gravity separation, dry and wet procedures, and auxiliary separation technologies such as magnetic and eddy current separation have all been utilized extensively in manual sorting processes. However, the future of plastic sorting lies in advanced AI systems such as CNNs, which can analyze and sort plastics with high precision. These AI systems, coupled with sensor-based technologies such as spectroscopy and robotics, promise to revolutionize plastic sorting by improving efficiency and accuracy. As we move forward, AI-driven solutions are expected to play a pivotal role in addressing the challenges posed by plastic waste and enhancing the sustainability of our planet.

KEYWORDS

- **mixed plastic waste**
- **plastic segregation**
- **sustainability**
- **image-based classification**

REFERENCES

1. Agarwal, S.; Gudi, R.; Saxena, P. Image Classification Approaches for Segregation of Plastic Waste Based on Resin Identification Code. *Trans. Indian Natl. Acad. Eng.* **2022,** *7* (3), 739–751. https://doi.org/10.1007/s41403-022-00324-4
2. Andrady, A. L.; Neal, M. A. Applications and Societal Benefits of Plastics. *Philos. Trans. Royal Soc. London B Biol. Sci.* **2009,** *364* (1526), 1977–1984. https://doi.org/10.1098/rstb.2008.0304
3. Annex A. Compose the OECD Global Plastics Outlook Database.

4. Anzano, J.; Bonilla, B.; Montull-Ibor, B.; Casas-González, J. Plastic Identification and Comparison by Multivariate Techniques with Laser-Induced Breakdown Spectroscopy. *J. Appl. Polym. Sci.* **2011,** *121* (5), 2710–2716. https://doi.org/10.1002/app.33801
5. Anzano, J.; Casanova, M. E.; Bermúdez, M. S.; Lasheras, R. J. Rapid Characterization of Plastics using Laser-Induced Plasma Spectroscopy (LIPS). *Polym. Test.* **2006,** *25* (5), 623–627. https://doi.org/10.1016/j.polymertesting.2006.04.005
6. Anzano, J.; Lasheras, R. J.; Bonilla, B.; Casas, J. Classification of Polymers by Determining of C1:C2:CN:H:N:O Ratios by Laser-Induced Plasma Spectroscopy (LIPS). *Polym. Test.* **2008,** *27* (6), 705–710. https://doi.org/10.1016/j.polymertesting.2008.05.012
7. Aschenbrenner, D.; Gros, J.; Fangerow, N.; Werner, T.; Colloseus, C.; Taha, I. Recyclebot–Using Robots for Sustainable Plastic Recycling. *Procedia CIRP* **2023,** *116*, 275–280. https://doi.org/10.1016/j.procir.2023.02.047
8. Bae, J. S.; Oh, S. K.; Pedrycz, W.; Fu, Z. Design of Fuzzy Radial Basis Function Neural Network Classifier Based on Information Data Preprocessing for Recycling Black Plastic Wastes: Comparative Studies of ATR FT-IR and Raman Spectroscopy. *Appl. Intell.* **2019,** *49* (3), 929–949. https://doi.org/10.1007/s10489-018-1300-5
9. Bakker, E. J.; Rem, P. C.; Fraunholcz, N. Upgrading Mixed Polyolefin Waste with Magnetic Density Separation. *Waste Manage.* **2009,** *29* (5), 1712–1717. https://doi.org/10.1016/j.wasman.2008.11.006
10. Beigbeder, J.; Perrin, D.; Mascaro, J. F.; Lopez-Cuesta, J. M. Study of the Physico-Chemical Properties of Recycled Polymers from Waste Electrical and Electronic Equipment (WEEE) Sorted by High Resolution Near Infrared Devices. *Resour. Conserv. Recy.* **2013,** *78*, 105–114. https://doi.org/10.1016/j.resconrec.2013.07.006
11. Bezati, F.; Froelich, D.; Massardier, V.; Maris, E. Addition of X-Ray Fluorescent Tracers into Polymers, New Technology for Automatic Sorting of Plastics: Proposal for Selecting Some Relevant Tracers. *Resour. Conserv. Recy.* **2011,** *55* (12), 1214–1221. https://doi.org/10.1016/j.resconrec.2011.05.014
12. Bhattacharya, P.; Lin, S.; Turner, J. P.; Ke, P. C. Physical Adsorption of Charged Plastic Nanoparticles Affects Algal Photosynthesis. *J. Phys. Chem. C* **2010,** *114* (39), 16556–16561. https://doi.org/10.1021/jp1054759
13. Blassino, M.; Solo-Gabriele, H.; Townsend, T. Pilot Scale Evaluation of Sorting Technologies for CCA Treated Wood Waste. *Waste Manage. Res.* **2002,** *20* (3), 290–301. https://doi.org/10.1177/0734242X0202000310
14. Bobulski, J.; Kubanek, M. Deep Learning for Plastic Waste Classification System. *Appl. Comput. Intell. Soft Comput.* **2021,** *2021*, 1–7. https://doi.org/10.1155/2021/6626948
15. Bonifazi, G.; Serranti, S. Imaging Spectroscopy Based Strategies for Ceramic Glass Contaminants Removal in Glass Recycling. *Waste Manage.* **2006,** *26* (6), 627–639. https://doi.org/10.1016/j.wasman.2005.06.004
16. Bonifazi, G.; Capobianco, G.; Serranti, S. A Hierarchical Classification Approach for Recognition of Low-Density (LDPE) and High-Density Polyethylene (HDPE) in Mixed Plastic Waste Based on Short-Wave Infrared (SWIR) Hyperspectral Imaging. *Spectrochim. Acta A Mol. Biomol. Spectrosc.* **2018,** *198*, 115–122. https://doi.org/10.1016/j.saa.2018.03.006
17. Bonifazi, G.; Palmieri, R.; Serranti, S. Hyperspectral Imaging Applied to End-of-Life (EOL) Concrete Recycling. *Tm – Technisches Messen* **2015,** *82* (12), 616–624. https://doi.org/10.1515/teme-2015-0044

18. Bonifazi, G.; Palmieri, R.; Serranti, S. Evaluation of Attached Mortar on Recycled Concrete Aggregates by Hyperspectral Imaging. *Constr. Build. Mater.* **2018,** *169*, 835–842. https://doi.org/10.1016/j.conbuildmat.2018.03.048
19. Bradley, D. Historical Development. In *Hydrocyclone*, 1965; pp 356–393.
20. Brunner, S.; Fomin, P.; Kargel, Ch. Automated Sorting of Polymer Flakes: Fluorescence Labeling and Development of a Measurement System Prototype. *Waste Manage.* **2015,** *38*, 49–60. https://doi.org/10.1016/j.wasman.2014.12.006
21. Buchan, R.; Yarar, B. Recovering Plastics for Recycling by Mineral Processing Techniques. *JOM* **1995,** *47* (2), 52–55. https://doi.org/10.1007/BF03221410
22. Chen, L.; Jin, S.; Li, W. In *Rapid Identification of Plastics Based on Raman Spectroscopy with the Combination of Support Vector Machine*, 2017 16th International Conference on Optical Communications and Networks (ICOCN); IEEE Publications, 2017; pp 1–3. https://doi.org/10.1109/ICOCN.2017.8121214
23. Chen, X.; Kroell, N.; Althaus, M.; Pretz, T.; Pomberger, R.; Greiff, K. Enabling Mechanical Recycling of Plastic Bottles with Shrink Sleeves through Near-Infrared Spectroscopy and Machine Learning Algorithms. *Resour. Conserv. Recy.* **2023,** *188*, 106719. https://doi.org/10.1016/j.resconrec.2022.106719
24. Cheng, Y.; Ekici, E.; Yildiz, G.; Yang, Y.; Coward, B.; Wang, J. Applied Machine Learning for Prediction of Waste Plastic Pyrolysis Towards Valuable Fuel and Chemicals Production. *J. Anal. Appl. Pyrolysis* **2023,** *169*, 105857. https://doi.org/10.1016/j.jaap.2023.105857
25. Chia, R. W.; Lee, J. Y.; Kim, H.; Jang, J. Microplastic Pollution in Soil and Groundwater: A Review. *Environ. Chem. Lett.* **2021,** *19* (6), 4211–4224. https://doi.org/10.1007/s10311-021-01297-6
26. Christensen, T. Ed.; In *Solid Waste Technology and Management*; John Wiley & Sons, 2011.
27. Cimpan, C.; Maul, A.; Jansen, M.; Pretz, T.; Wenzel, H. H. Central Sorting and Recovery of MSW Recyclable Materials: A Review of Technological State-of-the-Art, Cases, Practice and Implications for Materials Recycling. *J. Environ. Manage.* **2015,** *156*, 181–199. https://doi.org/10.1016/j.jenvman.2015.03.025
28. Conti, G. O.; Ferrante, M.; Banni, M.; Favara, C.; Nicolosi, I.; Cristaldi, A.; Fiore, M.; Zuccarello, P. Micro- and Nano-Plastics in Edible Fruit and Vegetables. The First Diet Risks Assessment for the General Population. *Environ. Res.* **2020,** *187*, 109677.
29. Cook, C. R.; Halden, R. U. Ecological and Health Issues of Plastic Waste. In *Plastic Waste and Recycling*; Academic Press, 2020; pp 513–527.
30. Costa, V. C.; Aquino, F. W. B.; Paranhos, C. M.; Pereira-Filho, E. R. Identification and Classification of Polymer e-waste Using Laser-Induced Breakdown Spectroscopy (LIBS) and Chemometric Tools. *Polym. Test.* **2017,** *59*, 390–395. https://doi.org/10.1016/j.polymertesting.2017.02.017
31. Cremers, D. A.; Radziemski, L. J. In *Handbook of Laser-Induced Breakdown Spectroscopy*; John Wiley & Sons, 2013.
32. Dall'Ara, A.; Bonoli, A.; Serranti, S. An Innovative Procedure to Characterize Properties from Tailored Composts. *Environ. Eng. Manage. J.* **2012,** *11* (10).
33. Dharmana, M. M.; Mahathi, L. U.; Indarapu, A.; Reddy, B. H. In *Plastic Segregation Machine using Audio Signal Analysis*, 2019 3rd International conference on Electronics, Communication and Aerospace Technology (ICECA); IEEE Publications, 2019; pp 982–985. https://doi.org/10.1109/ICECA.2019.8821842
34. Dodbiba, G.; Shibayama, A.; Miyazaki, T.; Fujita, T. Electrostatic Separation of the Shredded Plastic Mixtures Using a Tribo-Cyclone. *Magn. Electr. Separat.* **2002,** *11* (1–2), 63–92. https://doi.org/10.1080/07313630290002626

35. Duan, Q.; Li, J. Classification of Common Household Plastic Wastes Combining Multiple Methods Based on Near-Infrared Spectroscopy. *ACS ES&T Eng.* **2021,** *1* (7), 1065–1073. https://doi.org/10.1021/acsestengg.0c00183
36. Elton, D. C.; Boukouvalas, Z.; Butrico, M. S.; Fuge, M. D.; Chung, P. W. Applying Machine Learning Techniques to Predict the Properties of Energetic Materials. *Sci. Rep.* **2018,** *8* (1), 9059. https://doi.org/10.1038/s41598-018-27344-x
37. Erkinay Ozdemir, M.; Ali, Z.; Subeshan, B.; Asmatulu, E. Applying Machine Learning Approach in Recycling. *J. Mater. Cycles Waste Manage.* **2021,** *23* (3), 855–871. https://doi.org/10.1007/s10163-021-01182-y
38. Fang, B.; Yu, J.; Chen, Z.; Osman, A. I.; Farghali, M.; Ihara, I.; Hamza, E. H.; Rooney, D. W.; Yap, P. S. Artificial Intelligence for Waste Management in Smart Cities: A Review. *Environ. Chem. Lett.* **2023,** 1–31. https://doi.org/10.1007/s10311-023-01604-3
39. Feng, J.; Tang, X.; Jiang, X.; Chen, Q. Garbage Disposal of Complex Background Based on Deep Learning with Limited Hardware Resources. *IEEE Sens. J.* **2021,** *21* (18), 21050–21058. https://doi.org/10.1109/JSEN.2021.3100636
40. Fraunholcz, N. Separation of Waste Plastics by Froth Flotation – A Review, Part I. *Miner. Eng.* **2004,** *17*(2), 261–268. https://doi.org/10.1016/j.mineng.2003.10.028
41. Gandhi, R. R-CNN, Fast R-CNN, Faster R-CNN, YOLO—Object Detection Algorithms. *Towar* **2018**. https://towardsdatascience.com/r-cnn-fast-r-cnn-faster-r-cnn-yolo-object-detection-algorithms-36d53571365e (accessed Sept 19, 2023).
42. Geyer, R.; Jambeck, J. R.; Law, K. L. Production, Use, and Fate of All Plastics Ever Made. *Sci. Adv.* **2017,** *3* (7), e1700782. https://doi.org/10.1126/sciadv.1700782
43. Gondal, M. A.; Hussain, T.; Yamani, Z. H.; Baig, M. A. The Role of Various Binding Materials for Trace Elemental Analysis of Powder Samples using Laser-Induced Breakdown Apectroscopy. *Talanta* **2007,** *72* (2), 642–649. https://doi.org/10.1016/j.talanta.2006.11.039
44. Greenpeace. Plastic Debris in the World's Oceans [Online] 2006. https://www.greenpeace.org/archive-international/en/publications/reports/plastic_ocean_report/ (accessed Sept 4, 2023).
45. Gruber, F.; Grählert, W.; Wollmann, P.; Kaskel, S. Classification of Black Plastics Waste Using Fluorescence Imaging and Machine Learning. *Recycling* **2019,** *4* (4), 40. https://doi.org/10.3390/recycling4040040
46. Guo, H. N.; Wu, S. B.; Tian, Y. J.; Zhang, J.; Liu, H. T. Application of Machine Learning Methods for the Prediction of Organic Solid Waste Treatment and Recycling Processes: A Review. *Bioresour. Technol.* **2021,** *319*, 124114. https://doi.org/10.1016/j.biortech.2020.124114
47. Hori, K.; Tsunekawa, M.; Ueda, M.; Hiroyoshi, N.; Ito, M.; Okada, H. Development of a New Gravity Separator for Plastics—A Hybrid-jig—. *Mater. Trans.* **2009,** *50* (12), 2844–2847.
48. Hu, B.; Fraunholz, N.; Rem, P. Wetting Technologies for High-Accuracy Sink–Float Separations in Water-Based Media. *Open Waste Manage. J.* **2010,** *3* (1), 71–80. https://doi.org/10.2174/1876400201003010071
49. Hu, B.; Serranti, S.; Fraunholcz, N.; Di Maio, F.; Bonifazi, G. Recycling-Oriented Characterization of Polyolefin Packaging Waste. *Waste Manage.* **2013,** *33* (3), 574–584. https://doi.org/10.1016/j.wasman.2012.11.018
50. Huang, J.; Zhang, H.; Xiao, X.; Huang, J.; Xie, J.; Zhang, L.; Hu, H.; Dai, S.; Li, M.; Xu, Y. Method for C/N Ratio Estimation using Mask R-CNN and a Depth Camera for Organic Fraction of Municipal Solid Wastes. *Int. J. Agric. Biol. Eng.* **2021,** *14* (5), 222–229. https://doi.org/10.25165/j.ijabe.20211405.6382

51. Jahnavi, M. Introduction to Neural Networks, Advantages and Applications. *Mediu. Corp.* **2017**. https://towardsdatascience.com/introduction-to-neural-networks-advantages-and-applications-96851bd1a207 (accessed Sept 19, 2023).
52. Jähne, B. Concepts, Algorithms, and Scientific Applications. In *Digital Image Processing*, 1995; pp 321–.
53. Jaikumar, P.; Vandaele, R.; Ojha, V. In *Transfer Learning for Instance Segmentation of Waste Bottles using Mask R-CNN Algorithm*, International Conference on Intelligent Systems Design and Applications; Springer International Publishing, 2020; pp 140–149.
54. Jansen, M.; Feil, A.; Pretz, T. Recovery of Plastics from Household Waste by Mechanical Separation. *Waste Manage.* **2012,** *3*, 169–176.
55. Ji, T.; Fang, H.; Zhang, R.; Yang, J.; Fan, L.; Hu, Y.; Cai, Z. Low-Value Recyclable Waste Identification Based on NIR Feature Analysis and RGB-NIR Fusion. *Infrared Phys. Technol.* **2023,** *131*, 104693. https://doi.org/10.1016/j.infrared.2023.104693
56. Junjuri, R.; Gundawar, M. K. Femtosecond Laser-Induced Breakdown Spectroscopy Studies for the Identification of Plastics. *J. Anal. At. Spectrom.* **2019,** *34* (8), 1683–1692. https://doi.org/10.1039/C9JA00102F
57. Kadaous, W. What are the Advantages and Disadvantages of Deep Learning? Can you Compare it with the Statistical Learning Theory?—Quora. *Quora* [Online] 2016. https://www.quora.com/What-are-the-advantages-and-disadvantages-of-deep-learning-Can-you-compare-it-with-the-statistical-learning-theory#:~:text=Statistical%20learning-%20theory%20isn't,is%20a%20machine%20learning%20framework.&text=Has%20best%2Din%2Dclass%20performance,other%20solutions%20in%20multiple%20domains (accessed Sept 19, 2023).
58. Karthikeyan, M.; Subashini, T. S.; Jebakumar, R. SSD-Based Waste Separation in Smart Garbage using Augmented Clustering NMS. *Autom. Softw. Eng.* **2021,** *28*, 1–7.
59. KD. Top 4 Advantages and Disadvantages of Support Vector Machine or SVM. [Online] 2019. https://dhirajkumarblog.medium.com/top-4-advantages-and-disadvantages-of-support-vector-machine-or-svm-a3c06a2b107#:~:text=Advantages%20of%20Support%20Vector%20Machine,SVM%20is%20relatively%20memory%20efficient (accessed Sept 19, 2023)
60. Kibria, M. G.; Masuk, N. I.; Safayet, R.; Nguyen, H. Q.; Mourshed, M. Plastic Waste: Challenges and Opportunities to Mitigate Pollution and Effective Management. *Int. J. Environ. Res.* **2023,** *17* (1), 20. https://doi.org/10.1007/s41742-023-00507-z
61. Kilimci, Z. H.; Akyuz, A. O.; Uysal, M.; Akyokus, S.; Uysal, M. O.; Atak Bulbul, B. B.; Ekmis, M. A. An Improved Demand Forecasting Model using Deep Learning Approach and Proposed Decision Integration Strategy for Supply Chain. *Complexity* **2019,** *2019*, 1–15. https://doi.org/10.1155/2019/9067367
62. Kim, H.; Kim, S. In *Band Selection for Plastic Classification using NIR Hyperspectral Image*, 16th International Conference on Control., Automation and Systems (ICCAS); IEEE Publications, 2016; pp 302–304. https://doi.org/10.1109/ICCAS.2016.7832335
63. Kim, S. W.; Jeong, S. W.; An, Y. J. Microplastics Disrupt Accurate Soil Organic Carbon Measurement Based on Chemical Oxidation Method. *Chemosphere* **2021**, *276*, 130178. https://doi.org/10.1016/j.chemosphere.2021.130178
64. Ku, Y.; Yang, J.; Fang, H.; Xiao, W.; Zhuang, J. Deep Learning of Grasping Detection for a Robot used in Sorting Construction and Demolition Waste. *J. Mater. Cycles Waste Manage.* 2021, *23* (1), 84–95. https://doi.org/10.1007/s10163-020-01098-z

65. Kumar, S.; Yadav, D.; Gupta, H.; Verma, O. P.; Ansari, I. A.; Ahn, C. W. A Novel yolov3 Algorithm-Based Deep Learning Approach for Waste Segregation: Towards Smart Waste Management. *Electronics* **2020,** *10* (1), 14. https://doi.org/10.3390/electronics10010014
66. Lee, S. H.; Yeh, C. H.; Hou, T. W.; Yang, C. S. In *A Lightweight Neural Network Based on AlexNet-SSD Model for Garbage Detection*, 3rd High Performance Computing and Cluster Technologies Conference, June 22, 2019; pp 274–278. https://doi.org/10.1145/3341069.3341087
67. Li, J.; Chen, J.; Sheng, B.; Li, P.; Yang, P.; Feng, D. D.; Qi, J. Automatic Detection and Classification System of Domestic Waste Via Multimodel Cascaded Convolutional Neural Network. *IEEE Trans. Ind. Inf.* **2021,** *18* (1), 163–173. https://doi.org/10.1109/TII.2021.3085669
68. Li, J.; Li, C.; Liao, Q.; Xu, Z. Environmentally Friendly Technology for Rapid On-Line Recycling of Acrylonitrile-Butadiene-Styrene, Polystyrene and Polypropylene using Near-Infrared Spectroscopy. *J. Clean. Prod.* **2019,** *213*, 838–844. https://doi.org/10.1016/j.jclepro.2018.12.160
69. Liu, K.; Tian, D.; Wang, H.; Yang, G. Rapid Classification of Plastics by Laser-Induced Breakdown Spectroscopy (LIBS) Coupled with Partial Least Squares Discrimination Analysis Based on Variable Importance (VI-PLS-DA). *Anal. Methods* **2019,** *11* (9), 1174–1179. https://doi.org/10.1039/C8AY02755B
70. Liu, L.; Ouyang, W.; Wang, X.; Fieguth, P.; Chen, J.; Liu, X.; Pietikäinen, M. Deep Learning for Generic Object Detection: A Survey. *Int. J. Comput. Vision* **2020,** *128* (2), 261–318. https://doi.org/10.1007/s11263-019-01247-4
71. Liu, Y.; Zhao, T.; Ju, W.; Shi, S. Materials Discovery and Design using Machine Learning. *J. Materiomics* **2017,** *3* (3), 159–177. https://doi.org/10.1016/j.jmat.2017.08.002
72. Lu, Y.; Yang, B.; Gao, Y.; Xu, Z. An Automatic Sorting System for Electronic Components Detached from Waste Printed Circuit Boards. *Waste Manage.* **2022,** *137*, 1–8. https://doi.org/10.1016/j.wasman.2021.10.016
73. Lubongo, C.; Alexandridis, P. Assessment of Performance and Challenges in use of Commercial Automated Sorting Technology for Plastic Waste. *Recycling* **2022,** *7* (2), 11. https://doi.org/10.3390/recycling7020011
74. Luciani, V.; Bonifazi, G.; Rem, P.; Serranti, S. Upgrading of PVC Rich Wastes by Magnetic Density Separation and Hyperspectral Imaging Quality Control. *Waste Manage.* **2015,** *45*, 118–125. https://doi.org/10.1016/j.wasman.2014.10.015
75. Ma, W.; Wang, X.; Yu, J. A Lightweight Feature Fusion Single Shot Multibox Detector for Garbage Detection. *IEEE Access* **2020,** *8*, 188577–188586. https://doi.org/10.1109/ACCESS.2020.3031990
76. Mao, W. L.; Chen, W. C.; Fathurrahman, H. I. K.; Lin, Y. H. Deep Learning Networks for Real-Time Regional Domestic Waste Detection. *J. Clean. Prod.* **2022,** *344*, 131096. https://doi.org/10.1016/j.jclepro.2022.131096
77. Mattsson, K.; Johnson, E. V.; Malmendal, A.; Linse, S.; Hansson, L. A.; Cedervall, T. Brain Damage and Behavioural Disorders in Fish Induced by Plastic Nanoparticles Delivered through the Food Chain. *Sci. Rep.* **2017,** *7*(1), 11452. https://doi.org/10.1038/s41598-017-10813-0
78. Mauruschat, D.; Plinke, B.; Aderhold, J.; Gunschera, J.; Meinlschmidt, P.; Salthammer, T. Application of Near-Infrared Spectroscopy for the Fast Detection and Sorting of Wood–Plastic Composites and Waste Wood Treated with Wood Preservatives. *Wood Sci. Technol.* **2016,** *50* (2), 313–331. https://doi.org/10.1007/s00226-015-0785-x

79. Meikle, J. L. *American Plastic: A Cultural History*; Rutgers University Press, 1995.
80. Musu, W.; Tsuchida, A.; Kawazumi, H.; Oka, N. Quantitative Evaluation for the Peak Selectionin the Raman Spectroscopy Classification of Plastics Based on the Support Vector Machine. *Joho Chishiki Gakkaishi* **2020,** *30* (3), 361–369. https://doi.org/10.2964/jsik_2020_034
81. Ncube, L. K.; Ude, A. U.; Ogunmuyiwa, E. N.; Zulkifli, R.; Beas, I. N. An Overview of Plastic Waste Generation and Management in Food Packaging Industries. *Recycling* **2021,** *6* (1), 12. https://doi.org/10.3390/recycling6010012
82. Neo, E. R. K.; Low, J. S. C.; Goodship, V.; Debattista, K. Deep Learning for Chemometric Analysis of Plastic Spectral Data from Infrared and Raman Databases. *Resour. Conserv. Recy.* **2023,** *188*, 106718. https://doi.org/10.1016/j.resconrec.2022.106718
83. Neo, E. R. K.; Yeo, Z.; Low, J. S. C.; Goodship, V.; Debattista, K. A Review on Chemometric Techniques with Infrared, Raman and Laser-Induced Breakdown Spectroscopy for Sorting Plastic Waste in the Recycling Industry. *Resour. Conserv. Recy.* **2022,** *180*, 106217. https://doi.org/10.1016/j.resconrec.2022.106217
84. Nowakowski, P.; Pamuła, T. Application of Deep Learning Object Classifier to Improve e-waste Collection Planning. *Waste Manage.* **2020,** *109*, 1–9. https://doi.org/10.1016/j.wasman.2020.04.041
85. Peng, X.; Xu, B.; Xu, Z.; Yan, X.; Zhang, N.; Qin, Y.; Ma, Q.; Li, J.; Zhao, N.; Zhang, Q. Accuracy Improvement in Plastics Classification by Laser-Induced Breakdown Spectroscopy Based on a Residual Network. *Opt. Express* **2021,** *29* (21), 33269–33280. https://doi.org/10.1364/OE.438331
86. Picon, A.; Ghita, O.; Iriondo, P. M.; Bereciartua, A.; Whelan, P. F. In *Automation of Waste Recycling using Hyperspectral Image Analysis*, 15th Conference on Emerging Technologies y Factory Automation (ETFA 2010); IEEE Publications, 2010; pp 1–4. https://doi.org/10.1109/ETFA.2010.5641201
87. Plastic Europe. Waste Collection, Pretreatment and Sorting. [Online] 2023. https://legacy.plasticseurope.org/en/focus-areas/circular-economy/zero-plastics-landfill/waste-collection-pre-treatment-and-sorting (accessed Sept 4, 2023).
88. Plastics Europe. [Online] 2021. https://plasticseurope.org/knowledge-hub/plastics-the-facts-2022/ (accessed Sept 4, 2023).
89. Plug and Play Tech Center. *How Machine Learning and Robotics are Solving the Plastic Sorting Crisis,* 2020. *https://www.plugandplaytechcenter.com/resources/how-ai-and-robotics-are-solving-plastic-sorting-crisis/#: ~:text=Robotic%20recycle%20sorting%20uses%20artificial,come%20along%20with%20human%20labor* (accessed Sept 13, 2023).
90. Ragaert, K.; Huysveld, S.; Vyncke, G.; Hubo, S.; Veelaert, L.; Dewulf, J.; Du Bois, E. Design from Recycling: A Complex Mixed Plastic Waste Case Study. *Resour. Conserv. Recy.* **2020,** *155*, 104646. https://doi.org/10.1016/j.resconrec.2019.104646
91. Rahimi, A.; García, J. M. Chemical Recycling of Waste Plastics for New Materials Production. *Nat. Rev. Chem.* **2017,** *1* (6). https://doi.org/10.1038/s41570-017-0046
92. Rani, M.; Marchesi, C.; Federici, S.; Rovelli, G.; Alessandri, I.; Vassalini, I.; Ducoli, S.; Borgese, L.; Zacco, A.; Bilo, F.; Bontempi, E.; Depero, L. E. Miniaturized Near-Infrared (MicroNIR) Spectrometer in Plastic Waste Sorting. *Materials* **2019,** *12* (17), 2740. https://doi.org/10.3390/ma12172740
93. Reinsch, E.; Frey, A.; Albrecht, V.; Simon, F.; Peuker, U. A. Continuous Electric Sorting in the Recycling Process of Plastics. *Chemie Ingenieur Technik* **2014,** *86* (6), 784–796. https://doi.org/10.1002/cite.201400010

94. Rem, P.; Di Maio, F. D.; Hu, B.; Houzeaux, G.; Baltes, L.; Tierean, M. Magnetic Fluid Equipment for Sorting Secondary Polyolefins from Waste. *Environ. Eng. Manage. J.* **2013,** *12* (5), 951–958. https://doi.org/10.30638/eemj.2013.118
95. Roh, S. B.; Oh, S. K.; Park, E. K.; Choi, W. Z. Identification of Black Plastics Realized with the Aid of Raman Spectroscopy and Fuzzy Radial Basis Function Neural Networks Classifier. *J. Mater. Cycles Waste Manage.* **2017,** *19* (3), 1093–1105. https://doi.org/10.1007/s10163-017-0620-6
96. Rozenstein, O.; Puckrin, E.; Adamowski, J. Development of a New Approach Based on Midwave Infrared Spectroscopy for Post-Consumer Black Plastic Waste Sorting in the Recycling Industry. *Waste Manage.* **2017,** *68*, 38–44. https://doi.org/10.1016/j.wasman.2017.07.023
97. Rudin, C. Stop Explaining Black Box Machine Learning Models for High Stakes Decisions and Use Interpretable Models Instead. *Nat. Mach. Intell.* **2019,** *1* (5), 206–215. https://doi.org/10.1038/s42256-019-0048-x
98. Sakr, S.; Elshawi, R.; Ahmed, A. M.; Qureshi, W. T.; Brawner, C. A.; Keteyian, S. J.; Blaha, M. J.; Al-Mallah, M. H. Comparison of Machine Learning Techniques to Predict All-Cause Mortality Using Fitness Data: The Henry Ford Exercise Testing (FIT) Project. *BMC Med. Inform. Decis. Mak.* **2017,** *17* (1), 1.
99. Sampedro, G. A.; Kim, R. G.; Aruan, Y. J.; Kim, D. S.; Lee, J. M. In *Smart E-waste Bin Development Based on yolov4 Model*, 2021 1st International Conference in Information and Computing Research (iCORE). IEEE Publications, 2021; pp 125–128.
100. Santos, G.; Esmizadeh, E.; Riahinezhad, M. Recycling Construction, Renovation, and Demolition Plastic Waste: Review of the Status Quo, Challenges and Opportunities. *J. Polym. Environ.* **2023,** 1–31. https://doi.org/10.1007/s10924-023-02982-z
101. Sarc, R.; Curtis, A.; Kandlbauer, L.; Khodier, K.; Lorber, K. E.; Pomberger, R. Digitalisation and Intelligent Robotics in Value Chain of Circular Economy Oriented Waste Management–A Review. *Waste Manage.* **2019,** *95*, 476–492. https://doi.org/10.1016/j.wasman.2019.06.035
102. Sattler, H. P.; Yoshida, T. In *New Sorting System for Recycling of Magnesium and its Alloys After Use*, First International Conference on Processing Materials for Properties, 1993; pp 861–864.
103. Serranti, S.; Bonifazi, G. Hyperspectral Imaging Applied to End-of-Life Concrete Recycling. *Inimage Sensors Imag. Syst.* **2014,** *9022.*
104. Serranti, S.; Bonifazi, G. Techniques for Separation of Plastic Wastes. In *Use of Recycled Plastics in Eco-Efficient; Concrete.* Woodhead Publishing, 2019; pp 9–37.
105. Serranti, S.; Gargiulo, A.; Bonifazi, G. Characterization of Post-Consumer Polyolefin Wastes by Hyperspectral Imaging for Quality Control in Recycling Processes. *Waste Manage.* **2011,** *31* (11), 2217–2227. https://doi.org/10.1016/j.wasman.2011.06.007
106. Serranti, S.; Gargiulo, A.; Bonifazi, G. Classification of Polyolefins from Building and Construction Waste using NIR Hyperspectral Imaging System. *Resour. Conserv. Recy.* **2012,** *61*, 52–58. https://doi.org/10.1016/j.resconrec.2012.01.007
107. Serranti, S.; Gargiulo, A.; Bonifazi, G. Hyperspectral Imaging for Process and Quality Control in Recycling Plants of Polyolefin Flakes. *J. Near Infrared Spec.* **2012,** *20* (5), 573–581. https://doi.org/10.1255/jnirs.1016
108. Serranti, S.; Luciani, V.; Bonifazi, G.; Hu, B.; Rem, P. C. An Innovative Recycling Process to Obtain Pure Polyethylene and Polypropylene from Household Waste. *Waste Manage.* **2015,** *35*, 12–20. https://doi.org/10.1016/j.wasman.2014.10.017

109. Shapiro, M.; Galperin, V. Air Classification of Solid Particles: A Review. *Chem. Eng. Process.* **2005,** *44* (2), 279–285. https://doi.org/10.1016/j.cep.2004.02.022
110. Shen, H.; Pugh, R. J.; Forssberg, E. Floatability, Selectivity and Flotation Separation of Plastics by Using a Surfactant. *Coll. Surf. A Physicochem. Eng. Asp.* **2002,** *196* (1), 63–70. https://doi.org/10.1016/S0927-7757(01)00706-3
111. Singh, B. P. Wetting Mechanism in the Flotation Separation of Plastics. *Filtr. Separat.* **1998,** *35* (6), 525–527. https://doi.org/10.1016/S0015-1882(98)80027-X
112. Sitawarin, C.; Wagner, D. In *On the Robustness of Deep k-nearest Neighbors,* 2019 IEEE Security and Privacy Workshops (SPW); IEEE Publications, 2019; pp 1–7.
113. SL Recycling Limited. What is the Waste Sorting Process? [Online] 2023. https://www.slrecyclingltd.co.uk/what-is-the-waste-sorting-process/#:~:text=As%20the%20name%20suggests%2C%20the,to%20businesses%20separating%20recyclable%20materials (accessed Sept 4, 2023)
114. Stanley, K. O.; Clune, J.; Lehman, J.; Miikkulainen, R. Designing Neural Networks through Neuroevolution. *Nat. Mach. Intell.* **2019,** *1* (1), 24–35. https://doi.org/10.1038/s42256-018-0006-z
115. Subashini, M. M.; Vignesh, R. S. Thermoplastic Waste Segregation Classification System using Deep Learning Techniques. *Multimed. Tools Appl.* **2023,** *83* (29-30),1–7. https://doi.org/10.1007/s11042-023-16237-5
116. Svoboda, J.; Fujita, T. Recent Developments in Magnetic Methods of Material Separation. *Miner. Eng.* **2003,** *16* (9), 785–792. https://doi.org/10.1016/S0892-6875(03)00212-7
117. Svoboda, J. In *Magnetic Techniques for the Treatment of Materials*; Springer Science + Business Media, 2004.
118. Takoungsakdakun, T.; Pongstabodee, S. Separation of Mixed Post-Consumer PET–POM–PVC Plastic Waste Using Selective Flotation. *Sep. Purif. Technol.* **2007,** *54* (2), 248–252. https://doi.org/10.1016/j.seppur.2006.09.011
119. Tan, W.; Duan, Q.; Yao, L.; Li, J. A Sensor Combination Based Automatic Sorting System for Waste Washing Machine Parts. *Resour. Conserv. Recy.* **2022,** *181*, 106270. https://doi.org/10.1016/j.resconrec.2022.106270
120. Tatzer, P.; Wolf, M.; Panner, T. Industrial Application for Inline Material Sorting using Hyperspectral Imaging in the NIR Range. *Real Time Imag.* **2005,** *11* (2), 99–107. https://doi.org/10.1016/j.rti.2005.04.003
121. Tessarini, L.; Fileti, A. M. F. Audio Signals and Artificial Neural Networks for Classification of Plastic Resins for Recycling. *Digit. Chem. Eng.* **2022,** *5*, 100059. https://doi.org/10.1016/j.dche.2022.100059
122. Thokrairak, S.; Thibuy, K.; Jitngernmadan, P. In *Valuable Waste Classification Modeling Based on SSD-Mobilenet*, 5th International Conference on Information Technology (InCIT) IEEE Publications, 2020; pp 228–232. https://doi.org/10.1109/InCIT50588.2020.9310928
123. Thompson, R. C.; Swan, S. H.; Moore, C. J.; Vom Saal, F. S. Our Plastic Age. *Philos. Trans. Royal Soc. London B Biol. Sci.* **2009,** *364* (1526), 1973–1976. https://doi.org/10.1098/rstb.2009.0054
124. Ulrici, A.; Serranti, S.; Ferrari, C.; Cesare, D.; Foca, G.; Bonifazi, G. Efficient Chemometric Strategies for PET–PLA Discrimination in Recycling Plants using Hyperspectral Imaging. *Chemometr. Intell. Lab. Syst.* **2013,** *122*, 31–39. https://doi.org/10.1016/j.chemolab.2013.01.001

125. United Nations. Factsheet: Marine pollution [Online] 2020. https://sustainabledevelopment.un.org/content/documents/Ocean_Factsheet_Pollution.pdf(accessed on Sept 4, 2023)
126. United Nations Environmental Programm. [Online] 2023. https://www.unep.org/interactives/beat-plastic-pollution/#:~:text=Around%20the%20world%2C%20one%20million,once%20and%20then%20thrown%20away (accessed on Sept 4, 2023).
127. Van Der Valk, H. J. L.; Braam, B. C.; Dalmijn, W. L. Eddy-Current Separation by Permanent Magnets Part I: Theory. *Resour. Conserv.* **1986,** *12* (3–4), 233–252. https://doi.org/10.1016/0166-3097(86)90014-3
128. Walker, S.; Khan, W.; Katic, K.; Maassen, W.; Zeiler, W. Accuracy of Different Machine Learning Algorithms and Added-Value of Predicting Aggregated-Level Energy Performance of Commercial Buildings. *Energy Build.* **2020,** *209*, 109705. https://doi.org/10.1016/j.enbuild.2019.109705
129. Wang, C.; Liu, Y.; Chen, W. Q.; Zhu, B.; Qu, S.; Xu, M. Critical Review of Global Plastics Stock and Flow Data. *J. Ind. Ecol.* **2021,** *25* (5), 1300–1317. https://doi.org/10.1111/jiec.13125
130. Wang, C. Q.; Wang, H.; Fu, J. G.; Liu, Y. N. Flotation Separation of Waste Plastics for Recycling—A Review. *Waste Manage.* **2015,** *41*, 28–38. https://doi.org/10.1016/j.wasman.2015.03.027
131. Wang, Z.; O'Boyle, M. Machine Learning in Compiler Optimization. *Proc. IEEE* **2018,** *106* (11), 1879–1901. https://doi.org/10.1109/JPROC.2018.2817118
132. Wang, Z.; Li, H.; Yang, X. Vision-Based Robotic System for On-Site Construction and Demolition Waste Sorting and Recycling. *J. Build. Eng.* **2020,** *32*, 101769. https://doi.org/10.1016/j.jobe.2020.101769
133. Wang, Z.; Li, H.; Zhang, X. Construction Waste Recycling Robot for Nails and Screws: Computer Vision Technology and Neural Network Approach. *Autom. Constr.* **2019,** *97*, 220–228. https://doi.org/10.1016/j.autcon.2018.11.009
134. Wichai-Utcha, N.; Chavalparit, O. 3Rs Policy and Plastic Waste Management in Thailand. *J. Mater. Cycles Waste Manage.* **2019,** *21* (1), 10–22. https://doi.org/10.1007/s10163-018-0781-y
135. Wills, B. A.; Finch, J. In *Wills' Mineral Processing Technology: An Introduction to the Practical Aspects of Ore Treatment and Mineral Recovery*; Butterworth-Heinemann, 2015.
136. Wong, S. L.; Ngadi, N.; Abdullah, T. A. T.; Inuwa, I. M. Current State and Future Prospects of Plastic Waste as Source of Fuel: A Review. *Renew. Sustain. Energy Rev.* **2015,** *50*, 1167–1180. https://doi.org/10.1016/j.rser.2015.04.063
137. Workman, Jr., J.; Weyer, L. In *Practical Guide and Spectral Atlas for Interpretive Near-Infrared Spectroscopy*; CRC Press, 2012.
138. World Economic Forum. The New Plastics Economy: Rethinking the Future of Plastics; Ellen MacArthur Foundation, 2016. https://ellenmacarthurfoundation.org/the-new-plastics-economy-rethinking-the-future-of-plastics (accessed Sept 4, 2023)
139. Xia, J.; Huang, Y.; Li, Q.; Xiong, Y.; Min, S. Convolutional Neural Network with Near-Infrared Spectroscopy for Plastic Discrimination. *Environ. Chem. Lett.* **2021,** *19* (5), 3547–3555. https://doi.org/10.1007/s10311-021-01240-9
140. Yang, G.; Jin, J.; Lei, Q.; Wang, Y.; Zhou, J.; Sun, Z.; Li, X.; Wang, W. In *Garbage Classification System with Yolov5 based on Image Recognition*, 6th International

Conference on Signal and Image Processing (ICSIP); IEEE Publications, 2021; pp 11–18. https://doi.org/10.1109/ICSIP52628.2021.9688725

141. Zhang, Q.; Yang, Q.; Zhang, X.; Bao, Q.; Su, J.; Liu, X. Waste Image Classification based on Transfer Learning and Convolutional Neural Network. *Waste Manage.* **2021,** *135*, 150–157. https://doi.org/10.1016/j.wasman.2021.08.038
142. Zhang, Q.; Yang, Q.; Zhang, X.; Wei, W.; Bao, Q.; Su, J.; Liu, X. A Multi-Label Waste Detection Model Based on Transfer Learning. *Resour. Conserv. Recy.* **2022,** *181*, 106235. https://doi.org/10.1016/j.resconrec.2022.106235
143. Zhang, S.; Chen, Y.; Yang, Z.; Gong, H. Computer Vision Based Two-Stage Waste Recognition-Retrieval Algorithm for Waste Classification. *Resour. Conserv. Recy.* **2021,** *169*, 105543. https://doi.org/10.1016/j.resconrec.2021.105543
144. Zhang, S.; Rem, P. C.; Forssberg, E. The Investigation of Separability of Particles Smaller than 5 mm by Eddy Current Separation Technology. Part I: Rotating Type Eddy Current Separators. *Magnetic and Electrical Separation Magn. Electr. Separat.* **1999,** *9* (4), 233–251. https://doi.org/10.1155/1999/29832
145. Zhang, S.; Zhang, S.; Wang, B.; Habetler, T. G. Deep Learning Algorithms for Bearing Fault Diagnostics—A Comprehensive Review. *IEEE Access* **2020,** *8*, 29857–29881. https://doi.org/10.1109/ACCESS.2020.2972859
146. Zhang, Z.; Zhao, X.; Zhang, O.; Fu, G.; Xie, Y.; Liu, C. Residual Waste Quality Detection Method based on Gaussian-YOLOv3. In *Big Data Analytics for Cyber-Physical System in Smart City: BDCPS 2020, 28–29 December 2020*; Springer: Shanghai, China, 2021; pp 459–469. https://doi.org/10.1007/978-981-33-4572-0_67
147. Zhao, Y.; Li, J. Sensor-Based Technologies in Effective Solid Waste Sorting: Successful Applications, Sensor Combination, and Future Directions. *Environ. Sci. Technol.* **2022,** *56* (24), 17531–17544. https://doi.org/10.1021/acs.est.2c05874
148. Zheng, Y.; Bai, J.; Xu, J.; Li, X.; Zhang, Y. A Discrimination Model in Waste Plastics Sorting using NIR Hyperspectral Imaging System. *Waste Manage.* **2018,** *72*, 87–98. https://doi.org/10.1016/j.wasman.2017.10.015
149. Zhihong, C.; Hebin, Z.; Yanbo, W.; Binyan, L.; Yu, L. In *A Vision-Based Robotic Hrasping System using Deep Learning for Garbage Sorting*, 2017 36th Chinese Control. Conference, *CCC*. IEEE Publications, July 26, 2017; pp 11223–11226. https://doi.org/10.23919/ChiCC.2017.8029147
150. Zhu, H.; Basir, O. A Novel Fuzzy Evidential Reasoning Paradigm for Data Fusion with Applications in Image Processing. *Soft Comput.* **2006,** *10* (12), 1169–1180. https://doi.org/10.1007/s00500-005-0039-1
151. Zinchik, S.; Jiang, S.; Friis, S.; Long, F.; Høgstedt, L.; Zavala, V. M.; Bar-Ziv, E. Accurate Characterization of Mixed Plastic Waste using Machine Learning and Fast Infrared Spectroscopy. *ACS Sustain. Chem. Eng.* **2021,** *9* (42), 14143–14151. https://doi.org/10.1021/acssuschemeng.1c04281
152. Ziouzios, D.; Baras, N.; Balafas, V.; Dasygenis, M.; Stimoniaris, A. Intelligent and Real-Time Detection and Classification Algorithm for Recycled Materials using Convolutional Neural Networks. *Recycling* **2022,** *7* (1), 9. https://doi.org/10.3390/recycling7010009

Index

D

E

F

G

X

Z